MOTILITY IN CELL FUNCTION

JOHN M. MARSHALL
1920–1966

MOTILITY IN CELL FUNCTION

Proceedings of the First John M. Marshall Symposium in Cell Biology

edited by
Frank A. Pepe
Joseph W. Sanger
Vivianne T. Nachmias
Department of Anatomy
School of Medicine
University of Pennsylvania
Philadelphia, Pennsylvania

ACADEMIC PRESS
New York London Sydney Toronto San Francisco 1979
A Subsidiary of Harcourt Brace Jovanovich, Publishers

Academic Press Rapid Manuscript Reproduction

Proceedings of the First John M. Marshall Symposium in Cell Biology, Held December 2–3, 1977, in Philadelphia, Pennsylvania

ACADEMIC PRESS, INC.
111 Fifth Avenue, New York, New York 10003

United Kingdom Edition published by
ACADEMIC PRESS, INC. (LONDON) LTD.
24/28 Oval Road, London NW1 7DX

Library of Congress Cataloging in Publication Data

John M. Marshall Symposium in Cell Biology, 1st, Philadelphia, 1977.
Motility in cell function.

Symposium held Dec. 2-3, 1977, in Philadelphia, and sponsored by the Dept. of Anatomy, School of Medicine, University of Pennsylvania.
1. Cells—Motility—Congresses. 2. Muscle—Motility—Congresses. 3. Contractility (Biology)—Congresses. 4. Marshall, John Marvin, 1920-1966. 5. Biologists—United States—Biography. I. Pepe, Frank A. II. Sanger, Joseph W. III. Nachmias, Vivianne T. IV. Pennsylvania. University. Dept. of Anatomy. V. Title.
QH647.J64 1977 574.8′764 78-11880
ISBN 0-12-551750-5

PRINTED IN THE UNITED STATES OF AMERICA

79 80 81 82 9 8 7 6 5 4 3 2 1

Contents

MITOSIS

CONTRIBUTED PAPERS

List of Participants

Susan M. Abmayr, Carnegie–Mellon University, Department of Biological Sciences, 4400 Fifth Avenue, Pittsburgh, Pennsylvania 15213

Robert K. Abercrombie, University of Tennessee, Department of Zoology, Knoxville, Tennessee 37916

Robert S. Adelstein, National Institutes of Health, Bethesda, Maryland 20014

Seiel Aizu, Rutgers Medical School, Department of Anatomy, Piscataway, New Jersey 08854

Mark O. Aksoy, Carnegie–Mellon University, Department of Chemistry, 4400 Fifth Avenue, Pittsburgh, Pennsylvania 15213

Nina S. Allen, Dartmouth College, Department of Biological Science, Dartmouth, Hanover, New Hampshire 03755

Julian H. Alkon, University of Pennsylvania, Department of Physiology, Philadelphia, Pennsylvania 19104

Ernest W. April, Columbia University, College of Physicians, Department of Anatomy, 630 W 168th Street, New York, New York 10032

Debra K. Aromatorio, Carnegie–Mellon University, Department of Chemistry, 4400 Fifth Avenue, Pittsburgh, Pennsylvania 15213

John Aronson, Wistar Institute, Departments of Anatomy and Biology, 36th and Spruce/CB, Philadelphia, Pennsylvania 19104

Peggy J. Arps, Johns Hopkins University, Department of Biology, 34th and Charles Streets, Baltimore, Maryland 21218

Francis T. Ashton, University of Pennsylvania, Department of Anatomy, Philadelphia, Pennsylvania 19104

Roland M. Bagby, University of Tennessee, Department of Zoology, Knoxville, Tennessee 37916

Ruzena R. Bajcsy, University of Pennsylvania, Department of Computer and INF Science, 33rd and Walnut, Philadelphia, Pennsylvania 19104

R. A. Barlow, Jr., Haverford College, Department of Biology, Haverford, Pennsylvania 19041

Stephen M. Baylor, Yale University, Department of Physiology, 333 Cedar Street, New Haven, Connecticut 06510

David A. Begg, University of Virginia, Department of Biology, Charlottesville, Virginia 22901

Amy Behrman, Swarthmore College, Department of Biology, Swarthmore, Pennsylvania 19081

Pamela A. Benfield, Rosenstiel Basic Medical Sciences Research Center, Brandeis University, Waltham, Massachusetts 02154

Gudrun S. Bennett, University of Pennsylvania, Department of Anatomy, Philadelphia, Pennsylvania 19104

Malcolm Bersohn, Albert Einstein Medical College, Physiology/Cardiology, Department of Medicine, Montefiore Hospital and Medical Center, 111 East 210th Street, Bronx, New York 10467

Ashok K. Bhan, Montefiore Hospital and Medical Center, Cardiology Research Department, 111 East 210th Street, Bronx, New York 10467

Gopal Bhatnagar, Johns Hopkins University, Department of Dermatology, 601 North Broadway, Baltimore, Maryland 21205

Paul C. Bianchi, Jefferson University, Department of Pharmacology, 1020 Locust Street, Philadelphia, Pennsylvania 19107

Judith Biehl, University of Pennsylvania, Department of Anatomy, Philadelphia, Pennsylvania 19104

Stephen H. Blose, University of Pennsylvania, Department of Pathobiology, 3800 Spruce Street/H1, Philadelphia, Pennsylvania 19104

Krishna K. Bose, Carnegie–Mellon University, Department of Chemistry, 4400 Fifth Avenue, Pittsburgh, Pennsylvania 19217

J. Bowers, Haverford College, Department of Biology, Haverford, Pennsylvania 19041

Richard C. Brady, Temple University, Department of Biology, Philadelphia, Pennsylvania 19122

Philip W. Brandt, Columbia University of Physicians and Surgeons, Department of Anatomy, 630 West 168th Street, New York, New York 10032

Robert Brooks, Norwich–Eaton Pharmaceuticals, Norwich, New York

David L. Bruce, University of California, Department of Anesthesiology, Veterans Administration Hospital, 5901 E. Seventh Street, Long Beach, California 90801

Morris Burke, University of Maryland, Department of Physiology, 666 West Baltimore Street, Baltimore, Maryland 21201

Thomas M. Butler, Thomas Jefferson University, Department of Physiology, 1020 Locust Street, Philadelphia, Pennsylvania 19107

Jean-Paul Capony, Brookhaven National Laboratory of Biology, 50 Bell Avenue, Upton Long Island, New York 11973

Rita A. Carey, Temple University, Department of Cardiology, 3400 North Broad Street, Philadelphia, Pennsylvania 19140

Joan M. Caron, University of Connecticut Health Center, Department of Physiology, Farmington Avenue, Farmington, Connecticut 06032

Samuel K. Chacko, University of Pennsylvania, Department of Pathobiology, 305 Vet School/H1, Philadelphia, Pennsylvania 19104

Dorothy Chase, Boston University School of Medicine, Department of Physiology, 80 East Concord Street, Boston, Massachusetts 02118

Matthias Chiquet, Swiss Federal Institute of Technology, Institute of Cell Biology, ETH—Honggerberg CH-8093 Zurich, Switzerland

Prokash K. Chowrashi, University of Pennsylvania, Department of Anatomy, School of Medicine, Philadelphia, Pennsylvania 19104

Ronald F. Coburn, University of Pennsylvania, Department of Physiology, Philadelphia, Pennsylvania 19104

Amos Cohen, University of California at San Francisco, Department of Medicine, 1093 Moffitt, San Francisco, California 94143

David M. Cohen, University of Virginia Medical School, Department of Physiology, Charlottesville, Virginia

Robert A. Cohen, University of Pennsylvania, Department of Anatomy/G3, Philadelphia, Pennsylvania 19104

Richard H. Colby, Stockton State College of New Jersey, Faculty of Natural Sciences and Mathematics, Pomona, New Jersey 08240

Charles G. Connor, Temple University, Department of Biology, 128 Chatham Road, Upper Darby, Pennsylvania 19082

Mary Anne J. Conti, NIH—George Washington University, Cardiology Branch, Bethesda, Maryland 20014

Peter Cooke, University of Connecticut Health Center, Department of Physiology, Farmington, Connecticut 06032

George Cooper IV, University of Iowa, Department of Internal Medicine, University of Iowa Hospitals and Clinics, Iowa City, Iowa 52242

James Croop, University of Pennsylvania, Department of Anatomy/G3, School of Medicine, Philadelphia, Pennsylvania 19104

Roger Craig, A.R.C. Unit, Department of Zoology, South Parks Road, Oxford OX13PS England

Susan W. Craig, Johns Hopkins University, Department of Physiological Chemistry, 725 North Wolf Street, Baltimore, Maryland 21205

Renata Dabrowska, Carnegie–Mellon University, Department of Chemistry, 4400 Fifth Avenue, Pittsburgh, Pennsylvania 15213

Caroline H. Damsky, Wistar Institute, Departments of Anatomy and Biology, 36th and Spruce Streets, Philadelphia, Pennsylvania 19104

James L. Daniel, Temple University, Thrombosis Research Center, 3400 North Broad Street, Philadelphia, Pennsylvania 19140

Dandra J. Davidheiser, University of Pennsylvania, Department of Animal Biology, School of Vet Medicine/H1, Philadelphia, Pennsylvania 19104

Robert E. Davies, University of Pennsylvania, School of Vet Medicine/H1, Philadelphia, Pennsylvania 19174

M. W. Davis, Haverford College, Department of Biology, Haverford, Pennsylvania 19041

Joel S. Delfiner, University of Pennsylvania, Department of Anatomy/G3, School of Medicine, Philadelphia, Pennsylvania 19104

Adelaide M. Delluva, University of Pennsylvania, Department of Animal Biology, School of Vet Medicine/H1, 3800 Spruce Street, Philadelphia, Pennsylvania 19104

David J. DeRosier, Brandeis University, Rosenstiel Center, Waltham, Massachusetts 02154

Patricia A. Detmers, University of Pennsylvania, Department of Biology, Philadelphia, Pennsylvania 19104

Patrick F. Dillon, University of Virginia, Department of Physiology, Jordan Hall, Charlottesville, Virginia 22901

Nicholas W. DiTullio, Smith Kline and French Laboratories, Department of Biological Research, P.O. Box 7929, Philadelphia, Pennsylvania 19101

Paul Dolber, Duke University Medical Center, Department of Pathology, Durham, North Carolina 27710

Norman R. Dollahon, Villanova University, Department of Biology, Villanova, Pennsylvania 19085

Robert M. Dowben, University of Texas, Health Science Center, Department of Physiology, 5323 Harry Hines Blvd., Dallas, Texas 75235

Steven P. Driska, University of Virginia School of Medicine, Department of Physiology, Charlottesville, Virginia 22901

Abraham B. Eastwood, Columbia University, Department of Anatomy, 630 W. 168th Street, New York, New York 10032

John F. Eccleston, University of Pennsylvania, Departments of Biochemistry and Biophysics, Philadelphia, Pennsylvania 19104

Myra J. Elfvin, The Medical College of Pennsylvania, Department of Anatomy, 3300 Henry Avenue, Philadelphia, Pennsylvania 19129

Justin R. Fallon, University of Pennsylvania School of Medicine, Department of Anatomy/G3, Philadelphia, Pennsylvania 19104

Rose B. Fay, Princeton University, Department of Biology, Guyot Hall, Washington Road, Princeton, New Jersey 08540

Marcus Fechheimer, Johns Hopkins University, Department of Biology, Mergenthaler Hall, Baltimore, Maryland 21218

Steven A. Fellini, University of Pennsylvania, Department of Anatomy/G3, School of Medicine, Philadelphia, Pennsylvania 19104

Gregory Fischer, University of Virginia, Department of Biology, Charlottesville, Virginia 22901

Louis B. Flexner, University of Pennsylvania School of Medicine, Department of Anatomy/G3, Philadelphia, Pennsylvania 19104

Clara Franzini-Armstrong, University of Pennsylvania, Department of Biology/G5, Philadelphia, Pennsylvania 19104

John D. Friede, Villanova University, Department of Biology, Villanova, Pennsylvania 19085

Franklin Fuchs, University of Pittsburgh School of Medicine, Department of Physiology, Pittsburgh, Pennsylvania 15261

John Gergeley, Boston Biomedical Research Institute of Muscle Research, Boston, Massachusetts

Isidore Gersh, University of Pennsylvania School of Veterinary Medicine, Department of Anatomy, Philadelphia, Pennsylvania 19104

Anthony J. Giambalvo, Downstate Medical Center, Department of Biophysics, 450 Clarkson Avenue, Brooklyn, New York 11223

Jacques J. M. Gilloteaux, S.U.N.Y. at Syracuse, Department of Physiology, 766 Irving Avenue, Syracuse, New York 13210

William F. Gilly, Yale University, Department of Physiology, 333 Cedar Street, New Haven, Connecticut 06510

Peter S. Given, S.U.N.Y. College of Environmental Science and Forestry, Department of Chemistry, Syracuse, New York 13210

Kathryn Go, University of Pennsylvania School of Medicine, Ob–Gyn, Division of Reproductive Biology, 304 Med Labs, Philadelphia, Pennsylvania 19104

Patricia A. Gonnella, Temple University, Department of Biology, Broad Street and Montgomery Avenue, Philadelphia, Pennsylvania 19122

Gerald W. Gordon, University of Pennsylvania, Department of Biology, Philadelphia, Pennsylvania 19104

Linda M. Griffith, Harvard Medical School, Department of Anatomy, 25 Shattuck Street, Boston, Massachusetts 02115

Johanna G. Gwinn, University of Pennsylvania, Department of Anatomy/G3, Philadelphia, Pennsylvania 19104

William F. Harrington, Johns Hopkins University, Department of Biology, McCollum Pratt Institute, 34th and Charles Streets, Baltimore, Maryland 21218

David J. Hartshorne, Carnegie–Mellon University, Department of Biological Sciences, 4400 Fifth Avenue, Pittsburgh, Pennsylvania 15213

John C. Haselgrove, University of Pennsylvania, Johnson Research Foundation, C501 Richards Bldg./G4, Philadelphia, Pennsylvania 19104

Richard C. Haskell, Johns Hopkins University, Department of Biophysics, 34th and Charles Streets, Baltimore, Maryland 21218

Gyorgy Hegyi, Brookhaven National Laboratory, Department of Biology, 50 Bell Avenue, Upton Long Island, New York 11973

Robert S. Hikida, Ohio University, Department of Zoology, Athens, Ohio 45701

Robert S. Hilfer, Temple University, Biology Life Science Building, 12th and Norris Streets, Philadelphia, Pennsylvania 19122

Robert B. Hill, University of Rhode Island, Department of Zoology, Kingston, Rhode Island 02881

Sylvia Himmelfarb, Johns Hopkins University, Department of Biology, 34th and Charles Streets, Baltimore, Maryland 21218

Sarah E. Hitchcock, Carnegie–Mellon University, Department of Biological Sciences, 4400 Fifth Avenue, Pittsburgh, Pennsylvania 15213

Phyllis E. Hoar, University of Washington, Departments of Physiology and Biophysics, SJ-40, Seattle, Washington 98115

Holm A. Holmsen, Temple University Medical School, Thrombosis Research Center, 3400 North Broad Street, Philadelphia, Pennsylvania 19140

Howard Holtzer, University of Pennsylvania School of Medicine, Department of Anatomy, Philadelphia, Pennsylvania 19104

Sybil Holtzer, University of Pennsylvania School of Medicine, Department of Anatomy, Philadelphia, Pennsylvania 19104

John N. Howell, Ohio University, Departments of Zoology and Microbiology, Grosvenor Hall, Athens, Ohio 45701

Mary K. Hrapchak, Johns Hopkins University School of Medicine, Department of Physiological Chemistry, 725 North Wolfe Street, Baltimore, Maryland 21205

Hugh E. Huxley, Laboratory of Molecular Biology, Medical Research Council, Hill Road, Cambridge, England

Gabor Huszar, Department of Obstetrics and Gynecology, Yale Medical School, 333 Cedar Street, New Haven, Connecticut 06510

Shinya Inoué, University of Pennsylvania, Department of Biology, 217 Leidy Lab/G7, Philadelphia, Pennsylvania 19104

Mary J. Irish, CMDNJ—Rutgers Medical School, Department of Anatomy, Box 101, Piscataway, New Jersey 08854

Raja M. Iyengar, University of Pennsylvania, Department of Animal Biology, 3800 Spruce Street, Philadelphia, Pennsylvania 19104

Chungwha Iyengar, Bryn Mawr College, Department of Biology, Bryn Mawr, Pennsylvania 19010

James M. Jaeger, Temple University School of Medicine, Department of Physiology, 3420 North Broad Street, Philadelphia, Pennsylvania 19140

M. Mazher Jaweed, Jefferson University, Department of Rehabilitation Medicine, 11th and Sansom Street, Philadelphia, Pennsylvania 19104

Marilyn R. James, Presbyterian University of Pennsylvania Medical Center, Department of M.S.R.L., 51 North 39th Street, Philadelphia, Pennsylvania 19104

William H. Johnson, Rensselaer Polytech Institute, Department of Biology, Troy, New York 12181

James Junker, Duke University Medical Center, Department of Pathology, Durham, North Carolina 27710

Trudy Karr, Johns Hopkins University, Department of Biology, 34th and Charles Streets, Baltimore, Maryland 21218

Laura R. Keller, University of Virginia, Department of Biology, 14–5 Copeley Hill, Charlottesville, Virginia 22901

Thomas C. Keller, University of Virginia, Department of Biology, 14–5 Copeley Hill, Charlottesville, Virginia 22901

W. Glenn L. Kerrick, University of Washington, Departments of Physiology Biophysics, SJ-40, Seattle, Washington 98195

Dietrich Kessler, Haverford College, Department of Biology, Haverford, Pennsylvania 19104

Daniel P. Kiehart, University of Pennsylvania, Department of Biology, Philadelphia, Pennsylvania 19104

Harriet King, The Medical College of Pennsylvania, Department of Anatomy, 3300 Henry Avenue, Philadelphia, Pennsylvania 19129

Gretchen E. Klaussmann, Temple University, Department of Anatomy, 3400 North Broad Street, Philadelphia, Pennsylvania 19140

Peter J. Knight, The Johns Hopkins University, Department of Biology, Baltimore, Maryland 21218

Stephen J. Koons, S.U.N.Y. at Buffalo, Department of Biophysics, 4234 Ridge Lea Road, Buffalo, New York 14226

Edward D. Korn, National Heart, Lung, and Blood Institute, Laboratory of Cell Biology, Building 3, Room B1-22, Bethesda, Maryland 20014

George R. Kracke, Medical College of Pennsylvania, Departments of Physiology and Biochemistry, 3300 Henry Avenue, Philadelphia, Pennsylvania 19129

Meg A. Krilov, Temple University, Department of Biology, Philadelphia, Pennsylvania 19104

Joseph M. Krisanda, University of Virginia School of Medicine, Department of Physiology, Charlottesville, Virginia 22901

Robert R. Kulikowski, The University of Chicago, Department of Anatomy, 1025 East 57th Street, Chicago, Illinois 60637

Fredericka Kundig, Towson State University, Department of Biology, Towson, Maryland 21204

Richard M. Lane, Dickinson College, Department of Biology, Carlisle, Pennsylvania 17013

Barbara G. Langer, University of Pennsylvania School of Medicine, Department of Anatomy, Philadelphia, Pennsylvania 19104

Mara B. Lanterbach, Towson State University, Department of Biology, Towson, Maryland 21204

M. J. Lathwell, Haverford College, Department of Biology, Haverford, Pennsylvania 19041

Alan M. Laties, University of Pennsylvania, Department of Ophthalmology, Philadelphia, Pennsylvania 19104

Sherwin S. Lehrer, Department of Muscle Research, Boston Biomedical Research Institute, 20 Staniford Street, Boston, Massachusetts 02114

James L. Lessard, Children's Hospital Medical Center, Department of Fetal Pharmacology, Elland and Bethesda Avenues, Cincinnati, Ohio 45229

Robert M. Leven, University of Pennsylvania School of Medicine, Department of Anatomy, Philadelphia, Pennsylvania 19104

J. C. Rhea Levine, The Medical College of Pennsylvania, Department of Anatomy, 3300 Henry Avenue, Philadelphia, Pennsylvania 19129

Tsung-I Lin, University of Texas, Department of Biophysics, 5323 Harry Hines Blvd., Dallas, Texas 75235

Raye Z. Litten, University of Vermont, Departments of Physiology and Biophysics, Given Medical Building, Burlington, Vermont 05401

Ida J. Llewellyn-Smith, University of Toronto, Banting and Best Department of Medical Research, 112 College Street, Toronto, Ontario M5G IL6 Canada

Ariel G. Loewy, Haverford College, Department of Biology, Haverford, Pennsylvania 19041

Stephen J. Lovell, Johns Hopkins University, Department of Biology, 803B Dartmouth Road, Baltimore, Maryland 21212

Susan Lowey, Brandeis University, Department of Biochemistry, Rosenstiel Basic Science Research Center, Waltham, Massachusetts 02154

Roger C. Lucas, S.U.N.Y.—Downstate Medical Center, Department of Biochemistry, 450 Clarkson Avenue, Brooklyn, New York

Issei Mabuchi, University of Tokyo, College of General Education, Department of Biology, Komaba, Meguro-ku, Tokyo, 153 Japan

Susan D. MacLean, Harvard Medical School, Department of Anatomy, 25 Shattuck Street, Boston, Massachusetts 02115

Ashwani Malhotra, Montefiore Hospital and Medical Center, Department of Cardiology Research Laboratories, 111 East 210th Street, Bronx, New York 10467

David J. Marsh, Brandeis University, Rosenstiel Center, Waltham, Massachusetts 02154

Michael W. Marshall, Yale University, Department of Physiology, 333 Cedar Street, New Haven, Connecticut 06510

Joseph T. Martin, Temple University, Fels Research Institute, 3420 North Broad Street, Philadelphia, Pennsylvania 19140

Estela Maruenda, University of Pennsylvania, Department of Biology, Philadelphia, Pennsylvania 19104

Denis Martynowych, Haverford College, Department of Biology, Haverford, Pennsylvania 19041

Slavica S. Matacic, Haverford College, Department of Biology, Haverford, Pennsylvania 19041

Priscilla Mattson, Case Western Reserve University School of Medicine, Department of Anatomy, Cleveland, Ohio 44106

Rosemary M. Mazanet, University of Pennsylvania, Department of Anatomy, Philadelphia, Pennsylvania 19104

Grace M. Migliousi, Downstate Medical Center, Department of Cell Biology, 450 Clarkson Avenue, Brooklyn, New York 11236

Mark A. Milanick, University of Chicago, Department of Biophysics, 920 East 58th Street, Chicago, Illinois 60637

Barry Millman, University of Guelph, Department of Physics, Guelph, Ontario N1G 2WL, Canada

Vlasta Molak, Downstate Medical Center, Department of Biochemistry, 450 Clarkson Avenue, Brooklyn, New York 11203

Charles E. Montague, Johns Hopkins University, Department of Biophysics, 34th and Charles Streets, Baltimore, Maryland 21218

Robert Morell, University of Virginia, Department of Biology, Charlottesville, Virginia 22901

Diana J. Moss, University of Pennsylvania School of Medicine, Departments of Biophysics and Biochemistry, Philadelphia, Pennsylvania 19104

Ulrike Mrwa, Physiologisches Institute, Im Nevenheimer Feld 326, D-69 Heidelberg, Germany

Richard A. Murphy, University of Virginia School of Medicine, Department of Physiology, Charlottesville, Virginia 22901

George B. McClellan, University of Pennsylvania, Department of Physiology, C301 Richards Bldg./G4, Philadelphia, Pennsylvania 19104

J. Richard McIntosh, University of Colorado, Molecular, Cell, and Developmental Biology, Boulder, Colorado 80302

Vivianne T. Nachmias, University of Pennsylvania, Department of Anatomy, Philadelphia, Pennsylvania 19104

Chandrasekaran Nagaswami, University of Pennsylvania School of Medicine, Department of Anatomy, Philadelphia, Pennsylvania 19104

Leonard Nelson, Medical College of Ohio, Department of Physiology, Toledo, Ohio 43699

Jay Newmann, Johns Hopkins University, Department of Biophysics, 34th and Charles Streets, Baltimore, Maryland 21218

Richard Niederman, Veterans Administration Hospital, Dental Service, 1400 VFW Parkway, Boston, Massachusetts 02132

Benjamin Norris, Perth Amboy General Hospital, Department of Chemistry, 430B Annette Court, Somerset, New Jersey 08873

Charles P. Ordahl, Temple University Medical School, Department of Anatomy, 3400 North Broad Street, Philadelphia, Pennsylvania 19140

Joann J. Otto, University of Pennsylvania, Department of Biology, Leidy Labs/G7, Philadelphia, Pennsylvania 19104

Maurizio Pacifici, University of Pennsylvania School of Medicine, Department of Anatomy, Philadelphia, Pennsylvania 19104

Gail L. Pakstis, Temple University, Department of Biology, Broad Street and Montgomery Avenue, Philadelphia, Pennsylvania 19122

Lee D. Peachey, University of Pennsylvania, Department of Biology/G5, Philadelphia, Pennsylvania 19104

Suzanne M. Pemrick, S.U.N.Y. Downstate Medical Center, Department of Biochemistry, 450 Clarkson Avenue, Brooklyn, New York 11203

Frank A. Pepe, University of Pennsylvania, Department of Anatomy, Philadelphia, Pennsylvania 19104

Samuel V. Perry, University of Birmingham, Department of Biochemistry, P.O. Box 363, Birmingham B15 2TT, United Kingdom

Steven J. Phillips, Temple University School of Medicine, Department of Anatomy, 3420 North Broad Street, Philadelphia, Pennsylvania 19140

Thomas D. Pollard, Harvard University, Department of Anatomy, 25 Shattuck Street, Boston, Massachusetts 02115

Nancy Poste, Duke University Medical Center, Department of Pathology, Durham, North Carolina 27710

Maureen G. Price, University of Pennsylvania School of Medicine, Department of Anatomy, Philadelphia, Pennsylvania 19104

Dallas L. Pulliam, The Medical College of Pennsylvania, Department of Anatomy, 3300 Henry Avenue, Philadelphia, Pennsylvania 19129

Jack A. Rall, Ohio State University, Department of Physiology, 1645 Neil Avenue, Columbus, Ohio 43210

Lionel I. Rebhun, University of Virginia, Department of Biology, Gilmer Hall, Charlottesville, Virginia 22903

Alfred M. Reingold, University of Pennsylvania School of Medicine, Department of Anatomy, Philadelphia, Pennsylvania 19104

Stephen P. Remillard, Princeton University, Department of Biology, Guyot Hall Washington Road, Princeton, New Jersey 08540

Thomas F. Robinson, University of Pennsylvania School of Medicine, Department of Physiology, Philadelphia, Pennsylvania 19104

Michael E. Rodgers, Johns Hopkins University, Department of Biology, 34th and Charles Streets, Baltimore, Maryland 21218

Fred J. Roisen, CMDNJ—Rutgers Medical School, Department of Anatomy, P.O. 101, Piscataway, New Jersey 08854

Alburt M. Rosenberg, Swarthmore College, Department of Physics, Swarthmore, Pennsylvania 19081

Sharon Rosenberg, Downstate Medical Center, Department of Biochemistry, 450 Clarkson Avenue, Brooklyn, New York 11236

Neal A. Rubinstein, University of Pennsylvania School of Medicine, Department of Anatomy, Philadelphia, Pennsylvania 19104

Julie I. Rushbrook, Downstate Medical Center, Department of Biochemistry, 450 Clarkson Avenue, Brooklyn, New York 11203

Daniel Safer, University of Pennsylvania School of Medicine, Department of Anatomy, Philadelphia, Pennsylvania 19104

David W. Saffen, Johns Hopkins University, Department of Biology, 3301 St. Paul Street, Baltimore, Maryland 21218

Judith D. Saide, Boston University School of Medicine, Department of Physiology, 80 East Concord Street, Boston, Massachusetts 02118

Leon Salganicoff, Temple University Medical School, Department of Pharmacology, 3400 North Broad Street, Philadelphia, Pennsylvania 19104

Jean M. Sanger, University of Pennsylvania School of Medicine, Department of Anatomy, Philadelphia, Pennsylvania 19104

Joseph W. Sanger, University of Pennsylvania School of Medicine, Department of Anatomy, Philadelphia, Pennsylvania 19104

Satyapriya Sarkar, Boston Biomedical Research Institute, Department of Muscle Research, 20 Stanford Street, Boston, Massachusetts 02114

Harry L. Saunders, Smith Kline and French Laboratories, Department of Biological Research, P.O. Box 7929, Philadelphia, Pennsylvania 19101

Robert E. Savage, Swarthmore College, Department of Biology, Swarthmore, Pennsylvania 19081

David Sawyer, University of Connecticut Health Center, Department of Physiology, Farmington, Connecticut 06032

Vitaly Sawyna, The Medical College of Pennsylvania, Department of Anatomy, 3300 Henry Avenue, Philadelphia, Pennsylvania 19129

James Scheuer, Montefiore Hospital, Department of Medicine/Cardiology, 111 East 210th Street, Bronx, New York 10467

David Seiden, Rutgers Medical School, Department of Anatomy, P.O. Box 101, Piscataway, New Jersey 08854

Eileen M. Sharkey, Carnegie–Mellon University, Department of Biological Sciences, 4400 Fifth Avenue, Pittsburgh, Pennsylvania 15213

Charles R. Shear, University of Maryland School of Medicine, Department of Anatomy, 655 West Baltimore Street, Baltimore, Maryland 21201

Michael P. Sheetz, University of Connecticut Health Center, Department of Physiology, Farmington, Connecticut 06032

James M. Sherry, Carnegie–Mellon University, Department of Biological Sciences, 4400 Fifth Avenue, Pittsburgh, Pennsylvania 15213

Marion J. Siegman, Jefferson Medical College, Department of Physiology, 1020 Locust Street, Philadelphia, Pennsylvania 19106

Linda M. Siemankowski, S.U.N.Y. Downstate Medical Center, Department of Biochemistry, 450 Clarkson Avenue, Brooklyn, New York 11203

Raymond F. Siemankowski, S.U.N.Y. Downstate Medical Center, Department of Medicine, 450 Clarkson Avenue, Brooklyn, New York 11203

Bonnie F. Sloane, Pennsylvania Muscle Institute, and the University of Pennsylvania, Department of Physiology, and Presbyterian University of Pennsylvania Medical Center, 51 North 39th Street, Philadelphia, Pennsylvania 19104

Andrew P. Somlyo, Pennsylvania Muscle Institute, Presbyterian University of Pennsylvania Medical Center, 51 North 39th Street, Philadelphia, Pennsylvania 19104

Joachim R. Sommer, Duke University Medical Center, Department of Pathology, Durham, North Carolina 27710

M. Spady, Haverford College, Department of Biology, Haverford, Pennsylvania 19041

Sharon Speiser, S.U.N.Y. Downstate Medical Center, Departments of Anatomy and Cell Biology, 450 Clarkson Avenue, Brooklyn, New York 11236

Richard L. Stambaugh, University of Pennsylvania School of Medicine, Ob–Gyn, Division of Reproductive Biology, 304 Medical Laboratories, Philadelphia, Pennsylvania 19104

Marcia M. Steinberg, Temple University, Department of Biology, Broad and Montgomery, Philadelphia, Pennsylvania 19122

Peter R. Stewart, SGS Biochemistry, Australian National University, Canberra 1600, Australia

Gregory B. Stock, Johns Hopkins University, Jenkins Department of Biophysics, Baltimore, Maryland 21218

Karen H. Stump, Carnegie–Mellon University, Department of Chemistry, 4400 Fifth Avenue, Pittsburgh, Pennsylvania 15213

Andrew G. Szent-György, Brandeis University, Department of Biology, Waltham, Maryland 02154

Edwin W. Taylor, University of Chicago, Department of Biophysics, 928 East 58th Street, Chicago, Illinois 60637

D. Lansing Taylor, Harvard University, Department of Biological Laboratories, 16 Divinity Avenue, Cambridge, Massachusetts 02138

Lewis G. Tilney, University of Pennsylvania, Department of Biology, 233 Leidy Labs/G7, Philadelphia, Pennsylvania 19104

Paul Toselli, Boston University, Department of Biochemistry, 80 East Concord Street, Boston, Massachusetts 02332

Yoshiro Toyama, University of Pennsylvania School of Medicine, Department of Anatomy, Philadelphia, Pennsylvania 19104

Richard T. Tregear, Oxford University, ARC Unit, Department of Zoology, South Parks Road, Oxford OX1 3PS, England

David R. Trentham, University of Pennsylvania School of Medicine, Department of Biochemistry and Biophysics/G3, Philadelphia, Pennsylvania 19104

John A. Trotter, National Institutes of Health/NHLBI, Department of Molecular Cardiology, Bethesda, Maryland 20014

Joseph A. Truglia, S.U.N.Y. Downstate Medical Center, Department of Biochemistry, Box 450 Clarkson Avenue, Brooklyn, New York 10708

Kathleen M. Trybus, University of Chicago, Department of Biophysics, 920 East 58th Street, Chicago, Illinois 60637

K. C. Tsou, University of Pennsylvania, Department of Harrison Surgical Research, 1515 Ravdin Institute, Philadelphia, Pennsylvania 19104

David Turner, Institute of Cell Biology, Swiss Federal Institute of Technology, CH-8093 Zurich, Switzerland

Betty M. Twarog, S.U.N.Y. at Stony Brook, Department of Anatomical Sciences, Health Sciences Center, Stony Brook, New York 11790

Peter J. Vibert, Brandeis University, Rosenstiel Basic Medical Sciences Research Center, 415 South Street, Waltham, Massachusetts 02154

Phyllis R. Wachsberger, University of Pennsylvania School of Medicine, Department of Anatomy, Philadelphia, Pennsylvania 19104

Nancy Wallace, Duke University Medical Center, Department of Pathology, Durham, North Carolina 27710

Theo A. Wallimann, Brandeis University, Department of Biology, Waltham, Massachusetts 02154

Annemarie Weber, University of Pennsylvania School of Medicine, Departments of Biochemistry and Biophysics, 303 Anatomy–Chemistry Bldg./G3, Philadelphia, Pennsylvania 19104

Chris West, University of Pennsylvania School of Medicine, Department of Anatomy, Philadelphia, Pennsylvania 19104

Richard C. Whiting, United States Department of Agriculture Eastern Regional Research Center, 600 East Mermaid Lane, Philadelphia, Pennsylvania 19118

Frank J. Wilson, Rutgers Medical School, College of Medicine and Dentistry of New Jersey, Department of Anatomy, P.O. Box 101, Piscataway, New Jersey 08854

Saul S. Winegrad, University of Pennsylvania, Department of Physiology, Philadelphia, Pennsylvania 19104

John S. Wray, Brandeis University, Rosenstiel Basic Medical Sciences Research Center, Waltham, Massachusetts 02154

Jyh-jia W. Yang, Temple University, Department of Biology, Broad Street and Montgomery Avenue, Philadelphia, Pennsylvania 19122

Preface

This volume is based on the first of a series of symposia on anatomy sponsored by the Department of Anatomy, The School of Medicine, University of Pennsylvania. These will include symposia in cell biology, neurobiology, and developmental biology among others. It is a special pleasure to be involved in this symposium, which is cosponsored by the Pennsylvania Muscle Institute. I arrived at the University of Pennsylvania to work as a postdoctoral with John Marshall in the Department of Anatomy in 1957, and my career in muscle research started then. The encouragement and guidance provided by Dr. Marshall at that time were a major factor in my continuing interest in the molecular anatomy of myosin filaments. Dr. Marshall's interests spanned a variety of cell types from ameba to muscle.

This book is introduced by some brief recollections of Dr. Marshall by people who knew him and were enriched by interaction with him. A special lecture on myosin and actomyosin ATPase mechanisms was presented by Dr. David Trentham during the symposium and is included in the introduction. In Part 2, the structure of both muscle and nonmuscle filaments is considered. Part 3 covers regulation of myosin–actin interaction in skeletal muscle, cardiac muscle, smooth muscle, and nonmuscle systems and concludes with a review of comparative studies of regulation and evolution. Whole cells are treated in Part 4 again covering skeletal, cardiac and smooth muscle, and including catch muscle. In addition work with tissue culture cells, ameba, and the retina are included. Part 5 covers the functions of actin and microtubules in the mitotic spindle. Part 6 includes the contributed papers presented at the symposium.

The uniqueness of this symposium is that it covers the highlights of contractile function in diverse systems from nonmuscle cells and tissues to the variety of muscle tissues. The participants are leaders in their respective fields and recent rapid developments make this an opportune time for promoting these different approaches to the similar problems of contractility in diverse tissues.

The success of the symposium is due in major part to the efforts of the organizing committee which included Paul C. Bianchi (Professor and Chairman of Pharmacology, Thomas Jefferson University), Vivianne T. Nachmias (Associate Professor of Anatomy, School of Medicine, University of Pennsylvania), Frank A. Pepe (Professor and Chairman of Anatomy, School of Medicine, University of Pennsylvania), Joseph W. Sanger (Associate Professor of Anatomy, School of Medicine, University of Pennsylvania), Andrew P. Somlyo (Director, Pennsylvania Muscle Institute, Professor of Physiology, Presbyterian University of Pennsylvania Medical Center), and Saul Winegrad (Professor of Physiology, School of Medicine, University of Pennsylvania). A crucial element in making this symposium possible was the financial support provided by the National Science Foundation, Burroughs Wellcome Co., Ciba Geigy Corporation, ICI United States, Inc., McNeil Laboratories, Norwich Pharmacal Co., Smith Kline and French Laboratories, and William H. Rorer, Inc. We are grateful for their support. We are also grateful to those who chaired the various programs of the symposium and these included Dr. Paul C. Bianchi, Dr. Shinya Inoué, Dr. Vivianne T. Nachmias, Dr. Joseph W. Sanger, Dr. Andrew P. Somlyo, Dr. Andrew Szent-Györgyi, Dr. Annemarie Weber, and Dr. Saul S. Winegrad. Finally, we are indebted to the excellent administrative assistance provided by Mrs. Doerte Smith and the projection assistance provided by Mr. Robert Paltzman and Mr. Alfred Reingold.

Frank A. Pepe

INTRODUCTION

Motility in Cell Function
Proceedings of the First John M. Marshall Symposium in Cell Biology

JOHN M. MARSHALL: SCIENTIST AND HUMANIST 1920--1966

L. B. Flexner

John Marshall was killed April 29, 1966 in an automobile accident in Uganda while taking a short vacation with his family. He had gone to Makerere Medical College, Kampala, Uganda, nine months before on a Guggenheim Fellowship and as Visiting Professor of Anatomy to pursue his own research interests and to participate in the research training of staff and students. Such was the tragic end at the age of 46 of a man treasured by his family, friends, and colleagues, who had proved himself to be an extraordinarily gifted scientist, as well as a man of effective concern in his community.

Dr. Marshall was born in Kingsville, Texas, on April 15, 1920. He first attended the Texas College of Arts; then transferred to Rice Institute where he majored in chemistry, received the B.A. degree at the age of 21 and spent an additional year as a graduate Fellow in Colloid Chemistry. Three years later he entered the Medical College, University of Illinois, receiving the M.S. degree in physiology in 1948 and the M.D. degree cum laude in 1950. During these years at Illinois he was sequentially a Life Insurance Medical Research Student Fellow, a Markle Foundation Fellow, and Bristol Laboratories Fellow. At about the time that he had been awarded the M.S. degree he decided to qualify for the Ph.D. degree in pathology. He had finished all but a few formal requirements for the degree by the end of his internship in 1951. From 1951 to 1953 he worked as a postdoctoral Fellow of the American Cancer Society at the Carlsberg Laboratory in Denmark with Dr. H. Holter. In 1953, at the age of 33, he joined the staff of the Department of Anatomy at the University of Pennsylvania, where he held the rank of Professor of Anatomy at the time of his death. He had spent 13 years with us.

One of the striking qualities of Dr. Marshall was the breadth of his scientific interest and competence, characterized once by a colleague in biochemistry as awe-inspiring. Evidence of these qualities is to be found in his creative

ISBN 0-12-551750-5

attainments during his predoctoral and early postdoctoral years. While at Rice Institute, he collaborated in a publication on x-ray diffraction studies of copper sulphides. His work in the Department of Physiology at Illinois led to a series of papers on experimental renal hypertension. While in the midst of these studies, Dr. Marshall took a course in histochemistry, given by my friend and colleague, Dr. Isidore Gersh. This experience was critical. It led him abruptly away from his physiological problems and into histochemistry. The first published evidence of this new and abiding interest is to be found in his doctoral thesis entitled, "Localization of adrenocorticotrophic hormone by histochemical and immunochemical methods"; one of the first reported applications of the fluorescent antibody method of Coons to the localization of native antigens in tissues. Dr. Marshall's experience at the Carlsberg Laboratory was responsible for a second enduring area of concentration, for here with Dr. Holter he started his studies on pinocytosis in amoebae.

In his 13 years at Pennsylvania, continuing work on pinocytosis led to increasingly profound investigations on its mechanisms and implications. In a publication, shortly before his death, on intracellular transport in amoebae, Dr. Marshall pointed out: "It is often assumed that the transport mechanisms which maintain the internal environment of the cell and provide the materials required for growth and activity are localized in the cell membrane....We conclude that the amoeba is entirely dependent on membrane uptake to supply its requirement for chloride ion. The situation with regard to PO_4 is probably similar....The general concept suggested by the studies we have described is that all active transport processes in the amoeba are intracellular and depend upon dynamic transformations of membrane into cytoplasm, and of cytoplasm into membrane, rather than upon the operation of localized structures within an otherwise static 'cell membrane'." This point of view, although still controversial, was and is provocative in a fruitful way.

While at Pennsylvania, Dr. Marshall's histochemical interests were focused upon the protein constituents of skeletal muscle. At Illinois, he had adapted the fluorescent antibody method to the localization of the adrenocorticotrophic hormone in the cells of the adenohypophysis. Using the same method at the Carlsberg Laboratory, he had reported on the distribution of four enzymes or proenzymes in the acinar cells of the pancreas. Shortly after coming to Pennsylvania, he chose with several colleagues to work on skeletal muscle as a model system because it offers a highly organized, regularly repeating structure made up of several distinct proteins. In early work with anti-myosin, it was found thay myosin could be localized accurately in the sarcomere structure of relaxed muscle fibers and that in the differentiation of myoblasts the antibody method detected the synthesis of myosin at a stage much earlier than could be found by other methods. These results were followed by a study of the distribution of other proteins in the myofibril with the plan of mapping out the complete structure of the sarcomere. Beginning in 1961, emphasis was placed on the immunochemical analysis of the substructure of the myosin molecule using antisera of sharply defined specificity for particular regions of the molecule. This last problem was of the greatest interest to Dr. Marshall but unfortunately was not brough to fruition before his death.

In spite of the heavy demands made upon him by his creative work in the laboratory, Dr. Marshall was sensitive and responsive to what he considered to be his civic responsibilities. This can be exemplified by two of his numerous activities. He played a pioneering and leading role in combating urban blight and achieving racial integration in the neighborhood of Philadelphia in which he lived, Powelton Village, and was Chairman of the Powelton Neighbors Group from 1959 to 1960. His interest in public education led to his membership on the Board of the Citizens Committee on Public Education in Philadelphia of which he was Chairman from 1963 to 1965. To these tasks, as to all others, Dr. Marshall brought a very rare combination of dedication, high standards, keen insight, and felicity in dealing with people of diverse interests and backgrounds.

It was my good fortune to have been a colleague of Dr. Marshall throughout his years at Pennsylvania. Like many others, I valued his friendship, deeply respected his complete integrity, sought his advice and opinions, and marvelled at his knowledge, his imagination and originality, his clarity and incisiveness. To those, in earlier days, who might ask me what I thought of Dr. Marshall I would invariably reply: "It is a privilege to know him." This will make clear the depth of our loss.

Motility in Cell Function
Proceedings of the First John M. Marshall Symposium in Cell Biology

INTIMATIONS OF THINGS TO COME--JOHN MARSHALL AS A GRADUATE STUDENT

Isidore Gersh
Emeritus Professor of Anatomy

The word "intimations" in the title of this talk was selected to indicate the dynamic state of the development of John Marshall while he was a graduate student in my laboratory at the University of Illinois which was shared with and later taken over by Dr. H. R. Catchpole, who had a great deal to do with the development of John Marshall. I have tried to look back on those days, wondering how far one could have predicted his subsequent scientific career and social interests.

I think his outstanding quality then and later on was that he had a mind of his own. His actions as a conscientious objector during World War II mark him as a future dissenter. He was the only CO in a large family, and his behaviour as a CO was incomprehensible to the rest of his family. I would conclude that he learned early to work actively and consistently toward the realization of rational points of view. These were expressed in later life in his involvement in neighborhood integration, civil rights of all peoples in this country and elsewhere, sound education for change, and peace. His ardent pursuit of these aims in Philadelphia can be attested to by all who knew him.

Together with his rational political belief went a certain frugality in his life style and a certain modesty. These qualities, manifest as a graduate student, extended in his later years. I suppose his respect for others, no matter what their origins or position, was in a sense also an extension of these traits.

When it comes to science, it is more difficult to trace the exact expression in his mature days of his prior training and personality as a graduate student. He was trained in college as a chemist, and it is no surprise that he was inclined to look at cell products chemically and to measure the products of their activity quantitatively. He preferred working toward chemical purification of cell components, and this quality per-

ISBN 0-12-551750-5

sisted throughout his mature life. This step seemed to him essential for the specific identification of a cell product or property. Once identified, the question arose of sensitivity of the test procedure, and this led him into fluorescence microscopy. Suitable fixation concerned him greatly as a graduate student, and he tried to reduce to a minimum displacement, extraction, or chemical alteration of proteins in his microscopic preparation. In preparation for this brief talk, I reread his thesis of 1954. It still reads well and still feels like a fresh breeze. All the scientific qualities he showed then continued to be displayed in later years.

Once the decision was made to foresake the practice of medicine (which he made in Copenhagen and in Paris, while he was a Fellow of the American Cancer Society) it was clear that he would do research in the basic sciences. However, it was impossible to predict in exactly which manner or direction these properties would express themselves in later life. It was predictable that he would try to reduce cell activity to some chemically acceptable and measurable parameter, whether it were an endocrine or exocrine cell, a muscle or nerve cell, or a unicellular organism. His research might concern such general cellular activities as contractility, conductivity, permeability, pinocytosis, or secretion. As a graduate student, he flirted with all of these properties; there was no way to predict whether one of them, or some others, would next claim his attention.

This is not to imply that John Marshall was diletante or superficial. Far from it! He read widely and worked long and hard. He was, in addition, a technical wizard. He was the only graduate student I ever had for whom it was not necessary to demonstrate laboratory procedures. Simply describing the steps in any past or projected procedure was enough for him.

I have often wondered if scientific creativity is related to a certain political and social rationality in some people. Certainly in people like Bernal, Haldane, Pauling, and others the two traits seem blended. John Marshall was an admirable scientist and humanist, and I believe these qualities were clearly perceived in him when he was a graduate student, as intimations of things to come. The specific expression of these properties in science were not exactly clear, but the general features were clear for all to see.

JOHN M. MARSHALL: THE MAN

David L. Bruce

Your have heard a lot, today, about John Marshall the scientist, but not enough about John Marshall, the man. Many of you, and certainly all of the young members of this audience, never met John. What sort of person was he?

First of all, who am I? I am an academic physician engaged in the practice of, and research into, anesthesiology. In 1961, I came to the University of Pennsylvania to take my residency training. My chief, the late Dr. Robert Dripps, encouraged me to spend a postdoctoral year in the lab to learn how to do research. John Marshall was a neighbor and one night, sitting on his front porch, he agreed to take me as a postdoctoral student to study cell physiology. He didn't mind my lack of training in the more rigorous aspects of basic science, as long as I was motivated to learn. John was an open minded man.

Today, I heard a lot of papers that I didn't understand. My thoughts drifted back to the day that John and I took a train to New York and sat through a session of high-powered papers at the Biophysical Society meetings. At the coffee break I confided to him that I hadn't understood most of what I'd heard that morning. He replied, "if you *want* to, bad enough, you will." John was a "can do" man.

In his laboratory, I learned to blow glass. John said it was "good for the soul" to be able to make simple laboratory accessories. I saw a system for growing 100 grams of amebas a week, fashioned from a handcranked cream separator, a bunch of baker's bread pans, and the like. We had homemade ion-exchange columns on the water taps, filling giant water bottles which John hoisted up to storage shelves with surprising ease. John was a self-reliant man.

John didn't publish a great deal. Our one paper represented almost a year and a half of hard work, vast stores of data, numerous quality control checks of each and every analytic technique we used, and many rewrites before he'd agree to submit it. John was a completely honest, scientifically scrupulous man.

I could go on and on, and perhaps some of you think I already

ISBN 0-12-551750-5

have. I'd tell you about his inventiveness, exemplified by his unfinished studies in which he used living amebas as the resin in an ion-exchange column, but my time is up. I hope you have a better idea of a magnificent man who himself had better ideas than most of us ever will. John Marshall changed my life, for which I will always be grateful. I miss him.

Motility in Cell Function
Proceedings of the First John M. Marshall Symposium in Cell Biology

FROM AMEBA TO MUSCLE: ON SOME WORK BY AND WITH JOHN M. MARSHALL

*Vivianne T. Nachmias**

Department of Anatomy
School of Medicine
University of Pennsylvania

John Marshall would have liked this title, I think, for two reasons. One you may already know, is that not only did he work both on amebae and on muscle, but in the late 1950s when theories of the mechanisms of ameboid movement outnumbered the available pieces of evidence by a wide margin and monographs even appeared on mechanisms based on jet propulsion or changes in surface tension, he believed, as did a few others, that amebae contained a contractile system basically similar to that of muscle, an idea favored by 19th century biologists. He encouraged me through many early failures to continue to search for structural evidence for such a system in single amebae and himself used mass cultures to look for the contractile material in cytoplasmic fractions. Several chapters in this symposium on cell contractility show how far this belief in the basic similarity of the contractile processes has been fulfilled.

But, he would also have liked the title because it would have been an ideal one to "stand on its head" as he liked to term it, and to point out that the direction of scientific progress has been largely from muscle to amebae. Especially at the beginning, the availability of muscle proteins as well as methods and ideas developed from the muscle field made many types of experiments possible. More recently cellular contractility has developed its own separate identity, especially with the discovery of new proteins associated with actin and myosin.

Although John Marshall contributed new approaches and ideas and developed techniques relevant to both muscle and nonmuscle contractility, some of you may not know much about these contri-

**Current address: 15 Robinson Street, Cambridge, Massachusetts 02138.*

ISBN 0-12-551750-5

butions. He did not publish a great deal--partly because he was very busy not only with his own experiments but with major efforts to bring about social change both in the area where he lived and in the Philadelphia school system. Also, his philosophy of work included constant revision and ferment--which meant it was hard to stop the experiments, to sit down and codify and conclude in the written word. There are nevertheless, a number of basic papers from conferences which include the main results and these can be consulted for further details. What I would like to do here is to describe for you a few areas of his work that I knew closely and which in retrospect show how modern many of his ideas were. These few examples may also give you a feel for some of his approaches, enthusiasms, and cautions. Necessarily, this description is incomplete, leaving out both areas and people, but may still serve the purpose.

The first modern view that I remember is, however, not scientific at all, but was the first one encountered when entering his laboratory. John Marshall was very advanced in his abhorrence of coffee. Instead a large--2 liter--Erlenmeyer was seen heating in the lab early in the morning into which he shook a few handfuls of some vigorous brand of tea. This continued to brew as the day went on, becoming progressively darker and darker. It was offered (with a flourish) to all the many visitors. It was best to visit early in the day! Howard Holtzer came by frequently for famously energetic arguments on how sarcomeres grow, or how myosin aggregates, or how best to use antibodies to study development. As I remember, he would sit with an untouched beaker of tannic brew while Marshall would down a few. On at least one occasion David Bruce, by dint of getting to the lab early enough (ca: 7:00) had the pleasure of greeting Marshall with ripened tea and beaker by 7:30 A.M.

But to consider scientific areas, John Marshall in the 1950s was pursuing the localization of proteins in cells by fluorescent antibody techniques, using both pancreatic enzymes and pituitary hormones as well as myosin as antigens. Myosin was localized to the A band of muscle in this way. Such methods are giving most important results at the present time in cell contractile systems. These endeavors, carried out prior to gel electrophoresis involved tremendous amounts of work purifying proteins, preparing antibodies, coupling the markers and using gentle methods of tissue preservation, especially freeze dry procedures. He was a prodigious worker, keeping long hours and delegating little. He was then working on a new method for purifying myosin using lithium chloride which Susan Lowey came down to discuss and subsequently found valuable before the ion exchange method came into widespread use. He was also beginning to consider various models of myosin packing into thick filaments. This interest and the discussions it raised, stimulated Frank Pepe, a postdoctoral fellow at the time, who later developed the first detailed model of the thick filament (see this volume). He stimulated and helped Howard Holtzer and William Telfer to use fluorescent antibodies for developmental problems. In relation to the antibody work, Marshall was much concerned with problems of antigenic purity frequently pointing out that a trace contaminant could be a more potent antigen than the major component of a preparation. This has turned out to be a very real hazard, and has only been resolved by the sensitivity of acrylamide gels for both analysis and preparation.

John liked to be unconventional in approach. In a second line of work, he was trying to find new ways of obtaining plasma membranes. He used the giant amoeba which then bore the picturesque name *Chaos chaos*, now more subdued as *Chaos carolinensis*. In doing this, he prepared cytoplasmic fractions by centrifuging prechilled amebae without any homogenization at all. This resulted in the cells releasing a cytoplasmic "sol" fraction. This approach was taken up by Wolpert and collaborators (Wolpert *et al.*, 1964) to obtain the cytoplasmic fraction which would stream *in vitro* when warmed with ATP, and was later used by Pollard and Ito (1970) in their classic extension of that work. John had done some elegant quantitative work with H. Holter in Denmark on cell "drinking" or pinocytosis using fluorescent labelled proteins (cf. Holter, 1959). Subsequently, he and his student, Philip Brandt, used the same fluorescent labeled gamma globulin to demonstrate the elaborate channels formed during the process and the fate of the vesicles (Marshall *et al.*, 1959). I found an old quotation from this article which seems quite relevant now.

> "It appears that the drawing in of channels and the pinching off of channels or droplets from the plasmalemma depends on the attachment of the plasmalemma at many points to the underlying gel. Engulfment seems to follow directly from the normal process of gel contraction and sol extrusion. If this interpretation is correct, the crucial point in the pinocytosis mechanism occurs in the first stage, when the binding of proteins or other ions alters the physical state of the plasmalemma." (Marshall *et al.*, 1959)*

Some years later, Lucy Comly (1973) showed elegant examples of actin abutting on the plasma membrane in briefly glycerinated *Chaos* treated with heavy meromyosin, suggesting that these could represent the attachments of membrane to gel. How actin associates with membranes is now an active current problem.

**Quotation with the permission of the New York Academy of Sciences.*

A third area where Marshall's views are relevant now is in electron microscopy. He saw its tremendous potential, but oscillated from great enthusiasm to skepticism because of the possibility of artifacts, especially artifacts of fixation or movement of material during the fixation process. He emphasized the necessity for using tracers whenever the directionality of a process was being examined. He encouraged Carl Feldherr to explore and use colloidal gold treated with protein as a tracer to study nuclear pores. For the study of endocytosis, we decided to use ferritin as a protein tracer because individual molecules can be seen by electron microscopy, and to use frozen-dried amebae to minimize distortion and displacement during fixation. Unfortunately, the watery free-living amebae are refractory to many methods, including freeze drying. The structural results were disappointing: one could not make out any details. However, Alan Laties was often there working on freeze drying methods of preserving retinas for cytological work, and happily the method was of great value for his studies as described elsewhere in this volume. In our work, reluctantly on Marshall's side, and I have to admit more enthusiastically on mine, we turned to chemical fixation and obtained micrographs of amebae fixed briefly in osmium at pH 6.0, or in glutaraldehyde and osmium. We began to see some evidence for the presence of cytoplasmic filaments in these cells (Fig. 1), although the strands were delicate and were only seen in a small percentage of sections, especially those from the tails or uroid. Because Marshall was concerned with fixation artifacts, problems which are most relevant in the fixation of thin filaments today, he was only convinced that the structures we were seeing in sections had some relation to reality when we used the same fixatives on ferritin treated amebae in the following comparison approach.

Ferritin, as is well known, is a visible red protein with an iron dense core which makes it electron opaque, and it has an isoelectric point of about 4.4. Amebae do not object to pH 4.0, at least for short times. If one treats the cells with ferritin at pH 4 and 5°C, ferritin will adhere to the extensive negatively charged surface coat first shown by Pappas (1959), (Fig. 2) and if the amebae are rinsed at pH 4, much of the ferritin remains bound to the coat. When the amebae are warmed they remain in the rounded state and endocytose the bound ferritin or may form "caps" of ferritin. But if the pH is raised to 7 while the cells are still cold, the ferritin diffuses off the coat or glycocalyx into the solution. If such a cell is warmed it exhibits normal movement and does not endocytose or "cap." We prepared a modified derivative, methylated ferritin, by esterifying the carboxyls (Nachmias and Marshall, 1961). This modified protein is very insoluble above pH 4.5, remaining bound to the ameba surface even when the cells are washed at

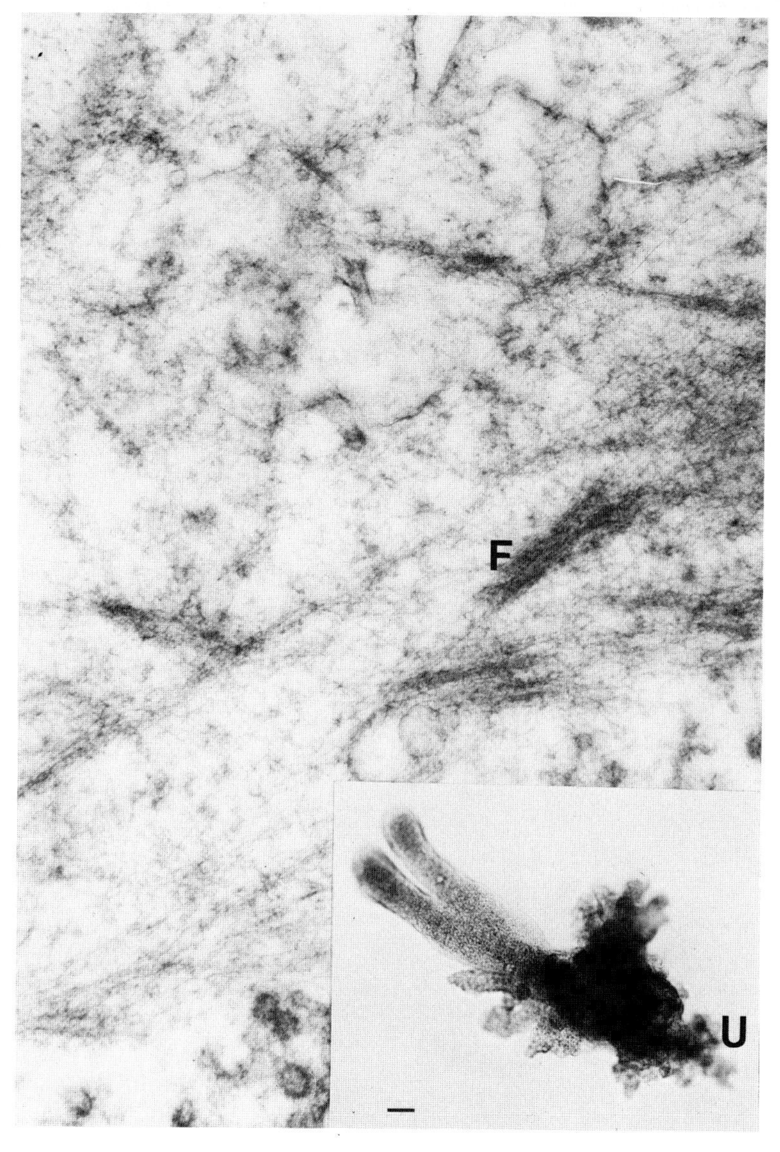

FIGURE 1. Cytoplasmic filaments (F) in a normally locomoting specimen of Chaos carolinesis. *Fixed in 2.5% glutaraldehyde buffered at pH 7 with 0.2* M *phosphate and postfixed in 2% osmium tetroxide for 10 minutes. The cell surface is at the upper left corner. (Reprinted from the* Journal of Cell Biology, 38, *p. 43, 1968 by courtesy of the Rockefeller University Press). x50,000. Mark is 1 mm. The insert is a light micrograph of a locomoting ameba fixed in 2% osmium with no distortion of shape; the block used for Fig. 1 was taken from a similar specimen from part of the uroid (U). x40. Mark is 0.1 µm.*

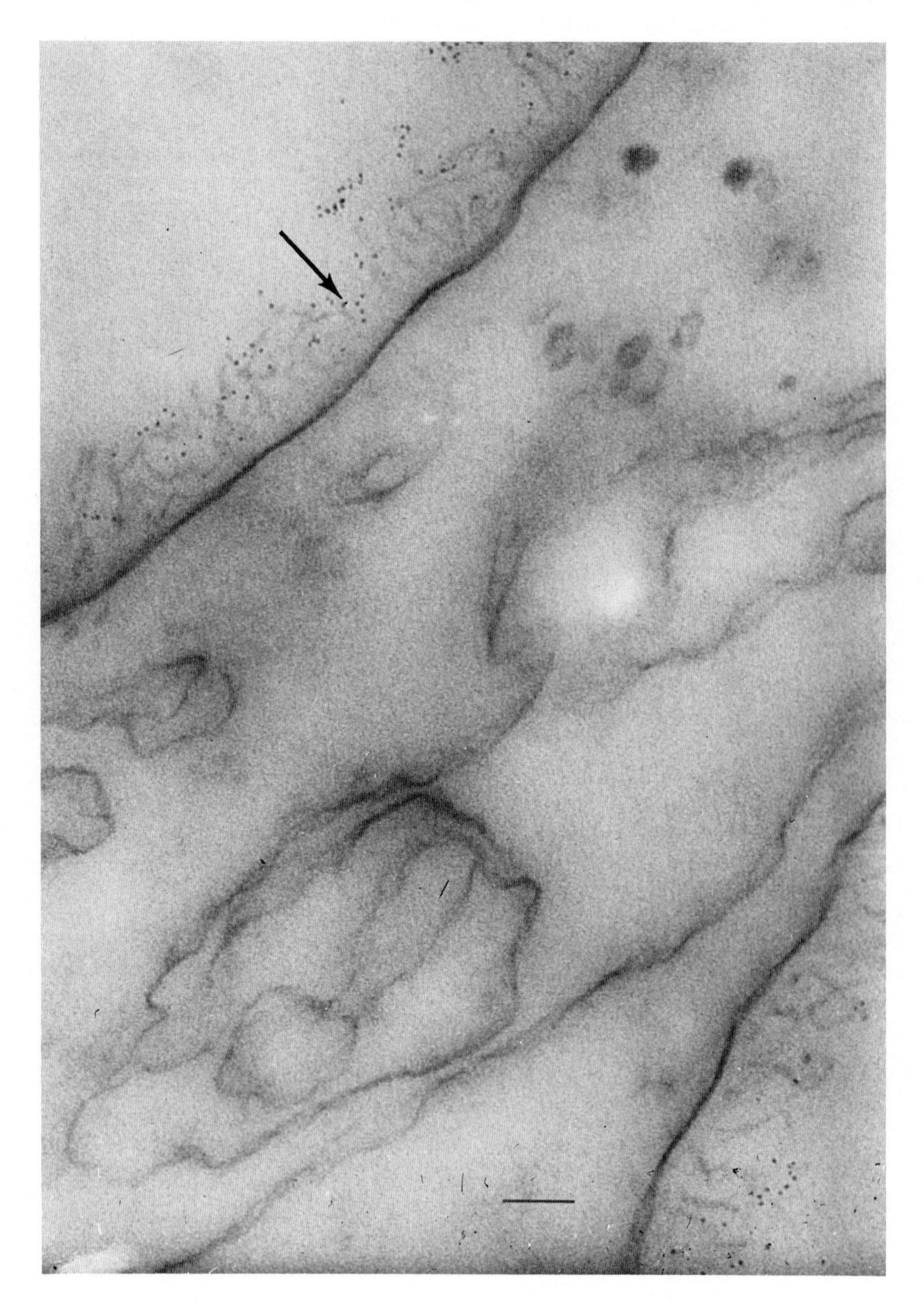

FIGURE 2. Portion of a narrow extension of Chaos to show some ferritin (arrow) still attached to the cell coat after a very brief washing at pH 7.0. x100,000. Mark is 0.1 μm.

pH 7 in the cold. On subsequent warming the protein is ingested (Fig. 3). Amebae were immersed in each protein solution at the same concentrations at pH 4.0 and 5°C, washed at the same pH so that only bound protein remained, and then warmed to room temperature at pH 4. Endocytosis occurred, and cells were fixed for eletron microscopy. Ingested ferritin was found to come free from the surface coat within the endocytic vacuole, (Fig. 4) while the methylated protein remained attached to the released surface material and eventually formed remarkable condensed structures of protein still bound to cell coat within the vesicle (Nachmias and Marshall, 1961). Thus the behavior of internalized protein as observed after fixation paralleled the external behavior of the protein assuming, as was reasonable from other data, that the pH inside the vesicle rose above pH 4.0 after ingestion. Neither ferritin nor methylated ferritin escaped from the vesicle into the cytoplasm nor into many small vesicles that appeared to bud off from (or to fuse with) the main vacuole (Fig. 4). This comparison of tracers and the cell's discrimination, shown after fixation, helped to convince us that electron microscope images of fixed cells could indeed be taken seriously even in the watery free-living amebae. Therefore, Marshall was willing to believe that the more substantial images of thick and thin filaments which I later found in endocytosing amebae treated with various inducers (Nachmias, 1964; Marshall and Nachmias, 1964) had in fact a close relation to actual assemblages.

There were, however, two problems in going directly to the view that the thick, short filaments and the thin long ones (Fig. 5) were similar to myosin and actin filaments of muscle. One was that from the mass cultures of amebae Marshall had isolated what looked like microtubule structures (Fig. 6)--although the widths were greater than the usual microtubules, being 33 nm. Microtubules were at that time the most popular candidates for a role in cytoplasmic streaming. Ours even altered in Mg ATP, exhibiting swelling. We reported on the presence of these structures (Marshall and Nachmias, 1964) but found it difficult to see how they related to the thick and thin filaments seen in individual cells. Was there some kind of structural conversion? Had we missed the tubules in the single cells? Dr. Myron Ledbetter examined our micrographs and said that they were not true microtubules. He was right. Puzzling over these remarkable "macro" tubules in subsequent years, in which I did not find anything resembling them in cells fixed in a variety of ways, including double fixation, I finally found the answer in a report on bacterial rhapidosomes (Baechler and Berk, 1972). Our structures were identical in size and shape, including the characteristic rounded ends, to the structures they isolated from *Pseudomonas aeruginosa*. The amebae fed on paramecia which fed on bacteria. According to Baechler and Berk, the rhapidosomes have a

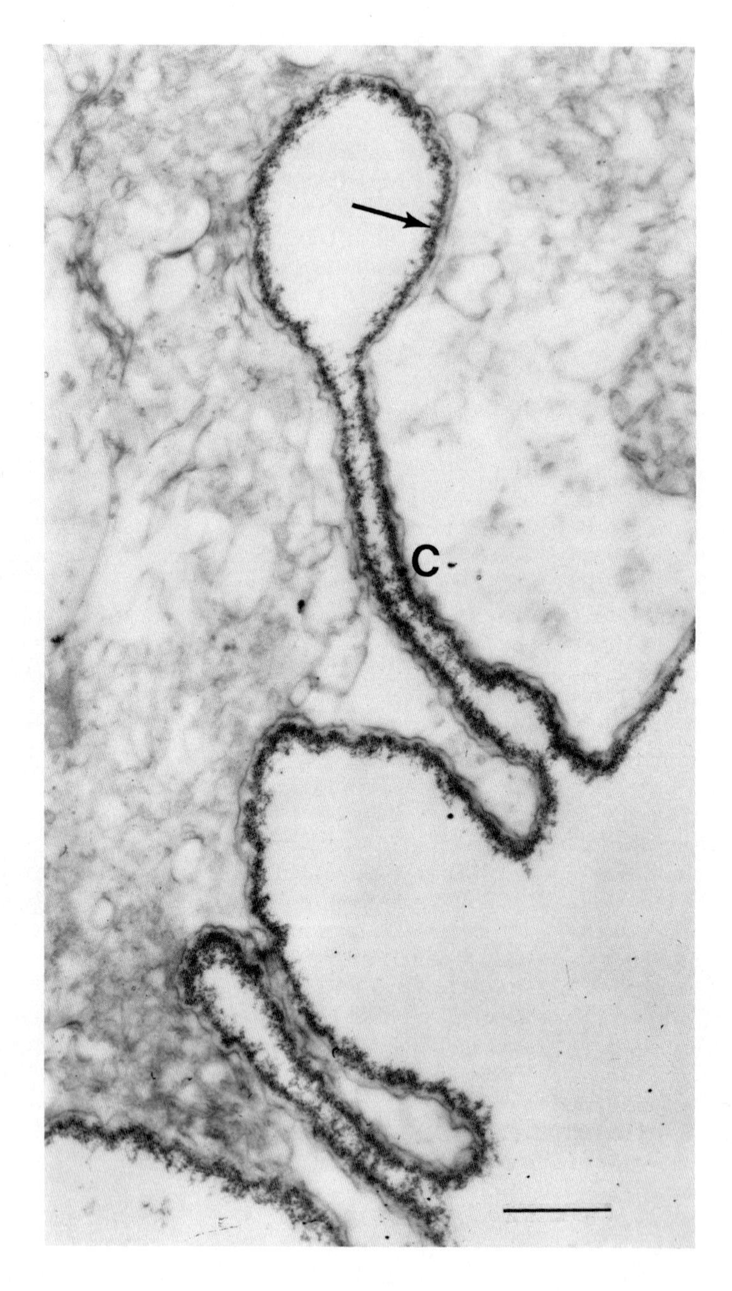

FIGURE 3. Ameba beginning to take methylated ferritin bound to the cell coat (arrow). Note the channel (C) about 4 μm long. Micrograph by John Marshall. x6,000. Mark is 1 μm.

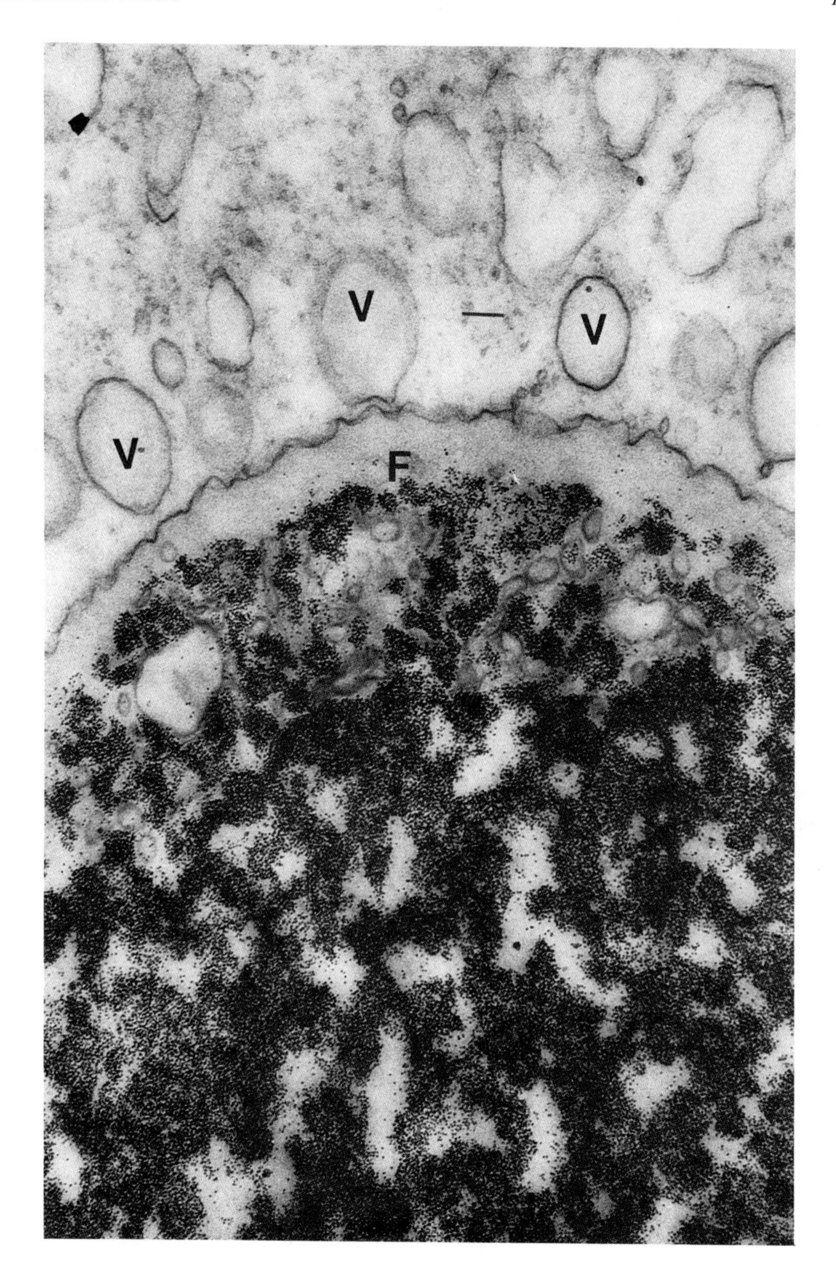

FIGURE 4. *Section through ferritin containing vacuole (F) 30 min after ingestion of ferritin. Coat filaments are released from the membrane. (Vesicles (V) close to the major vacuole uniformly exclude ferritin). Approximately ×60,000. Mark is 0.1 μm.*

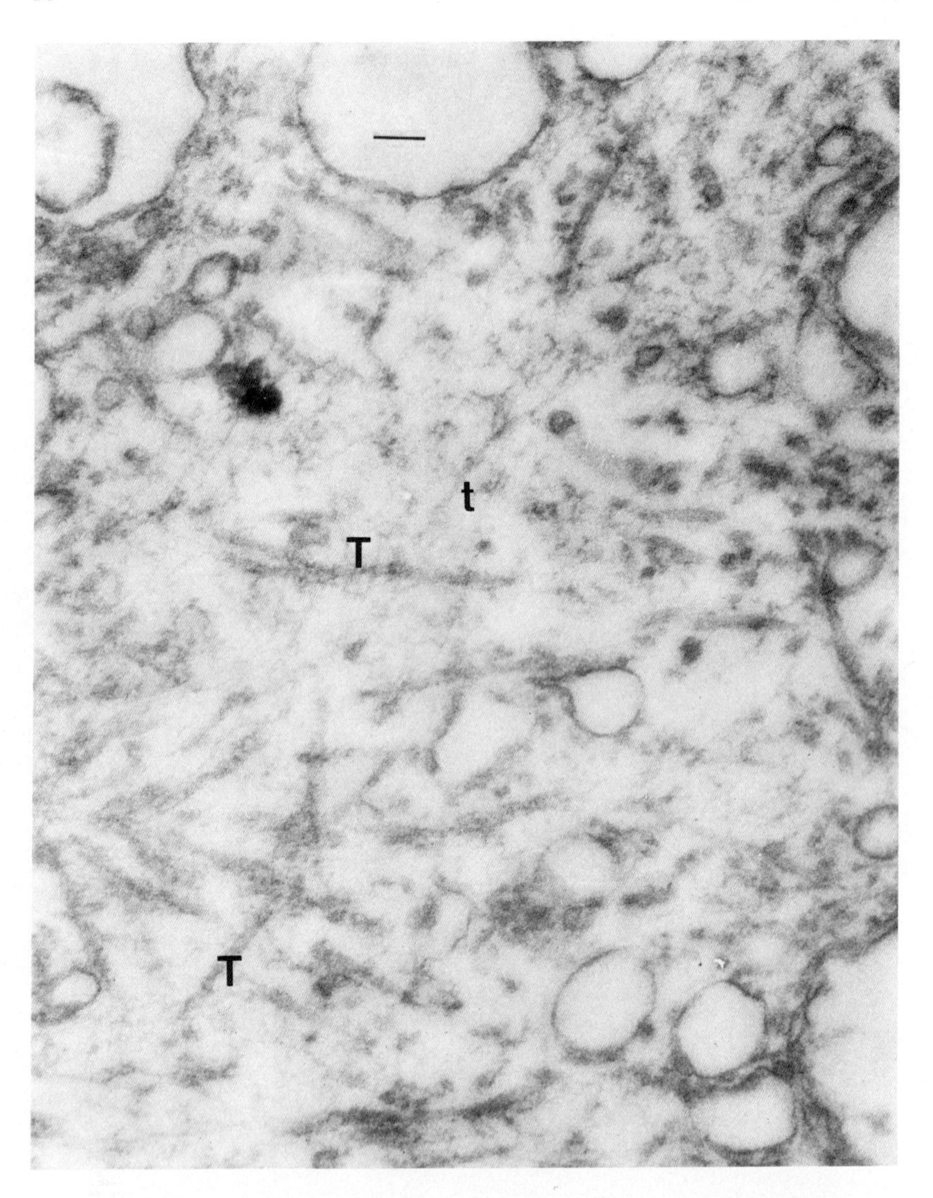

FIGURE 5. Section through region of ameba cytoplasm close to the surface after induction of endocytosis with protamine. Both thick (T) and thin (t) filaments are present. ×80,000. Mark is 0.1 μm.

tendency to stick to lipopolysachrides, and so the most likely explanation is binding during harvesting of the mass cultures. The giant microtubules were in fact red herrings.

The second reason for uncertainty about the thin and thick filaments was that we saw images of thick filaments which appeared to condense as if they were forming, perhaps *from* the thin filaments (Fig. 7). A few years later, Pollard and Ito (1970) demonstrated that the streaming which occurred *in vitro* in crude preparations of amebae cytoplasmic sols, did not take place when the thick filaments were removed by centrifugation. They concluded that the thick and thin filaments had to interact to cause streaming. Although this has not been proven directly, it is still the most fruitful and best interpretation of the data, and pointed directly to the conclusion that the two types of filaments probably represented two different proteins. This deduction seemed compelling, but I was not totally convinced of the filament composition until I succeeded in purifying myosin from another organism with vigorous cytoplasmic

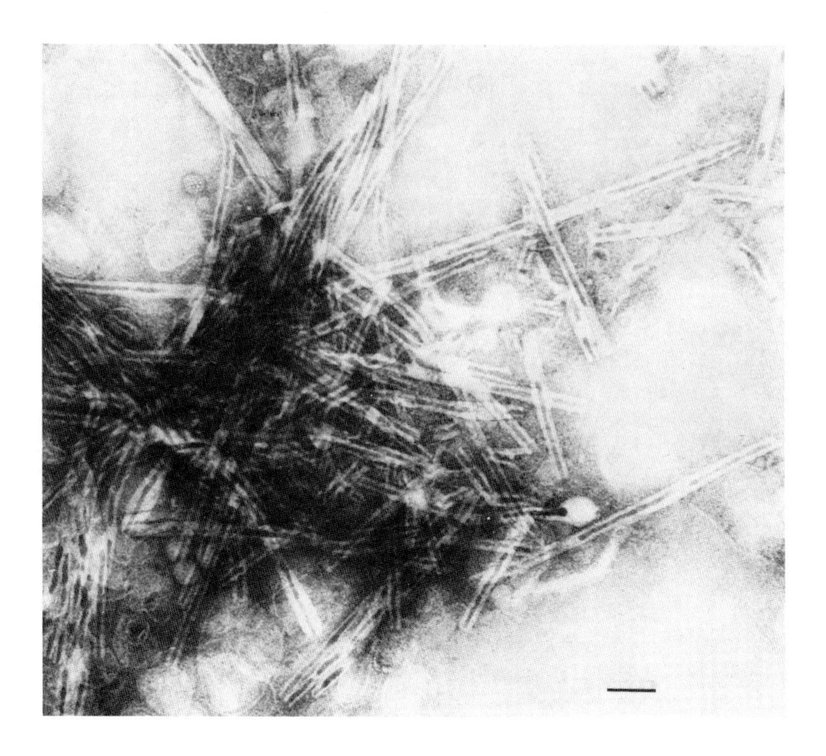

FIGURE 6. Negatively stained sample of tubular structures seen isolated from mass cultures of Chaos carolinensis. *Width of tubules is 33 nm, internal dimension about 10 nm, 1% sodium phosphotungstate, pH 7.5. ×68,000. Mark is 0.1 μm.*

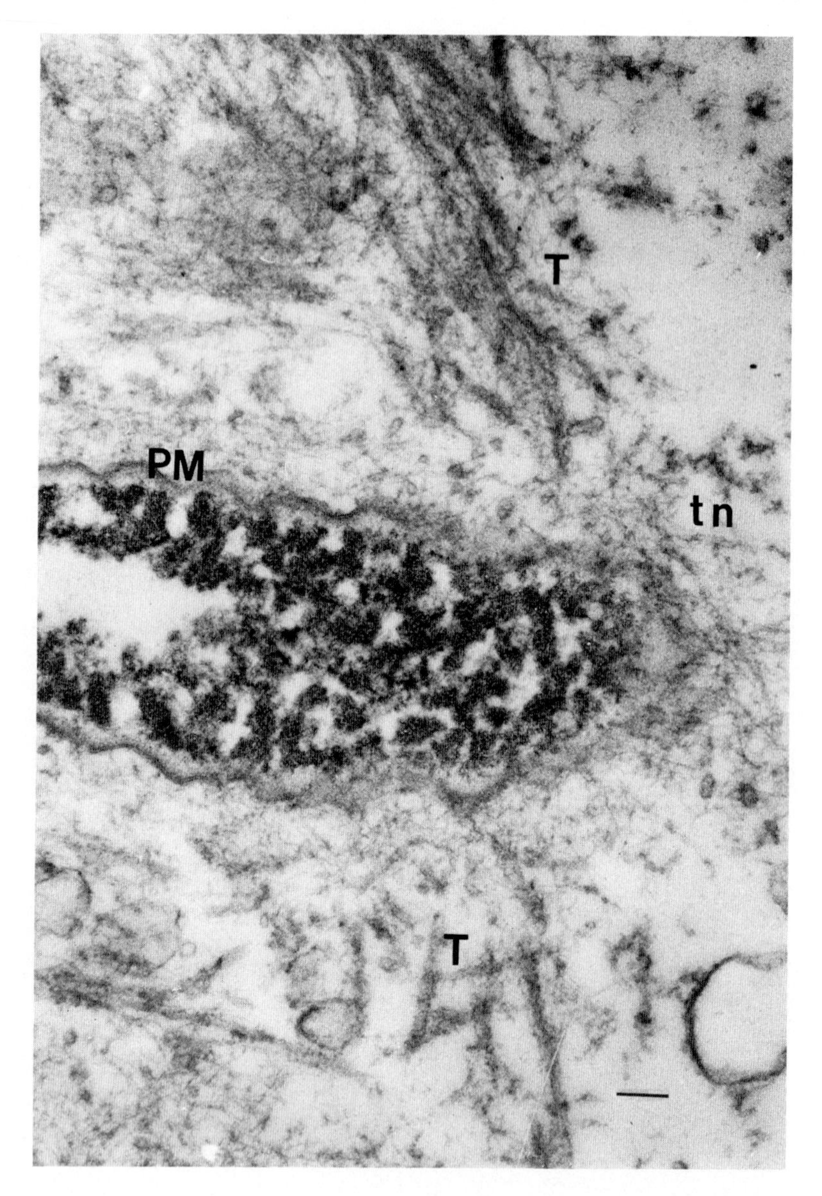

FIGURE 7. Cross section of a region of cytoplasm just subjacent to the plasma membrane (PM). Ameba has been treated with 1% Alcian blue to induce pinocytosis (endocytosis). Next to this channel a thin filament network is found (tn) while deeper in the cytoplasm arrays of filaments can be seen which appear to be condensing into thick (T) filaments. ×80,000. Reprinted from J. Histochemistry and Cytochemistry 13, *p. 100, 1965 by courtesy of the Williams & Wilkins Co. Mark is 0.1 μm.*

streaming and found that it could form bipolar filaments of the same dimensions seen in the fixed amebae specimens some years before (Nachmias, 1972a).

The thick filaments that we observed then appear to be very similar to those formed by nonmuscle myosins isolated from mammalian cells as well as other amebae, eggs and even sperm (Adelstein *et al.*, 1972; Stossel and Pollard, 1973; Hinssen and d'Haese, 1974; Niederman and Pollard, 1975; Clarke and Spudich, 1974; Condeelis, 1977; Mabuchi, 1976a,b; Kobayashi *et al.*, 1977) although the myosin from some of the cells form filaments *in vitro* which are thinner and shorter than 0.45-0.5 μm, the length found in fixed amebae. But lengths of such filaments depend on ionic conditions (Clarke and Spudich, 1974; Condeelis, 1977) and the filaments from different cell types appear all to have a basic bipolar structure. (In some instances assembly into long filaments with axially shifted bipolar units is found, Hinssen and d'Haese, 1974). Recently a myosin forming bipolar filaments has also been found and purified from the soil ameba *Acanthamoeba* and named myosin II (Maruta and Korn, 1977 and Pollard, this volume). This ameba earlier appeared to have a quite different form of myosin (myosin I) that is soluble even at low ionic strength and can interact with actin and have its enzymatic activity increased in the presence of a cofactor (Pollard and Korn, 1973) or when phosphorylated (Korn, this volume). Quite recently a report has appeared of a myosin from *Nitella* which also forms bipolar thick filaments (Kato and Tonomura, 1977). This is most interesting since the type of cytoplasmic streaming demonstrated by this alga is a continuous rotation in one direction unlike that found in amebae or other animal cells. The bipolar myosins when mixed with actin in solution form arrowhead structures showing a polar attachment as Huxley first found with muscle myosin (1963) and this has been found in all cases which have been tested (cf. Nachmias, 1972b; Clarke and Spudich, 1974; Niederman and Pollard, 1975; Mabuchi, 1976a; Condeelis, 1977). *Acanthamoeba* myosin I is thus far unique in forming nonpolar attachments, while myosin II does form polar attachments (Pollard, this volume). It will be most interesting to understand why *Acanthamoeba* has two types of myosin, whether both are involved in tension production, and whether other cells contain a type I myosin.

Although the nature of the filaments has been considerably clarified, another question which concerned John Marshall in the 1960s is still puzzling. If the contractile mechanisms of amebae (and other cells) are basically similar to those in muscle, to what is the labile nature of nonmuscle movement due? Is it a consequence of lack of stable sarcomere structures alone, such as Z-lines and M proteins? We can now ask more specifically than he could whether filaments (of myosin or actin) polymerize and depolymerize during movements or stimulation of pinocytosis. Can one end of a myosin filament be active while the other end is inactive and can this give rise to streaming? Fountain streaming can be seen in both amebae and in caffeine derived membrane bound sacs of cytoplasm from *Physarum* and yet both *Physarum* and amebae can be induced to make threads which contract (Kamiya, 1968; Taylor *et al.*, 1973). Recently Ms. C. Meyers in my laboratory has also observed pure rotation of contracted endoplasm within an ectoplasmic shell in caffeine droplets reactivated with Mg ATP, low levels of calcium and a lytic agent (Fig. 8). The rotation is accompanied by protrusions of ectoplasm which are seen more commonly than the rotation itself. Although the movements are not vigorous, they show that even in an organism whose cytoplasm is capable of contracting when oriented as a thread, a movement similar to rotational streaming can be observed when it is prepared in another way.

After this digression to events which at least have clarified the nature of the filaments, let me return finally to two more of John Marshall's ideas. He was both old and new fashioned in considering the role of calcium central, and ionic gradients as relevant. Work done with David Bruce showed that potential and presumably, therefore, ionic gradients existed between different parts of the amebae cytoplasm. These were thought to be related to the gel and sol states. Marshall thought that calcium was a trigger for the endocytic process as well as playing a central role in the control of movement, an idea for which Taylor *et al.* (1973) have provided direct and fascinating evidence in their demembranated amebae models in which calcium levels control the resting, contracting and "flare" states, while Hatano (1970) and Matthews (1977) have presented evidence for calcium control of motility in the caffeine derived droplets of *Physarum*. Marshall thought that endocytosis at least in amebae resulted from a stimulus that allowed attachment of the membrane to the underlying gel, and that this was also a calcium effect; we do not yet know if this is correct. Since calcium appears to break down gels made *in vitro* from cytoplasmic sols a different, current view is that such breakdown leads to contractility by releasing structural rigidity to allow actomyosin interactions to occur.

Finally, alluded to in Dr. Flexner's talk was Marshall's speculation that membranes could form or unfold from the cytoplasm without vesicles fusing with the cell membrane. My attempt to show this in amebae with labelled surfaces was suggestive (Nachmias, 1966). Lewis Tilney has good evidence in favor of this idea when membrane forms in the extrusion of the acrosomal process (Tilney, 1977).

As we continue through this conference honoring this intense and idealistic scientist, these selected examples show that Marshall could be here and discover that though some

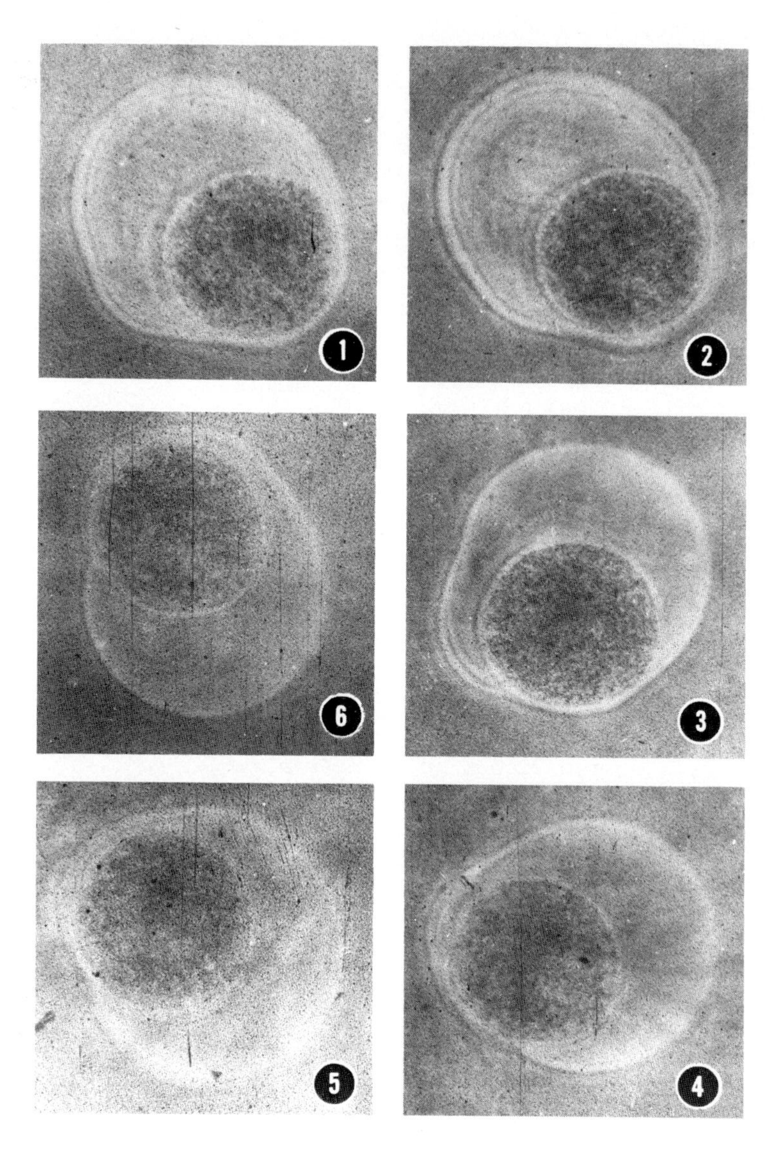

FIGURE 8. Individual frames from a 16 mm film taken at ×100 magnification. Frames show stages in the rotation of contracted endoplasm within the ectoplasmic shell of a caffeine droplet from plasmodium of Physarum. *The membrane bound droplet was suspended in an "activation solution" containing 27mM KCl, 3 mM NaCl, 5 mM EGTA, 5 mM $MgCl_2$, 2.5 mM $CaCl_2$, 0.5 mM Mg ATP, 5 mM HEPES buffer pH 7.0, 0.02 mM A 23187 ionophore, 0.015% Saponin. Movement began within 2-3 min after suspension. The contracted endoplasm moved 0.68 μ/sec around the circumference taking 43 sec for the complete route.*

answers are clearer, debate still continues on many points that he thought about 15 and 20 years ago. It continues with more specific questions about the proteins that seem to be involved in cell motility. Pursuing these questions, and the fact that this conference is taking place are the best tributes we can pay John Marshall and the influence he had on several of us. Those who knew him cannot forget his enthusiasm, ingenuity, and originality and his deep interest in the mechanisms cells use to move themselves or their environment.

ACKNOWLEDGMENT

I would like to thank Ms. Cynthia Meyers for her work with caffeine droplets and Mr. Jeffrey S. Sullender for expert help with photography. A part of this work was supported by Grant #AM-17492 from the National Institutes of Health.

REFERENCES

Adelstein, R. S., Conti, M. A., Johnson, G. S., Pastan, I., and Pollard, T. D. (1972). Isolation and characterization of myosin from cloned mouse fibroblasts, *Proc. Nat. Acad. Sci. U.S.A. 69*, 3693-3697.

Baechler, C. A., and Berk, R. S. (1972). Ultrastructural observations of *Pseudomonas aeruginosa:* rhapidosomes, *Microstructures 3*, 24-31.

Clarke, M., and Spudich, J. A. (1974). Isolation and characterization of myosin from amoebae of *Dictyostelium discoideum, J. Mol. Biol. 86*, 209-222.

Comly, L. T. (1973). Microfilaments in *Chaos carolinensis*, *J. Cell Biol. 58*, 230-237.

Condeelis, J. S. (1977). The self assembly of synthetic filaments of myosin isolated from *Chaos carolinensis* and *Amoeba proteus, J. Cell Sci. 25*,387-402.

Hatano, S. (1970). Specific effect of calcium on movement of plasmodial fragment obtained by caffeine treatment, *Expt. Cell Res. 61*, 199-203.

Hinssen, H., and D'Haese, J. (1974). Filament formation by slime mold myosin isolated at low ionic strength, *J. Cell Sci. 15*, 113-129.

Holter, H. (1959). Problems of pinocytosis, with special regard to amebae, *Ann. N. Y. Acad. Sci. 78*, 524-437.

Huxley, H. E. (1963). Electron microscope studies on the structure of natural and synthetic protein filaments from striated muscle, *J. Mol. Biol. 7*, 281-308.

Kamiya, N. (1968). The mechanism of cytoplasmic movement in a myxomycete plasmodium *in* Aspects of Cell Motility, *Symposia Soc. Exp. Biol. 23,* 199-214.

Kato, T., and Tonomura, Y. (1977). Identification of myosin in *Nitella flexilis, J. Bioch. 82,* 777-782.

Kobayashi, R., Goldman, R. D., Hartshorne, D. J., and Field, J. B. (1977). Purification and characterization of myosin from bovine thyroid, *J. Biol. Chem. 252,* 8285-8290.

Mabuchi, I. (1976a). Myosin from starfish egg: properties and interaction with actin, *J. Mol. Biol. 100,* 569-582.

Mabuchi, I. (1976b). Isolation of myosin from starfish sperm heads, *J. Bioch. Tokyo 80,* 413-415.

Marshall, J. M., Schumaker, V. N., and Brandt, P. W. (1959). Pinocytosis in amoebae, *Ann. N. Y. Acad. Sci. 78,* 515-523.

Marshall, J. M. and Nachmias, V. T. (1964). Cell surface and pinocytosis, *J. Histochem. Cytochem. 13,* 92-104.

Maruta, H., and Korn, E. D. (1977). Purification from *Acanthamoeba castellanii* of proteins that induce gelation and syneresis of F-actin, *J. Biol. Chem. 252,* 399-402.

Matthews, L. M. (1977). Ca^{++} regulation in caffeine-derived microplasmodia of *Physarum polycephalum, J. Cell Biol. 72,* 50 -50 .

Nachmias, V. T. (1964). Fibrillar structures in the cytoplasm of *Chaos chaos, J. Cell Biol. 23,* 183-188.

Nachmias, V. T. (1966). A study by electron microscopy of the formation of new surface by *Chaos chaos, Exp. Cell Res. 43,* 583-601.

Nachmias, V. T. (1972a). Filament formation by purified *Physarum* myosin, *Proc. Nat. Acad. Sci. U.S.A. 69,* 2011-2014.

Nachmias, V. T. (1972b). Electron microscope observations on myosin from *Physarum polycephalum, J. Cell Biol. 52,* 648-663.

Nachmias, V. T., and Marshall, J. M. Jr. (1961). Protein uptake by pinocytosis in amoebae: Studies on ferritin and methylated ferritin, *In* "Biological Structure and Function Proceedings of the First IUB/IUBS International Symposium" (T. W. Goodwin and O. Lindberg, eds.), Vol. 11, pp. 605-616. Academic Press, New York.

Niederman, R., and Pollard, T. D. (1975). Human platelet myosin II. *In vitro* assembly and structure of myosin filaments, *J. Cell Biol. 67,* 72-92.

Pappas, G. D. (1959). Electron microscope studies on amoebae, *Ann. N. Y. Acad. Sci. 78,* 448-473.

Pollard, T. D., and Ito, S. (1970). Cytoplasmic filaments of *Amoeba proteus* 1. The role of filaments in consistency changes and movement, *J. Cell Biol. 46,* 267-289.

Pollard, T. D., and Korn, E. D. (1973). *Acanthamoeba* myosin. Interaction with actin and with a new cofactor protein required for actin activation of Mg^{2+} adenosine triphosphatase activity, *J. Biol. Chem. 248*, 4691-4697.

Stossel, T. P., and Pollard, T. D. (;073). Myosin in polymorphonuclear leukocytes, *J. Biol. Chem. 248*, 8288-8294.

Taylor, D. L., Condeelis, J. S., Moore, P. L., and Allen, R. D. (1973). The contractile basis of amoeboid movement I. The chemical control of mobility in isolated cytoplasm, *J. Cell Biol. 59*, 378-394.

Tilney, L. G. (1977). Actin: its association with membranes and the regulation of its polymerization, *In* "International Cell Biology" (B. R. Brinkley and K. R. Porter, eds.), pp. 388-402. Rockefeller University Press, New York.

Wolpert, L., Thompson, C. M., and O'Neill, C. H. (1964). Studies on the isolated membrane and cytoplasm of *Amoeba proteus* in relation to ameboid movement, *In* "Primitive Motile Systems in Cell Biology" (R. D. Allen and N. Kamiya, eds.), pp. 143-171 Academic Press, New York.

Motility in Cell Function
Proceedings of the First John M. Marshall Symposium in Cell Biology

RECENT ADVANCES IN THE MYOSIN AND ACTOMYOSIN ATPase MECHANISMS

Michael A. Geeves,[1] *C. Fred Midelfort,*[2]
David R. Trentham, and Paul D. Boyer[3]

Department of Biochemistry and Biophysics
University of Pennsylvania
School of Medicine
Philadelphia, Pennsylvania

INTRODUCTION

In recent years considerable advances have been made in elucidation of the Mg^{2+}-dependent myosin and actomyosin ATPase mechanisms using transient and steady-state kinetic methods (Trentham *et al*, 1976; Taylor 1978). In the case of frog muscle there is good correlation between the myosin and actomyosin ATPase steady-state activities and the rates of ATP hydrolysis in resting and maximally working muscle, respectively, so that the generally accepted physiological function of the ATPases appears correct (Ferenczi *et al*, 1978).

The Mg^{2+}-dependent ATPase can be described by a seven-step mechanism eq. (1).

$$\mathrm{M + ATP} \overset{1}{\rightleftharpoons} \mathrm{M.ATP} \overset{2}{\rightleftharpoons} \mathrm{M^*.ATP} \overset{3}{\rightleftharpoons} \mathrm{M^{**}.ADP.P_{\mathit{i}}} \overset{4}{\rightleftharpoons} \mathrm{M^*.ADP.P_{\mathit{i}}} \overset{5}{\rightleftharpoons} \mathrm{M^*.ADP + P_{\mathit{i}}} \overset{6}{\rightleftharpoons} \mathrm{M.ADP} \overset{7}{\rightleftharpoons} \mathrm{M + ADP} \tag{1}$$

[1]*Present address: Division of Natural Sciences, Thimann Laboratories, University of California, Santa Cruz, California 95064.*

[2]*Present address: Department of Biochemistry, The Albert Einstein School of Medicine, Bronx, New York.*

[3]*Present address: Molecular Biology Institute, University of California, Los Angeles, California 90024.*

ISBN 0-12-551750-5

k_{+i} and k_{-i} are forward and reverse rate constants and K_i $(=k_{+i}/k_{-i})$ the equilibrium constant of the ith step. Relevant constants for the purpose of this paper are shown in Table I. The rate determining step of the ATPase is controlled by k_{+4}. Based on the work of Taylor and his colleagues (White and Taylor, 1976) the actomyosin ATPase in its simplest terms appears to be an extension of the myosin ATPase with actin involved in each step except for that in which ATP is cleaved (eq. (2)).

$$\mathrm{AM + ATP} \underset{5}{\overset{1}{\rightleftharpoons}} \mathrm{AM.ATP} \overset{2}{\rightleftharpoons} \mathrm{A + M^{*}.ATP} \overset{3}{\rightleftharpoons} \mathrm{M^{**}.ADP.P_{\mathit{i}} + A} \overset{4}{\rightleftharpoons}$$

$$\mathrm{AM.ADP.P_{\mathit{i}}} \rightleftharpoons \mathrm{AM + ADP + P_{\mathit{i}}} \quad (2)$$

With the exception of step 3, equilibrium constants of successive steps in the two mechanisms differ due to the presence of actin which has a high affinity for myosin. The rate limiting step of the myosin ATPase is bypassed in the actomyosin ATPase leading to three orders of magnitude greater catalytic center activity of the latter. Lymn and Taylor (1971) proposed a scheme (Fig. 1) in which elementary steps of the actomyosin ATPase are related to cycling of cross-bridges. This model can be extended to include more recently identified steps of

TABLE I. Equilibrium Constants, K_i, and Forward Rate Constants, k_{+i}, of the Myosin ATPase Mechanism in Terms of Equation 1

K_i		k_{+i} (sec^{-1})	
K_1	$4.5 \times 10^3\ M^{-1}$		--
K_2	7.2×10^7	k_{+2}	400
K_3	9	k_{+3}	$\geqslant 160$
		k_{+4}	0.06

The second-order association rate constant, k_{+1} is considered to be large, $> 10^7\ M^{-1}\ sec^{-1}$ (i.e., the bimolecular process is diffusion controlled). These data were obtained using both heavy meromyosin and subfragment 1 in a solvent (with minor variations in different laboratories) of 5 mM $MgCl_2$, 100 mM KCl, and 50 mM tris at pH 8.0 and 21°C (Trentham et al., 1976; Goody et al., 1977).

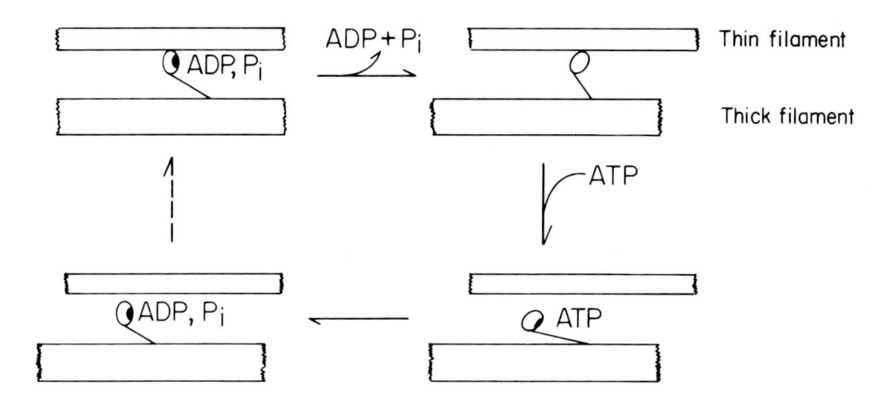

FIGURE 1. Biochemical events of the actomyosin ATPase linked to the cross-bridge cycle.

the actomyosin ATPase. In addition the work of Hill (1974, 1977) allows one to accommodate formally into this scheme continuous sliding of actin and myosin filaments past one another and energy transduction associated with performance of mechanical work.

As indicated in eqs. (1) and (2) the myosin and actomyosin ATPase mechanisms are postulated to have a common step--the cleavage of ATP. This hypothesis is supported by measurements of the equilibrium constants of the cleavage step in the two ATPases which were found to be equal (Eccleston *et al*, 1975). The work described here examines the cleavage step further. First, the rate constant of the cleavage step is measured, and secondly the nature of the cleavage reaction is explored. Experiments were carried out using myosin subfragment 1 prepared from rabbit skeletal muscle as described by Bagshaw and Trentham (1973).

THE KINETICS OF THE ATP CLEAVAGE REACTION

The difficulty in evaluating k_{+3} and k_{-3} in the case of myosin arises because the rate constant of the previous step, k_{+2}, is smaller than or of the same order as k_{+3}. Thus, in general, when ATP is mixed with myosin, the rate of appearance of products is controlled wholly or in part by k_{+2}. An important reason for evaluating k_{+3} is the need to find out to what extent ATP cleavage is rate limiting in the actomyosin ATPase. If cleavage is rate limiting, it provides an explanation as to why actin and myosin are predominantly dissociated during actomyosin ATPase activity and suggests that M*.ATP is wholly or in part the "refractory state" described by Eisenberg *et al.* (1972).

Early evidence that the rate of ATP cleavage was slower than previously thought was provided unexpectedly in transient ^{18}O-experiments designed to test whether hydrolysis as opposed to phosphoenzyme or metaphosphate formation occurs in the cleavage step (Trentham, 1977). The principle of the experiment is illustrated in Fig. 2. The key point is that kinetic measurements begin after the protein isomerization (step 2 of eq. (1), and hence the experiment is not affected by the value of k_{+2}. ATP was mixed with excess subfragment 1 in a double-mixer quenched-flow apparatus (a modification of the instrument described by Gutfreund (1969)). ^{18}O-water was added at 117 msec when essentially all the nucleotide was bound to the protein as M*.ATP or M**.ADP.P_i. The reaction was then quenched with acid at various times. As shown by the scheme in Fig. 2 unlabelled M**ADP.P_i will disappear during the incubation with ^{18}O-water and the rate of its disappearance will be controlled by k_{-3}. In practice measurements record ^{18}O-incorporation into M**.ADP.P_i and M*.ATP. The results show (Table 2) that even after 35 msec significantly less than 1 ^{18}O-atom is incorporated per M**.ADP.P_i. Since according to Fig. 2 unlabelled M**.ADP.P_i decays exponentially it follows that k_{-3}<15 sec^{-1} and, since K_3 = 9, k_{+3}< 135 sec^{-1}. The inequality sign is introduced because additional ^{18}O may have been incorporated into M**.ADP.P_i following rearrangement of the P_i moiety as outlined in steps 3 and 4 of Fig. 2. The presence of ^{18}O in M*.ATP will indicate the extent of this "intermediate" exchange, hence the inclusion of the labelling of M*.ATP in Table 2. The experiment is difficult technically, particularly the measurement of ^{18}O in M*.ATP since the amounts of material for analysis are small. It appears that the labelling of M**.ADP.P_i exceeds that of M*.ATP so $k_{+3} \simeq 135$ sec^{-1} and is less than k_{+2} (= 400 sec^{-1}) under these reaction conditions, though the accuracy of the experiment is not high.

Another approach which permitted evaluation of k_{+3} was to carry out a single turnover experiment as described in Fig. 3. In this 5 μM ATP was mixed with 19 μM subfragment 1 and the time courses of the concentrations of free ATP, M*.ATP and M**.ADP.P_i were followed. The striking feature of the results is that the ratio of M*.ATP to M**.ADP.P_i varies throughout the course of the experiment and this means that k_{+3} is not all that much greater than the pseudo first-order constant, k, controlling the binding of ATP. Under the reaction conditions $k = K_1 k_{+2}$[subfragment 1] and has an observed value of 17 sec^{-1} when measured from the rate of free ATP depletion. The ratio of M**.ADP.P_i to M*.ATP at 200 msec gives K_3 = 7.

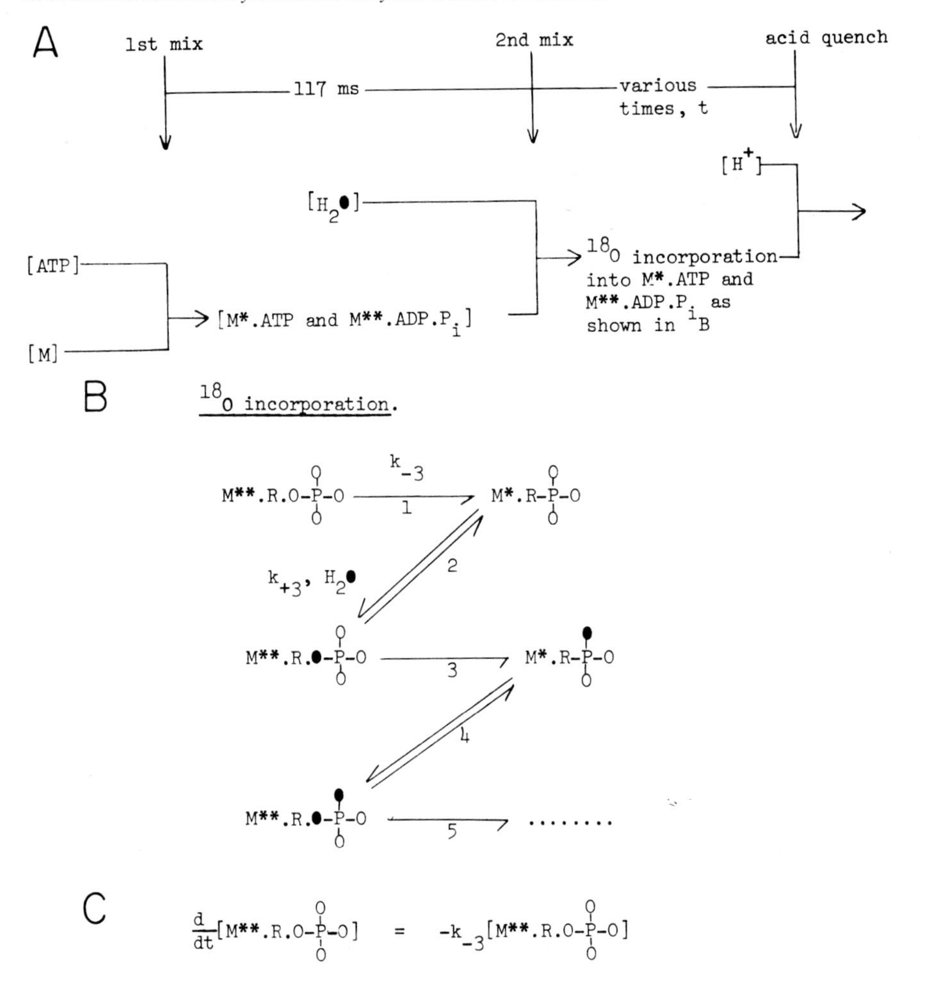

$$\frac{d}{dt}[M^{**}.R.O\text{-}P\text{-}O] = -k_{-3}[M^{**}.R.O\text{-}P\text{-}O]$$

FIGURE 2. Flow diagram showing the design of the ^{18}O-experiment to measure k_{-3} *together with a postulated mechanism showing how ^{18}O is incorporated into M*.ATP and M**.ADP.P_i. 172.5* μM *[$\gamma^{32}P$]ATP was added to 295 γM subfragment 1 in a double-mixer quenched-flow apparatus at 24°C in 50* mM *KCl, 2* mM *$MgCl_2$, and 20* mM *tris adjusted to pH 8 with HCl. At 117 msec 4.23 a.p.e. ^{18}O-water was added. The reaction mixture was quenched at various times into 0.4* M *$HClO_4$. The pH was adjusted to 4.8 with KOH/HOAc and carrier ATP and P_i were added. ATP and P_i were separated by the charcoal method, and the ratio of [$\gamma^{32}P$]ATP to [^{32}P]P_i was measured. The ^{18}O-content of P_i and of the γ-PO_3 group of ATP were analyzed as described by Bagshaw* et al. *(1975). Symbols: ATP, R-P-O; subfragment 1,* M.

*(A) The flow diagram. (B) Postulated mechanism of ^{18}O incorporation. (C) Rate equation showing that [M**.R.O-P-O] decays exponentially.*

TABLE 2. Extent of ^{18}O-incorporation into M.ATP and M**.ADP.P_i with time*

		Expected a.p.e. per oxygen atom labelled		Observed a.p.e.		Fraction of 1 oxygen atom labelled	
t, msec	K_3	M*.ATP	M**.ADP.P_i	M*.ATP	M**.ADP.P_i	M*.ATP	M**.ADP.P_i
4	5.5	0.048	0.162	0.000	0.010	0.00	0.06
15	5.6	0.048	0.168	0.004	0.028	0.08	0.17
35	6.0	0.043	0.154	0.003	0.030	0.07	0.19

*t records the time of quenching the reaction mixture after ^{18}O-water addition (Fig. 2). K_3 (= [M**.ADP.P_i]/([M*.ATP]) was evaluated from the ratio of ATP to P_i determined from ^{32}P analysis after charcoal separation and acid hydrolysis of ATP. 0.137, 0.140, and 0.119 μmol of ATP were isolated from M*ATP for ^{18}O-analysis in the three experiments. The limit of sensitivity in ^{18}O-measurements is 0.002 a.p.e., so the a.p.e. found in the ATP samples is barely significant above zero. (a.p.e. ≡ atom percent excess of ^{18}O above background of natural abundance of ^{18}O which is 0.2%.)*

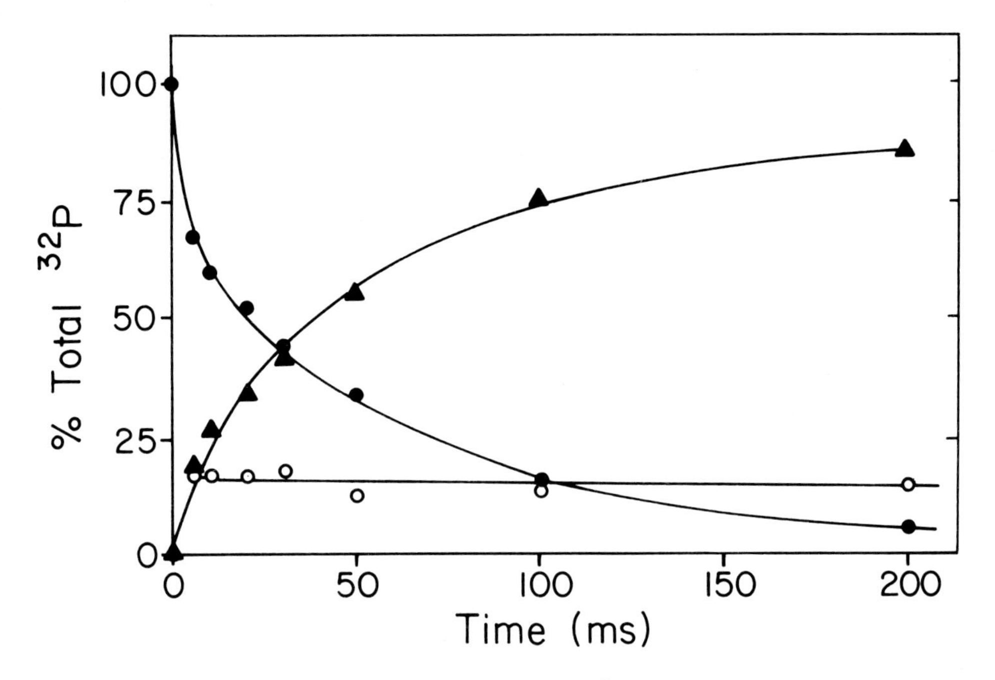

FIGURE 3. Intermediates of the Mg^{2+}-dependent subfragment 1 ATPase during a single turnover. 19 μM subfragment 1 was mixed with 5 μM [$\gamma^{32}P$]ATP at 20°C in 50 mM KCl, 5 mM $MgCl_2$, 50 mM tris adjusted to pH 8.0 with HCl in a quenched-flow apparatus. At the times indicated the reaction was quenched into 1N $HClO_4$ and the ratio of [$\gamma^{32}P$]ATP to [^{32}P]P_i measured as described by Bagshaw and Trentham (1973). From this the ratio of ([free ATP] + [M.ATP]) to M**.ADP.P_i was determined. The experiment was repeated except that at the indicated times 3.3 mM nonradioactive ATP was added to the reaction mixture. The reaction was allowed to continue for a further 60 sec and was then quenched with 1N $HClO_4$ and the ratio of [$\gamma^{32}P$]ATP to [^{32}P]P_i measured. M*.ATP present at the time of nonradioactive ATP addition was hydrolyzed in the 60 sec prior to $HClO_4$ addition and did not dissociate from the protein as ATP in view of the small value of k_{-2} (Table 1). Free [$\gamma^{32}P$]ATP was unable to bind to subfragment 1 because of competition by the 3.3 mM nonradioactive ATP, so measurement of [$\gamma^{32}P$]ATP and [^{32}P]P_i gave the ratio of [free ATP] to ([M*.ATP]+ [M**.ADP.P_i]). The combination of the two measurements enabled the ratios of free ATP to M*.ATP to M**.ADP.P_i to be determined at various times as shown in the figure. ●, free ATP; o, M.ATP and ▲, M**.ADP.P_i.*

$$\mathrm{M + ATP} \xrightarrow{k} \mathrm{M^*.ATP} \underset{}{\overset{k_{+3}}{\rightleftharpoons}} \mathrm{M^{**}.ADP.P_{\it i}} \qquad (3)$$

Computer simulation of the reaction scheme (eq. (3)) was performed to determine the best fit of the data to k_{+3}, where k_{+3} was given values from 50 to 500 sec^{-1}. A good fit to the data was obtained when k_{+3} = 150 sec^{-1}, though any value in the range 100-200 sec^{-1} was compatible with the experimental results.

In principle k_{+3} can be measured directly from the rate of M**.ADP.P_i formation when excess ATP is mixed with subfragment 1 provided the processes preceding ATP cleavage are sufficiently rapid. At low ionic strength this appears to be the case. When either 100 μ*M* or 200 μ*M* ATP is mixed with subfragment 1, presteady-state P_i formation occurs with an observed rate constant of 36 sec^{-1} for the exponential process (Fig. 4). This rate constant is 2.7-fold smaller than the observed rate constants (98 and 102 sec^{-1}) of the exponential protein fluorescence change when 100 and 200 μ*M* ATP, respectively, were mixed with 5 μ*M* subfragment 1. The kinetics of the fluorescence change were measured in a stopped-flow spectrofluorimeter. A complete analysis as described by Bagshaw *et al.* (1974) over the range 100-500 μ*M* ATP indicated $K_1 = 3.1 \times 10^4$ M^{-1} and k_{+2} = 122 sec^{-1} under these conditions. The fluorescence change probably reflects at least in part the protein isomerization (step 2 of eq. (1)) since nonhydrolysable substrates such as ADP and ATP (βγ-NH) also cause a protein fluorescence change on binding (Werber *et al.*, 1972). While this result is in line with those of Sleep and Taylor (1976) and Taylor (1977), there must be some reservation about its interpretation in view of the uncertainty as to the nature of the protein fluorescence change (Taylor, 1978). If the interpretation of the above result is correct, a lag phase in P_i formation should have occurred but it will not be readily observed unless data are collected in the first 10 msec of the reaction. Evidence for a lag phase or at least complex exponential character in P_i formation has been obtained by Bagshaw *et al.* (1972, Fig. 2) and by S. P. Chock and E. Eisenberg (personal communication).

To a first approximation the rate constant of the transient phase in Fig. 4 measures ($k_{+3} + k_{-3}$). K_3 was measured by the method of Bagshaw and Trentham (1973) and found to be 5 under those conditions, from which it follows that k_{+3} = 30 sec^{-1} 0.01 *M* ionic strength. However since the value of the rate constant controlling ATP binding, $k_{+2}K_1[\mathrm{ATP}]/(1+K_1[\mathrm{ATP}])$, is of the same order as k_{+3}, transient P_i formation is not a simple exponential process, though it looks like one (Fig. 4). The observed rate constant will be less than ($k_{+3} + k_{-3}$). The reaction sequence of eq. (3) was simulated with k = 102 sec^{-1} (the observed rate constant of the fluorescence change at 200 μ*M* ATP) and with

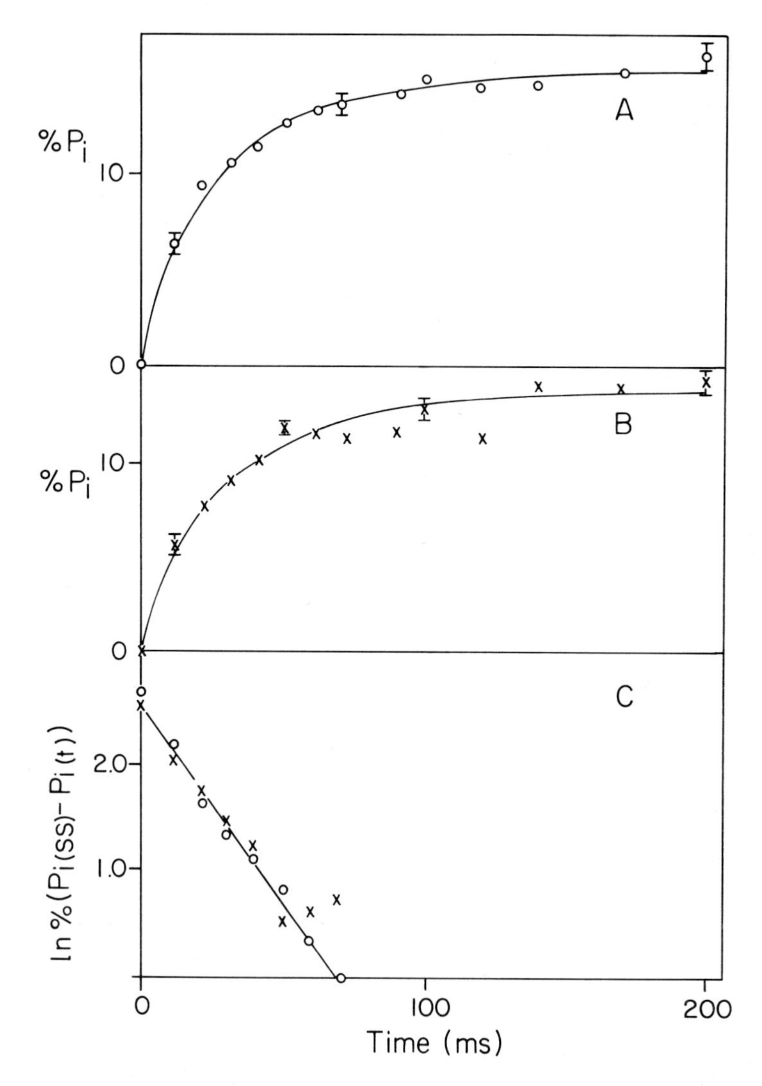

FIGURE 4. Presteady state P_i formation when ATP was mixed with subfragment 1 at low ionic strength. (A) 100 μM [$\gamma^{32}P$]ATP was mixed with 20 μM subfragment 1 in a solvent of 2 mM KCl, 2 mM $MgCl_2$, and 4 mM tris adjusted to pH 8.0 with HCl at 20°C. The reaction was quenched at the times indicated and the P_i formed plotted as a percentage of the total ^{32}P. (B) As A but with 200 μM instead of 100 μM [$\gamma^{32}P$]ATP. (C) A line drawn through steady-state P_i formation was extrapolated back to zero time. At times t ln %($P_{i(ss)}-P_{i(t)}$) was plotted against t, where %$P_{i(ss)}$ is read from the extrapolated steady-state line at time t. The observed rate constant of the exponential process in (A) and (B) equals the negative value of slope of the line in (C).

(k_{+3} + k_{-3}) taking various values. The rate profile of M**.ADP.P_i formation approximated to a single exponential with a rate constant of 36 sec^{-1} (the observed rate constant in Fig. 4) over the last 90% of the reaction when (k_{+3} + k_{-3}) = 42 sec^{-1}. It follows that k_{+3} = 35 sec^{-1} at 0.01 *M* ionic strength. k_{+3} therefore shows a marked ionic strength dependence (cf. Table 1).

The refractory state has only been characterized at low ionic strength. At an F-actin concentration (6 mg/ml) approaching saturation, the actin activated subfragment 1 ATPase activity was 29 sec^{-1} when measured at 0.01 *M* ionic strength in the presence of 2 m*M* MgATP at pH 8 and 20°C. It follows that, if the Lymn and Taylor scheme is correct, cleavage of M*.ATP is at least partially rate determining in the actin activated subfragment 1 ATPase. In which case M*.ATP is identified as one and possibly the only component of the refractory state.

THE NATURE OF THE ATP CLEAVAGE REACTION

Proposed mechanisms of the hydrolytic role of water in steps 2 to 4 of the ATPase fall into two general classes. On the one hand it has been postulated that water enters the reaction in step 3 so that this step is more explicitly written as:

$$M^*.ATP + H_2O \rightleftharpoons M^{**}.ADP.P_i \quad (4)$$

The failure to detect phospho-enzyme intermediates (Sartorelli *et al.*, 1966) together with transient ^{18}O-studies suggest that, if the class of reaction described by eq. (4) is correct, then the reaction is a simple hydrolysis in which the P_i is indeed inorganic phosphate (Trentham, 1977). Accordingly in this class of reaction, steps 2 and 4 (eq. (1)) both represent protein isomerizations with no change in the covalent bond structure of ATP, ADP or P_i.

On the other hand it has been proposed that water enters the reaction prior to step 3 (Young *et al.*, 1974). In terms of eq. (1) water addition occurs in step 2, so that M*.ATP represents protein-bound ATP in which the γ-phosphorus atom is pentacoordinated to oxygen atoms, step 3 becomes a pseudorotation in which the nature of the pentacoordinate structure changes, and step 4 involves bond cleavage of protein-bound ATP (eq. (5)).

$$M.ATP + H_2O \overset{2}{\rightleftharpoons} M^{*}.ATP \overset{3}{\rightleftharpoons} M^{**}.ADP.P_i \overset{4}{\rightleftharpoons} M^{*}.ADP.P_i \quad (5)$$

Thus according to this second class of mechanism $M^{**}.ADP.P_i$ is proposed to correspond to a hydrated state of protein-bound ATP and step 3 can be written more explicitly as eq. (6) (the protonation of the oxygen atoms is not intended to have mechanistic significance).

$$M^{*}.\ A\text{-}O\text{-}\underset{O}{\overset{O}{P}}\text{-}O\text{-}\underset{O(e)}{\overset{O}{P}}\text{-}O\text{-}P\big(OH(a)\big)\big(OH(e)\big)\big(OH(e)\big)\big(OH(a)\big) \rightleftharpoons M^{**}.\ A\text{-}O\text{-}\underset{O}{\overset{O}{P}}\text{-}O\text{-}\underset{O}{\overset{O}{P}}\text{-}O\text{—}P\big(OH(e)\big)\big(OH(a)\big)\big(HO(e)\big)\big(OH(e)\big) \quad (6)$$

The pentacoordinate structure around the γ-phosphorus atom can be drawn as a trigonal bipyramid (eq. (7)) whose corners mark the positions of the five oxygen atoms which form three equatorial bonds (Marked (*e*)) and 2 apical bonds (marked (*a*)) with the phosphorus atom. In a pseudorotation the two apical bonds shorten to become equatorial, and two of the equatorial bonds lengthen and become apical (eq. (7)).

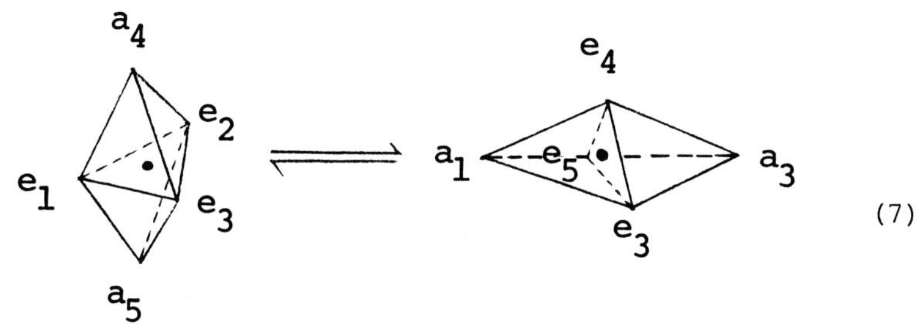

Empirical rules govern which bonds may take up apical coordination; the most important of these is that electron deficient atoms preferentially take up apical positions (e.g., the oxygen in -O-H is preferentially apical to that in $-O^-$ (Westheimer, 1968)). Isolation of the nucleotide from $M^{*}.ATP$ and $M^{**}.ADP.P_i$ involves quenching the reaction into protein denaturant--generally 1 N acid. When this happens an apical bond breaks so that $M^{*}.ATP$ can only yield ATP while $M^{**}.ADP.P_i$ can, and according to the Young *et al.* (1974) mechanism does, yield ADP and P_i.

It is the purpose of this work to distinguish which of these classes of mechanism occurs in the Mg^{2+}-dependent myosin ATPase. For ease of description the class of reaction described by eq. (4) will be called the "hydrolysis" mechanism and that by eqs. (5) and (6) the "pseudorotation" mechanism. In the hydrolysis mechanism protein-bound ATP is cleaved reversibly in step 3.

This means that the β,γ-bridging oxygen atom of protein-bound ATP may become equivalent to the β-nonbridging oxygen atoms. So, if the β,γ-bridging oxygen atom of ATP is labelled with ^{18}O, this ^{18}O-atom may "scramble" in step 3 and become a non-bridging β-oxygen atom in M*.ATP (eq. (8) where ● represents ^{18}O).

M*.Adenosine-O-P-O-P-●-P-O + H_2O ⟶ M**.Adenosine-O-P-O-P ●.P_i

↓ (8)

M*.Adenosine-O-P-O-P(●)-O-P-O ⟵ M**.Adenosine-O-P-O-P(●) O.P_i

On the other hand in the pseudorotation mechanism, there is no bond cleavage before the essentially irreversible step 4 and the β,γ-bridging oxygen of protein-bound ATP will not scramble.

However, scrambling will not necessarily be complete in the case of the hydrolysis mechanism because the β-phosphate group of ADP in M**.ADP.P_i may be so tightly coordinated to the protein that rotational freedom of the β-phosphate group is restricted. Thus an experimental result which shows evidence of scrambling rules out a pseudorotation mechanism as being a sufficient description of step 3 (eq. 1). If scrambling is incomplete, it provides evidence of rotational restraint of the β-phosphate group of ADP in M**.ADP.P_i.

The concept for carrying out this type of experiment was first developed by Midelfort and Rose (1976) in their studies of glutamine synthetase. Studies by Wimmer and Rose (1977) have shown that scrambling of ATP occurs in the presence of chloroplast lamellae and light indicating that ATP binds reversibly to the protein and is cleaved on the protein.

The ^{18}O-analytical problem differs from that presented earlier (Fig. 2) in that it is the distribution of ^{18}O in a molecular species rather than ^{18}O-incorporation that is of major concern. The solution to this problem is aided by the use of an ^{18}O-analytical technique in which the distribution of ^{18}O in P_i is measured and so the ratio of P_i containing 0,1,2 and three ^{18}O-atoms per molecule is found. The detailed enzymology and chemistry which follow are designed specifically to separate the β-nonbridging and β,γ-bridging oxygen atoms of ATP isolated from M*.ATP.

Analysis of this aspect of the myosin ATPase mechanism presents two further problems. First ATP binds tightly to the protein and once liganded does not dissociate reversibly to any significant extent (Goody *et al.*, 1977). Thus one is forced to work with very small quantities of nucleotides dictated by the amount of M*.ATP that can be isolated. Secondly the three nonbridging oxygen atoms bound to the γ-phosphorus of ATP undergo exchange. Since the starting material [$^{18}O_4$] ATP is labelled in these oxygen atoms as well as the β,γ-bridging oxygen atoms, this complicates the analytical procedure. On the other hand extensive studies have already been done on the myosin ATPase and the resulting knowledge of its mechanism means that the work is carried out on rather well defined intermediates, M*.ATP and M**.ADP.P_i, whose involvement in the myosin and actomyosin ATPases and cross-bridge cycle of muscle contraction is becoming increasingly clear.

Principle of the Experimental Approach

The ATP starting material is labelled on all four oxygen atoms of the γ-phosphate group. This [$^{18}O_4$]ATP is mixed with excess subfragment 1 and at 2 sec most of the nucleotide will exist as M*.ATP and M**.ADP.P_i effectively at equilibrium (K_3 = 9, Table 1). A small fraction of the nucleotide will exist as M*.ADP and ADP (k_{+4} = 0.06 sec^{-1}, k_{+6} = 2.3 sec^{-1} and k_{+5}, k_{+7} are large under the reaction conditions) and is of little further concern. If [ATP]>[subfragment 1] only a 1:1 stoichiometric amount of ATP will bind at the active site. The three oxygen atoms derived from the nonbridging oxygen atoms of the γ-phosphate group of protein-bound [$^{18}O_4$]ATP will undergo exchange with the solvent (Bagshaw *et al.*, 1975). At 2 sec, therefore, one can invisage the predominant oxygen labelling patterns for the two mechanisms as in Fig. 5.

Analysis of the ATP proceeds as follows. At 2 sec the reaction mixture is quenched into acid and the amount of ATP present is measured. ATP is isolated by column chromatography. 50% of the ATP is treated with dihydroxyacetone and glycerokinase, yielding dihydroxyacetone phosphate which is hydrolyzed in alkali. The resulting P_i is acidified and methylated yielding trimethyl phosphate which is analyzed in a mass-spectrometer (eq. (9)) (Midelfort and Rose, 1976). The analysis shows the labelling pattern of the γ-phosphate group and hence indicates the extent of oxygen exchange with the solvent.

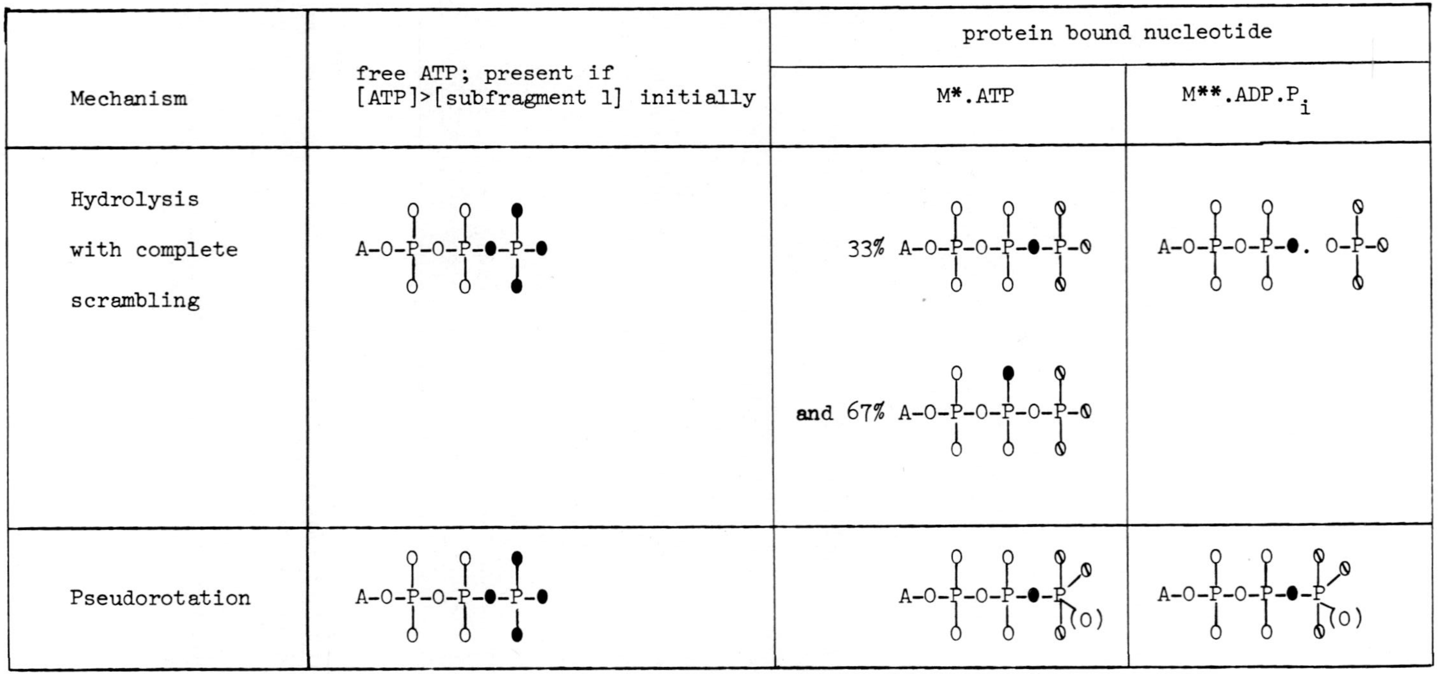

Mechanism	free ATP; present if [ATP]>[subfragment 1] initially	protein bound nucleotide	
		M*.ATP	M**.ADP.P_i
Hydrolysis with complete scrambling	A-O-P-O-P-●-P-●	33% A-O-P-O-P-●-P-⊘ and 67% A-O-P-O-P-O-P-⊘	A-O-P-O-P-●. O-P-⊘
Pseudorotation	A-O-P-O-P-●-P-●	A-O-P-O-P-●-P (0)	A-O-P-O-P-●-P (0)

FIGURE 5. Expected distribution of ^{18}O-label in nucleotides depending on the mechanism of subfragment 1 catalyzed ATP hydrolysis. ● represents ^{18}O and ⊘ partially enriched oxygen atoms. ⊘-oxygen atoms may undergo exchange with ^{16}O-water in the solvent, so that their level of ^{18}O-enrichment is expected to be low in ATP and P_i isolated from M.ATP and M**.ADP.P_i, respectively. (0) is a solvent oxygen atom.*

$$\text{A-O-}\overset{O}{\underset{O}{P}}\text{-O-}\overset{O}{\underset{O}{P}}\text{-}\bullet\Big|\overset{\bullet}{\underset{\bullet}{P}}\text{-}\bullet + \begin{matrix}CH_2OH\\ CO\\ CH_2OH\end{matrix} \underset{}{\overset{\text{glycerokinase}}{\rightleftharpoons}} \text{ADP} + \begin{matrix}CH_2\text{-O-}\overset{\bullet}{\underset{\bullet}{P}}\text{-}\bullet\\ CO\\ CH_2OH\end{matrix} \xrightarrow{OH^-} (C_3H_6O_3) + \text{O-}\overset{\bullet}{\underset{\bullet}{P}}\text{-}\bullet \xrightarrow{H^+,\,CH_2N_2} \text{trimethyl phosphate} \qquad (9)$$

(cleavage indicated between the β- and γ-phosphorus)

The remaining 50% ATP is analyzed to find out the extent of scrambling. The ATP is treated with acetyl-CoA synthetase, acetyl-CoA and a small amount of pyrophosphate. This allows the β- and γ-phosphates to interchange. The ATP is isolated and then analyzed by a reaction sequence as in eq. (9). Overall the sequence of reactions for the analysis of an ATP molecule with a β-nonbridging ^{18}O-atom is shown in eq. (10).

$$\text{A-O-}\overset{O}{\underset{O}{P}}\text{-O-}\overset{\bullet}{\underset{O}{P}}\text{-O-}\overset{\bullet}{\underset{\bullet}{P}}\text{-}\bullet \underset{}{\overset{\text{acetyl CoA, }PP_i\text{, synthetase}}{\rightleftharpoons}} \text{A-O-}\overset{O}{\underset{O}{P}}\text{-}\bullet\text{-}\overset{\bullet}{\underset{\bullet}{P}}\text{-O-}\overset{\bullet}{\underset{O}{P}}\text{-O}$$

$$\xrightarrow{\text{dihydroxyacetone, glycerokinase}} \text{ADP} + \underset{50\%}{\begin{matrix}CH_2O\text{-}\overset{\bullet}{\underset{\bullet}{P}}\text{-}\bullet\\ CO\\ CH_2OH\end{matrix}} + \underset{50\%}{\begin{matrix}CH_2O\text{-}\overset{\bullet}{\underset{O}{P}}\text{-O}\\ CO\\ CH_2OH\end{matrix}} \xrightarrow{1)\ OH^-\quad 2)\ H^+,\ CH_2N_2} \text{trimethyl phosphate} \qquad (10)$$

Chemical ionization of trimethyl phosphate yields a parent ion equal to the molecular weight of trimethyl phosphate +1, so that ions of m/e 141,143,145, and 147 (containing 0,1,2 and 3 ^{18}O-atoms, respectively) are detected. Scrambling is characterized by high 143 peak in trimethyl phosphate derived from ATP treated as described in eq. (10).

^{18}O-Analytical Methods and Reagents

The principle of the ^{18}O-analytical procedures used to identify the extent and distribution of ^{18}O-enrichment in $[^{18}O]$ATP has been described above. Experiment details are given by Midelfort and Rose (1976). It should be noted that small amounts of $[^{16}O]PP_i$ were added to promote the pyrophosphate exchange reaction with yeast acetyl-CoA synthetase (eq. (10)). The molar ratio of added PP_i to ATP was approximately 1 to 10. ^{18}O-content and distribution in P_i was determined after conversion into trimethyl phosphate by direct gas chromatography-mass spectral analysis on a Finnigan model 3300 instrument (Columbia University, Department of Chemistry). A 2-ft OV-1 column at 60-80°C was used to separate trimethyl phosphate from solvent. Trimethyl phosphate was ionized with chemical ionization using methane gas. The signal intensities at *m/e* 141,143,145, and 147 were monitored to produce continuous curves which were then integrated. Two sets of results were taken for each sample. In all cases the reproducibility was considered satisfactory. For example each pair of 141:143 peak ratios whose values lay in the range 10 to 100 had a mean variation of 6%. The results recorded in the tables are those in which the total peak areas were largest and hence likely to be most accurate. All results are corrected for the natural abundance of ^{18}O in the ^{16}O-water.

The synthesis of [^{18}O]ATP labelled in the four terminal oxygens of the γ-phosphoryl group ([$^{18}O_4$]ATP) was as described by Midelfort and Rose (1976). [$^{18}O_4$]ATP was stored in methanol at -20°C. A sample was cleaved using the glycerokinase reaction (eq. (9)) and mass spectral analysis showed that each of the nonbridging γ-oxygen atoms had a 91.3% probability of being ^{18}O (Table 5). Thus the β,γ-nonbridging oxygen atom will also be approximately 91.3% ^{18}O-enriched--a conclusion confirmed by the result of analysis III, Table 4.

Isolation of ATP and P_i from M.ATP and M**.ADP.P_i*

Two experiments involving analysis of M*.ATP are described here. Since the analysis depends critically on the experimental details and particularly on the relative initial concentrations of ATP and subfragment 1, the isolation of ATP and P_i will be described for each experiment. The general procedures are similar to those of Bagshaw *et al.* (1975). In experiment 1 [$^{18}O_4$]ATP was mixed with molar excess subfragment 1, and the products were separated by ion exchange chromatography.

The conditions of experiment 2 were similar except that [$^{18}O_4$]ATP was in 1.05-fold excess of subfragment 1. At first sight this means an added complication since some ATP will be analyzed which will not have reacted with subfragment 1. However, as is discussed below, this free ATP allows for controls to be done which test the validity of the analytical procedure.

Experiment 1: 12 μ*M* [$^{18}O_4$]ATP (1.5. μmol) plus trace amounts of [γ^{32}P]ATP were treated at 20°C with 18 μ*M* subfragment 1 in 50 m*M* KCl, 2 m*M* $MgCl_2$ and 50 m*M* tris adjusted to pH 8.0 with HCl. The reaction was carried out in several stages using 4 ml of reactants at a time in the quenched flow apparatus. Each sample was quenched at 2 sec into an equal volume of 7% $HClO_4$ at 0°C, and the pH raised to 4 with a solution containing 0.6 *M* potassium acetate and 3 *M* KOH. The solution was centrifuged to remove protein and $KClO_4$. The supernatant and washings were collected. Based on a ^{32}P count 66% of the initial radioactivity was recovered. Losses occur in the dead-space of the quenched-flow apparatus. To determine the ratio of ATP to P_i present an aliquot of the solution was chromatographed on polyethylene imine plates and analyzed for [$\gamma^{32}P$]ATP and [^{32}P]P_i as described by Bagshaw and Trentham (1973). A total of 0.106 μmol ATP was present. 1.5 μmol of carrier unlabelled ATP was added to the solution which was adjusted in batches to pH 7.6, diluted with water to a conductivity of 1 mmho cm^{-1} then chromatographed on a 2cm diam x 30 cm DE 52 (Whatman) column that had been washed with 1 *M* NH_4HCO_3 and then water. Some P_i washed straight through the column. The rest of the P_i and ATP was eluted with a 3 l 0.02 to 0.3 *M* gradient of tri-*n*-ethylammonium bicarbonate at pH 7.5. The column eluate was analyzed for ^{32}P and absorption at 260 nm. The P_i, ADP and ATP peaks were all well resolved. The ATP fraction was collected, concentrated on a rotary evaporator (the water bath being maintained at 30°C), repeatedly washed with methanol and evaporated to dryness.

Experiment 2: The same procedure was carried out as in Experiment 1 except that 20 μ*M* [$^{18}O_4$]ATP (4.00 μmol) and tracer [$\gamma^{32}P$]ATP were mixed with 19 μ*M* subfragment 1. P_i was concentrated and washed as for ATP. The P_i was then taken up in water, purified using the *iso*butane/benzene technique and precipitated as NH_4MgPO_4 (Bagshaw *et al.*, 1975).

Results

The magnitudes of the 143 peaks and of the 143/145 peak ratios of trimethyl phosphate isolated from [^{18}O]ATP provide evidence for or against scrambling. The fact that the 143 peaks and the 143/145 peak ratios (Tables 3 and 4) of samples II in both experiments are greater than those of samples I shows that scrambling of the β,γ-bridging O-atom of M*.ATP, and hence bond cleavage, occurs in step 3 of the myosin ATPase as in eq. (4).

Further analysis of the data permits insight into the nature of bound ADP in M**.ADP.P_i and calculation of the rate constant associated with scrambling of the β-phosphate group.

In experiment 1 the isolated [^{18}O]ATP arises solely from M*.ATP. Extensive intermediate exchange is indicated in I (Table 3). With no exchange the 147 peak would have been 5.23%. From I the relative peak intensities with complete and with no scrambling can be predicted (IV and V) and compared with the observed result (VI) after correction for the effect of ^{18}O dilution in the exchange reaction through addition of pyrophosphate. Scrambling has occurred but only to the extent of 28%.

TABLE 3. Experiment 1: Analysis of [$^{18}O_4$]ATP with and without Pyrophosphate Exchange after Reaction with Subfragment 1.

	m/e of trimethyl phosphate	141	143	145	147
		% of total peak intensity			
I	Derived from γ-PO_3 of ATP with no PP_i exchange	99.41	0.27	0.15	0.17
II	Derived from γ-PO_3 of ATP with PP_i exchange	99.29	0.63	0.03	0.05
III	Predicted II if no scrambling	99.72	0.13	0.07	0.08
IV	Predicted II if complete scrambling	97.63	2.15	0.07	0.08
V	II corrected for unlabelled PP_i dilution	99.22	0.69	0.03	0.06

I; ATP was isolated as described in the text after [$^{18}O_4$]ATP had reacted with subfragment 1 for 2 sec. The γ-PO_3 group was analyzed without PP_i exchange as outlined in eq. (9). II; as I except ATP was subjected to PP_i exchange (eq. (10)). III; calculated from I by dividing the 143, 145, and 147 peaks by 2. Since, if there is no enrichment of the β-nonbridging oxygen atoms, PP_i exchange will halve ^{18}O-enrichment of the γ-nonbridging oxygen atoms (eq. (10)). IV; calculated from II but assuming that one β-nonbridging oxygen atom per M.ATP (not free ATP) is 60.8% enriched (i.e., the β,γ-nonbridging oxygen atom, which was originally 91.3% enriched has scrambled). V; unlabelled PP_i (10% of the molarity of ATP) is added to promote PP_i exchange (see ^{18}O-analytical methods and reagents).*

TABLE 4. *Experiment 2: Analysis of* $[^{18}O_4]$*ATP with and without Pyrophosphate Exchange after Reaction with Subfragment 1*

	m/e of trimethyl phosphate	141	143	145	147
		% of total peak intensity			
I	Derived from γ-PO_3 of ATP with no PP_i exchange	96.05	0.29	0.73	2.93
II	Derived from γ-PO_3 of ATP with PP_i exchange	97.78	0.67	0.32	1.23
III[a]	Derived from β-PO_3 of ADP from ATP with no PP_i exchange	90.88	9.01	0.04	0.07
IV	Predicted II if no scrambling	98.03	0.14	0.36	1.47
V[b]	Predicted II if complete scrambling of M*.ATP	94.96	2.02	0.36	1.47
VI	II adjusted to 147 peak of IV and V which corrects for unlabelled PP_i dilution	97.35	0.80	0.38	1.47

I; as I Table 3. II; as II Table 3. III; ADP was isolated after the glycerokinase catalyzed reaction and thus originated from the ATP analyzed in I. IV; calculated from I by dividing the 143 145 and 147 peaks by 2. V; calculated from I assuming that one β-nonbridging O-atom per M.ATP is 60.8% enriched due to scrambling. VI; in this experiment (cf. experiment 1) unlabelled* PP_i *is corrected for by adjusting the 147 peak to the predeicted value which is independent of scrambling.*

[a]*The intensity of the 143 peak derived from this ADP shows 10.1-fold dilution of* $[^{18}O]$*ATP with carrier prior to the DEAE column and provides the most reliable estimate of the dilution. In addition the low 145 and 147 peaks show that* $[^{18}O_4]$*ATP had been synthesized with clean labelling in the β,γ-bridging O-atom and not in the β-nonbridging O-atoms.*

[b]*The 143 peak is predicted as follows. At the time of quenching trimethyl phosphate derived from* γ-PO_3 *of M*.ATP will have a practically zero 147 peak (cf. Table 3, I) due to intermediate exchange and that of free ATP will generate a 147 peak of 7.8% peak total based on the 10.1-fold dilution factor deduced from III. Thus the 2.93% 147 peak intensity in I indicates 38% of the isolated ATP was derived from free ATP and 62% was from M*.ATP. The β,γ-bridging O-atom of M*.ATP but not that of free ATP is then assumed to undergo complete scrambling.*

The data of experiment 2 (Table 4) differ from those of experiment 1 due to the presence of some free [$^{18}O_4$]ATP at the time of quenching the reaction. This free [$^{18}O_4$]ATP is not subject to intermediate exchange and hence has 145 and 147 peaks that are measurable with reasonable accuracy. The 145 to 147 peak ratios should be equal in samples I and II (Table 4) whether or not scrambling has occurred, and the fact that they are equal is a good check on the reliability of the experimental procedure. In addition the intensity of the 147 peak in II is predicted from that in I after account is taken of added pyrophosphate. The observed result again checks the reliability of the procedure. The significance of analyzing ADP derived from the glycerokinase catalyzed reaction is described in the table legend. The result of experiment 2 shows that scrambling has occurred to the extent of 35%.

From these two experiments we conclude that the pseudorotation mechanism as formulated in eq. (5) is untenable. The result is consistent with the simple hydrolysis mechanism (eq. (4)). However, the β-phosphate group of ADP in M**.ADP.P_i rearranges slowly so that even after 2 sec only 32% scrambling of the β-phosphate group has occurred, which gives a rate constant for the process of 0.19 sec^{-1}.*

Wimmer and Rose (1977) analyzing the interaction of ATP with the ATP synthase of chloroplast lamellae compared the scrambling rate of the β,γ-bridge oxygen to the rate of water oxygen exchange with the γ-PO_3 group of ATP. They found the two processes occurred at almost the same rate. In contrast analysis of M.*ATP in sample I, Table 3 shows that oxygen exchange between the γ-PO_3 group of M*.ATP and water occurs much more rapidly than the scrambling.

The oxygen exchange of the γ-PO_3 group of M*.ATP is apparently much more extensive than our own earlier work suggests (Bagshaw *et al.*, 1975). It was therefore of interest to analyze the extent of oxygen exchange between water and P_i isolated from M**.ADP.P_i. Again the results show (Table 5) that this intermediate exchange is much more extensive than most earlier work has suggested. Careful comparison is needed between the CO_2 (Boyer and Bryan, 1967) and trimethyl phosphate

Note added in Proof: In collaboration with Dr. M. R. Webb we have analyzed oxygen scrambling of M*.ATP following exchange of ^{18}O from the β-nonbridge to β,γ-bridge oxygen position of ATP. In this more sensitive experiment we find much more extensive (possibly complete) scrambling of the oxygen atoms suggesting that the β-phosphate group of myosin-bound ADP is freer to rotate in the M.ADP.P_i state than the complementary experiments described here indicate.*

*TABLE 5: Intermediate Exchange in P_i Isolated from M**.ADP.P_i Following the Reaction of [$^{18}O_4$]ATP with Subfragment 1*

	m/e of trimethyl phosphate	141	143	145	147
		% of total peak intensity			
I	From P_i of M**.ADP.P_i in experiment 1	99.15	0.51	0.11	0.23
II	From P_i of M**.ADP.P_i in experiment 2	98.0	1.28	0.32	0.40
III	From γ-PO_3 of [$^{18}O_4$]ATP using glycerokinase (eq.(9))	1.6	1.9	17.5	79.0

*In I and II P_i was isolated from M**.ADP.P_i without dilution. Spontaneous as opposed to subfragment 1 catalyzed hydrolysis of [$^{18}O_4$]ATP would give rise to P_i with an isotopic distribution corresponding to that in III. It is possible that the 147 peaks in I and II arise from spontaneous ATP hydrolysis, which will therefore be, at maximum, 0.5% of total ATP hydrolysis.*

methods for analyzing myosin catalyzed exchange reactions. The latter method has the advantage of showing the distribution as well as the amount of ^{18}O-label in P_i. There is no evidence from our data that exchange of one oxygen atom in the γ-PO_3 group of ATP is markedly slower than any other as suggested by the work of Shukla and Levy (1977).

CONCLUSIONS

1) The rate constant, k_{+3}, controlling ATP cleavage in the Mg^{2+}-dependent myosin ATPase has been shown by a variety of techniques to be in the range 35-150 sec^{-1} (pH 8, 20°C, I=0.01-0.08 *M*) and comparable in value to the rate constant, k_{+2}, controlling isomerization of the myosin-ATP binary complex, M*ATP.

2) k_{+3} is markedly ionic strength dependent, and has a sufficiently small value at low ionic strength that M*.ATP is likely to be partially or wholly the steady-state intermediate of the actomyosin ATPase mechanism. The extent to which a myosin-products complex (Chock *et al.*, 1976) is also a component of the steady-state intermediate and hence the refractory state remains to be established.

3) Hydrolysis of ATP is associated with the process M*.ATP $\rightleftharpoons$ M**.ADP.P_i in the Mg^{2+}-dependent myosin subfragment 1 ATPase. The β-phosphate group of ADP appears to be tightly coordinated to subfragment 1 in the M**.ADP.P_i state.*

4) Intermediate oxygen exchange seems to be more extensive than earlier work has suggested. The new results are in line with what might be expected from the myosin ATPase kinetic constants, since $k_{+4} \ll k_{-3}$, although other processes other than simple reversal of the hydrolysis step 3 may control the rate of intermediate exchange.

ACKNOWLEDGMENTS

Most of the experimental work described here was carried out in the Department of Biochemistry, University of Bristol. We thank Professor H. Gutfreund for helpful discussions and for the use of many facilities. We are grateful to the Science Research Council, U.K., the Muscular Dystrophy Association of America, and the U.S. National Science Foundation [Grant PCM 75-18884 (University of California)].

REFERENCES

Bagshaw, C. R. *et al.* (1974). *Biochem. J. 141,* 351-364.
Bagshaw, C. R., Eccleston, J. F., Trentham, D. R., Yates, D. W., and Goody, R. S. (1972). *Cold Spring Harbor Symp. Quant. Biol. 37,* 127-135.
Bagshaw, C. R., and Trentham, D. R. (1973). *Biochem. J. 133,* 323-328.
Bagshaw, C. R., Trentham, D. R., Wolcott, R. G., and Boyer, P. D. (1975). *Proc. Nat. Acad. Sci. U.S.A. 72,* 2592-2596.
Boyer, P. D., and Bryan, D. M. (1967). *Methods Enzymol. 10,* 60-71.
Chock, S. P., Chock, P. B., and Eisenberg, E. (1976). *Biochemistry 15,* 3244-3253.
Eisenberg, E., Dobkin, L., and Keilley, W. W. (1972) *Proc. Nat. Acad. Sci. U.S.A. 69,* 667-671.
Eccleston, J. F., Geeves, M. A., Trentham, D. R., Bagshaw, C. R., and Mrwa, U. (1975). *In* "The Molecular Basis of Motility" (L. Heilmeyer, ed.,), pp. 42-52, 26th Mosbach Symp. Springer-Verlag, Berlin.
Ferenczi, M. A., Homsher, E., Simmons, R. M., and Trentham, D. R. (1978). *Biochem. J. 171,* 165-175.
Goody, R. S., Hofmann, W., and Mannherz, H. G. (1977). *Eur. J. Biochem. 78,* 317-324.
Gutfreund, H. (1969). *Methods Enzymol. 16,* 229-249.
Hill, T. L. (1974). *Progr. Biophys. Mol. Biol. 28,* 267-340.
Hill, T. L. (1977). *In* "Free Energy Transduction in Biology--The Steady-State Kinetic and Thermodynamic Formalism," pp. 103-129. Academic Press, New York.
Lymn, R. W., and Taylor, E. W. (1971). *Biochemistry 10,* 4617-4624.
Midelfort, C. F., and Rose, I. A. (1976). *J. Biol. Chem. 251,* 5881-5887.
Sartorelli, L., Fromm, H. J., Benson, R. W., and Boyer, P. D. (1966). *Biochemistry 5,* 2877-2884.
Shukla, K. K., and Levy, H. M. (1977). *Biochemistry 16,* 132-136.
Sleep, J. A., and Taylor, E. W. (1976). *Biochemistry 15,* 5813-5817.
Taylor, E. W. (1977). *Biochemistry 16,* 732-740.
Taylor, E. W. (1978). *Critical Rev. Biochem.* in press.
Trentham, D. R. (1977). *Biochem. Soc. Trans. 5,* 5-22.
Trentham, D. R., Eccleston, J. F., and Bagshaw, C. R. (1976)/ *Quart. Rev. Biophys. 9,* 217-281.
Werber, M., Szent-Gyorgyi, A. G., and Fasman, G. (1972). *Biochemistry 11,* 2872-2883.
Westheimer, F. H. (1968). *Acc. Chem. Res. 1,* 70-78.

White, H. D., and Taylor, E. W. (1976). *Biochemistry 15*, 5818-5826.
Wimmer, M. J., and Rose, I. A. (1977). *J. Biol. Chem. 252*, 6769-6775.
Young, J. H., McLick, J., and Korman, E. F. (1974) *Nature 249*, 474-476.

MYOSIN FILAMENT STRUCTURE

Motility in Cell Function
Proceedings of the First John M. Marshall Symposium in Cell Biology

ISOLATION AND DISTRIBUTION OF MYOSIN ISOENZYMES

Susan Lowey, Laura Silberstein,**
*Geraldine F. Gauthier,** and John C. Holt†*

*Rosenstiel Basic Medical Sciences Research Center,
Brandeis University
Waltham, Massachusetts

**Laboratory of Electron Microscopy
Wellesley College
Wellesley, Massachusetts

†National Institute for Biological Standards and Control
London, NW3 6RB, England

INTRODUCTION

In a symposium on cellular motility, it seems appropriate to include an illustration of the myosin molecule, the central figure in contractility (Fig. 1). With the possible exception of *Acanthamoeba* myosin, all myosins isolated to date consist of the familiar two globular heads attached to an α-helical rod-like tail (Slayter and Lowey, 1967; Pollard and Weihing, 1974; Elliot *et al.*, 1976). The long dimension of the head is generally thought to be about 150 Å from a variety of measurements (Moore *et al.*, 1970; Mendelson *et al.*, 1973), whereas the rod, which forms the backbone of the thick filament, is about 1400-1500 Å long (Lowey *et al.*, 1969; Cohen *et al.*, 1970). Despite similarities in overall shape, enzymic activity, and aggregation properties, there are differences among myosins: The myosins of nonmuscle cells, and of smooth, cardiac, and skeletal muscles can have somewhat different regulatory mechanisms and modes of packing in the thick filament (Adelstein *et al.*, 1976; Clarke and Spudich, 1977). As will be described here, even the myosin of a relatively homogeneous "fast, white" skeletal muscle consists of several isoenzymic forms. The reason for the existence of so many diverse forms of myosin is

ISBN 0-12-551750-5

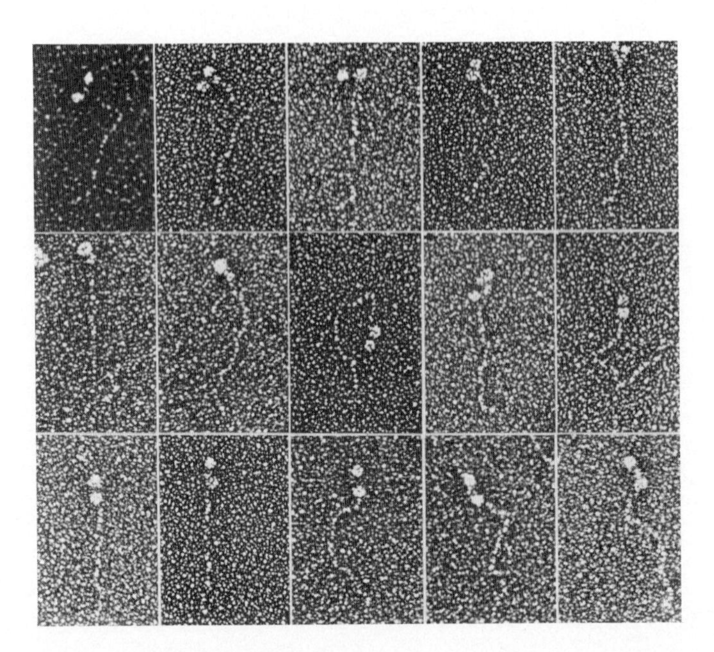

Fig. 1. A composite of rotary shadowed myosin molecules. (Reprinted from Slayter and Lowey, 1967).

not clear at present, but the isolation and localization of these isoenzymes may lead to a better understanding of their function.

RESULTS AND DISCUSSION

The substructure and nomenclature of the subunits of myosin is reviewed in Fig. 2. The two heavy chains of molecular weight 200,000 are associated with four moles of light chains of molecular weight about 20,000 (Lowey and Risby, 1971). There are two chemical classes of light chains: In vertebrate fast skeletal myosin we have called them the "alkali light chain" and the "DTNB light chain" due to the dissociating ability of high pH and the sulfhydryl reagent, DTNB, on these light chains (Weeds and Lowey, 1971). We stress that this terminology has no meaning for other myosins, where DTNB has no similar effect. It does appear, however, that all myosins have one class of light chains (analogous to the "alkali light chain") which cannot be removed without loss of ATPase; and another class (analogous to the "DTNB light chain") which is not essential for activity, is able to bind calcium, and can, in many muscles, be phosphoryl-

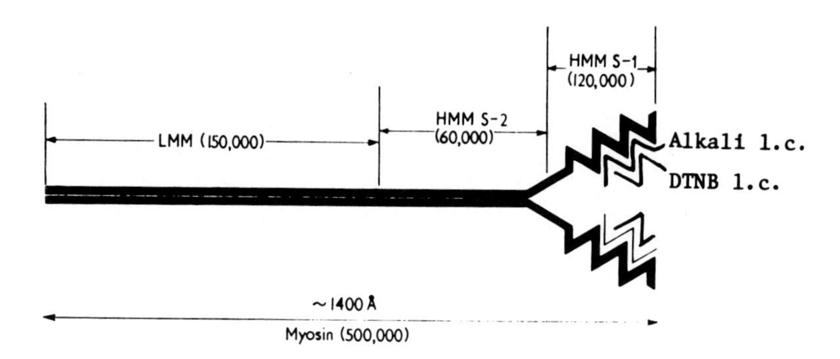

Fig. 2. Schematic representation of the myosin molecule. (Reprinted with modification from Lowey et al., *1969).*

ated by specific kinases. A number of papers in this volume will deal with the regulatory role of this latter light chain in certain types of muscle and nonmuscle cells. We will restrict our discussion to the "alkali light chains" of vertebrate fast skeletal myosin.

The light chain pattern of a number of muscles from the chicken is shown in Fig. 3 (Lowey and Risby, 1971). The group on the left represents myosins from fast muscles (pectoralis and posterior latissimus dorsi), whereas the slower myosins of the anterior latissimus dorsi and cardiac muscle are illustrated on the right. Differences in the light chain patterns of the two groups have given rise to the speculation that light chains contribute to determining the speed of contraction of a specific muscle. The alkali light chains in chicken pectoralis

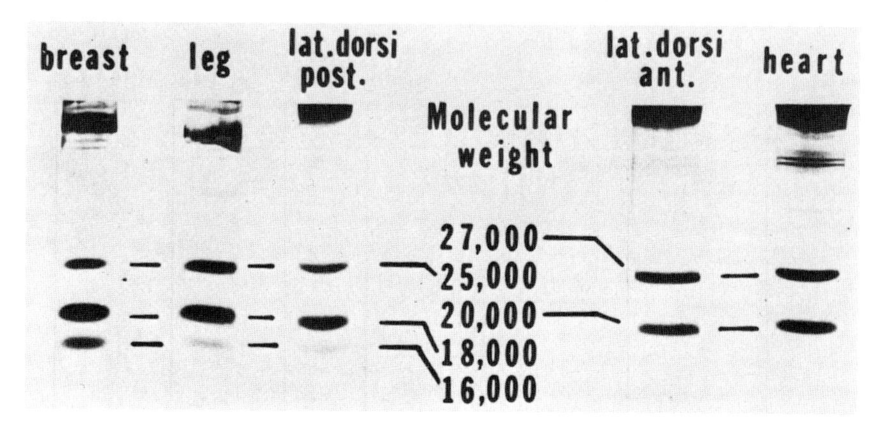

Fig. 3. Light chains from chicken muscles. (Reprinted from Lowey and Risby, 1971).

myosin consist of about 1 mole of alkali 1 (molecular weight about 21,000) and 1 mole of alkali 2 (molecular weight about 17,000). There are 2 moles of the DTNB light chain (molecular weight about 19,000). The heterogeneity in size of the alkali light chains cannot be explained by the proteolytic degradation of the larger A1 light chain to the smaller A2 light chain. The sequence of the rabbit A2 light chain is largely identical with that of the A1 light chain, with the exception of five residues in the N-terminal end of A2 not found in A1. This small difference in structure establishes that the light chains are the products of two related genes. The A1 light chain also contains an additional 41 residues at its N-terminal end; this so-called "difference peptide" has an unusual amino acid composition, being especially rich in proline, lysine, and alanine (Frank and Weeds, 1974). A schematic representation of the amino acid sequence of the A1 and A2 light chains from rabbit myosin is shown in Fig. 4.

The question arose whether the A1 and A2 light chains were located in each of the two heads of a single myosin molecule (a "heterodimer") or whether a myosin molecule contained only two moles of either A1 or A2 in a single molecule ("homodimers"). The unequal stoichiometry of the A1 and A2 light chains in rabbit myosin had suggested the existence of isoenzymes (Sarkar, 1972; Weeds *et al.*, 1975), but a determination of the relative amounts of heterodimer and homodimer requires the direct isolation of these molecules. Since all myosin molecules are so alike in their chemical properties, the normal fractionation procedures of ion-exchange chromatography and gel filtration are not applicable. Rather, one has to devise some type of affinity chromatography based on a unique feature of one of the isoenzymes not shared by the other. An immunological approach

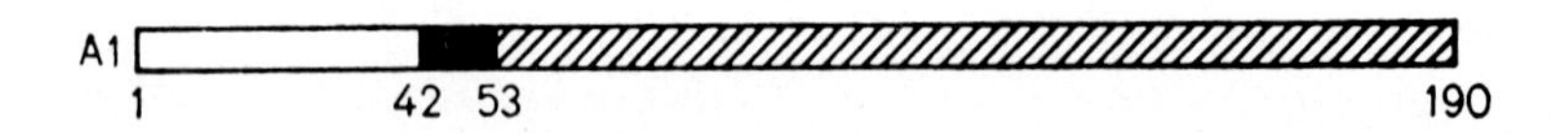

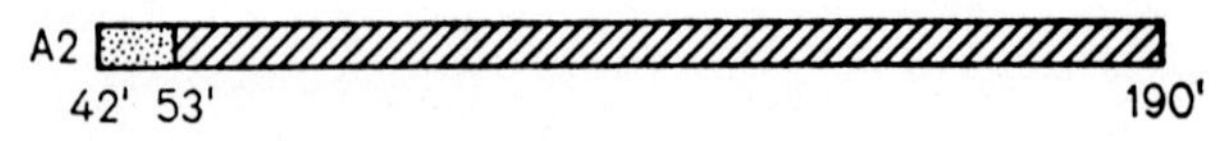

Fig. 4. Schematic representation of A1 and A2 sequences. Sequences identical in both proteins are shown with diagonal hatching; amino acid replacements are found in residues 42'-52' in A2 when compared with 42-52 in A1. Alkali 1 also contains an additional 41 residues at its N-terminal end. (Reprinted from Frank and Weeds, 1974).

was suggested by the finding that antibodies against the A1 light chain contain a subpopulation of antibodies which are unique for determinants of A1, presumably the "difference peptide," and which do not cross-react with A2 (Holt and Lowey, 1975). The isolation of these specific antibodies and their application in affinity chromatography is summarized in Fig. 5. A2 light chains coupled to Sepharose 4B are used as an immunoadsorbent to bind all antibodies in anti-A1 serum which recognize A2. The unretained antibodies are presumably specific for the "difference peptide" (Δ1). These anti-Δ1 antibodies are then coupled to Sepharose, and the question asked is whether this immunoadsorbent can distinguish between A1 and A2 myosin. The column was first tested with a mixture of the isolated heads of myosin (S1). As Fig. 6 shows, the column was very successful at fractionating S1(A1) from S1(A2): The S1 (A2) was not retained by the column, whereas S1(A1) could only be eluted with 4*M* guanidine hydrochloride. The column was also successful in fractionating tryptic HMM and, most important, native myosin, Fig. 7.

These experiments (Holt and Lowey, 1977) provided convincing evidence that the bulk of the myosin in chicken pectoralis consists of homodimers of A1 and A2 myosin. However, the bound A1 myosin fraction does contain at least 20% A2 light chain, and the possibility of 20% or more heterodimers in the myosin population cannot be excluded. Although it is quite likely that this contamination represents nonspecific binding of myosin, or aggregates of myosin retained by the column, further evidence for or against the existence of heterodimers would clearly be desirable. Recently we have succeeded in preparing an anti-Δ2 immunoadsorbent column which has the ability to bind A2 containing molecules specifically, while passing A1 molecules into the void volume. The rationale used to prepare this column was identical to that described in Fig. 5, except that antibodies to A2 light chain were passed through an immunoadsorbent of A1 coupled to Sepharose. Antibodies specific for A2 light chain, that is, anti-Δ2, were then coupled to Sepharose, and the efficiency of the column tested as before with myosin heads (S1), Fig. 8. These experiments showed that we now have two immunoadsorbents capable of fractionating myosin isoenzymes. When myosin is passed through these affinity columns, about equal amounts of myosin are passed and retained, the type of isoenzyme in the two fractions depending on the specificity of the columns, Fig. 9. The finding that the areas under the

PROBLEM: DOES MYOSIN CONSIST OF 2 POPULATIONS OF MOLECULES?

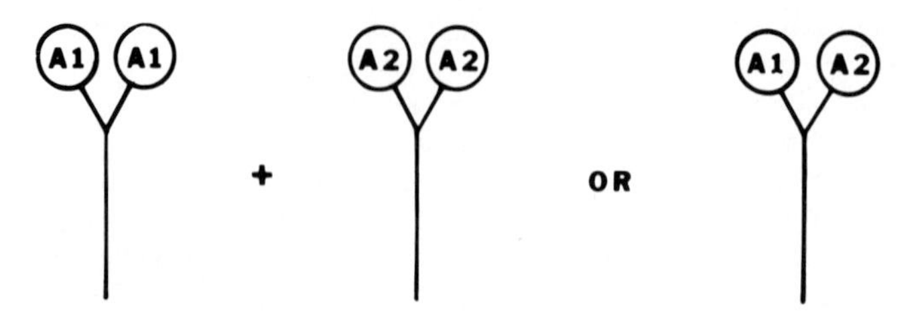

APPROACH: FRACTIONATE MYOSIN BY AFFINITY CHROMATOGRAPHY

1. MAKE ANTI-A1 ANTIBODIES

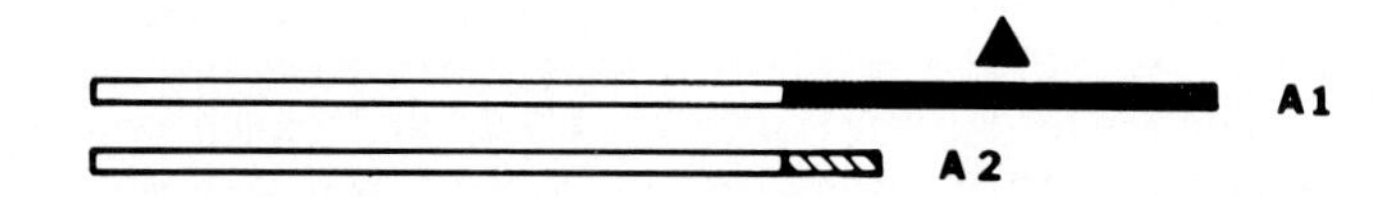

2. ISOLATE ANTI-Δ

3. PASS MYOSIN THROUGH ANTI-Δ COLUMN

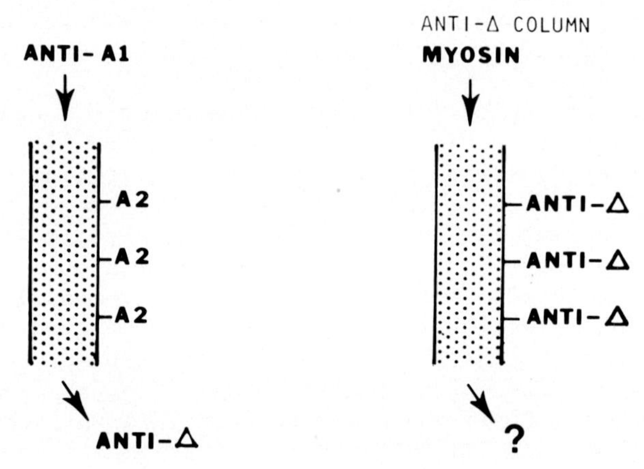

Fig. 5. Schematic representation of the procedure used to separate myosin isoenzymes by affinity chromatography. Antibodies prepared against A1 light chain (step 1) were passed through an immunoadsorbent consisting of A2 coupled to Sepharose 4B (step 2); antibodies in the serum that cross-react with A2 were bound to the column, whereas the unretained fraction presumably contained antibodies specific for the "difference peptide" (Δ) in A1. These anti-Δ antibodies were isolated from the serum by an immunoadsorbent composed of A1 (not shown), and then coupled to Sepharose to make the anti-Δ affinity column (step 3). In order to prepare an immunoadsorbent specific for A2 light chain, the same procedure was followed, except that the starting serum was anti-A2 and the column in step 2 consisted of Sepharose-coupled A1 light chain.

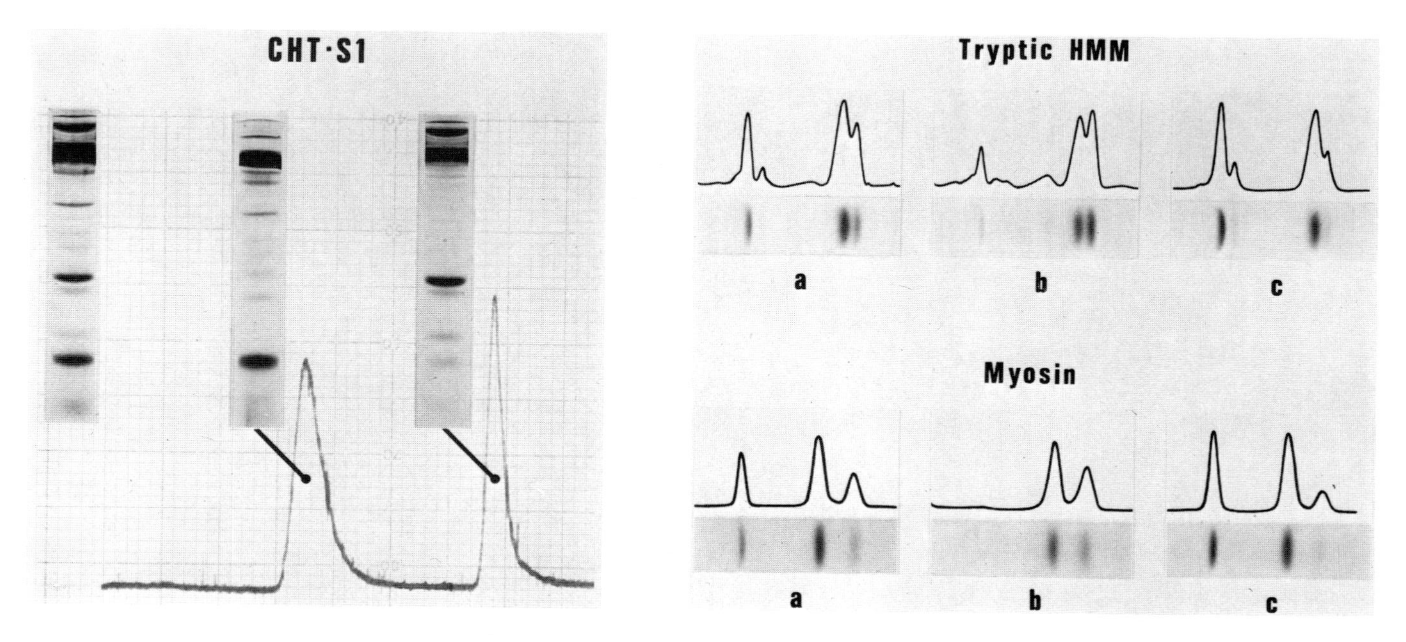

Fig. 6. *Separation of subfragment 1 isoenzymes by an affinity column specific for A1. From left to right: the starting material consisting of S1 prepared by chymotryptic digestion of myosin; the unretained alkali 2 fraction; and the 4 M guanidine·HCl eluted alkali 1 fraction. (Reprinted from Holt and Lowey, 1977).*

Fig. 7. *Densitometry of HMM and myosin before and after fractionation by an affinity column specific for A1. (a) Starting material. (b) Unretained alkali 2 fraction. (c) Alkali 1 fraction eluted with 4 M guanidine·HCl. (Reprinted from Holt and Lowey, 1977).*

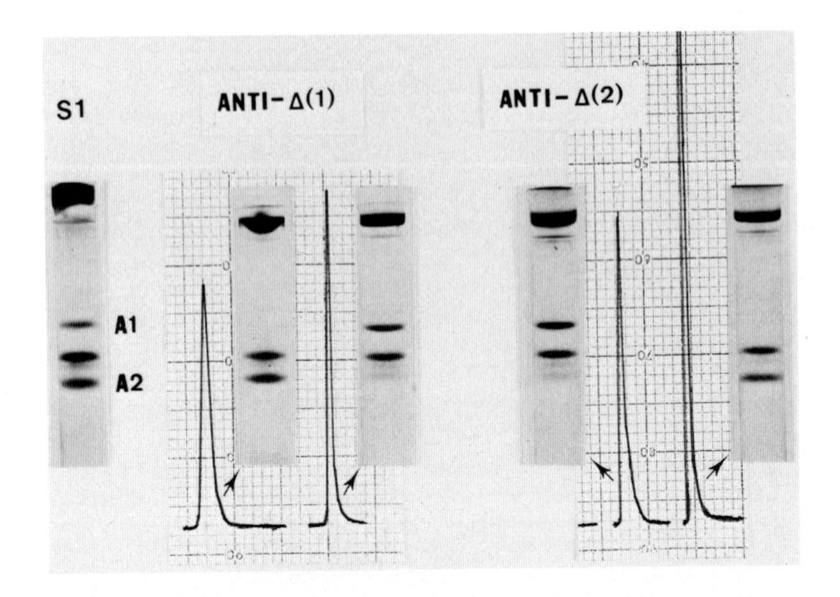

Fig. 8. Separation of subfragment 1 isoenzymes by affinity columns specific for A1 or A2. From left to right: the starting material consisting of S1 prepared by papain digestion of myosin; the unretained alkali 2 fraction and the 4 M guanidine·HCl eluted alkali 1 fraction from an anti-Δ1 affinity column. When the same material was applied to an anti-Δ2 column, alkali 1 was not retained, and alkali 2 was bound until eluted with 4 M guanidine·HCl.

traces are roughly equal implies that the amount of heterodimer, if any exists, is probably less than 25%, but further experiments are needed to resolve this ambiguity. As well as demonstrating the existence of myosin homodimers, and providing a means for their isolation, these affinity columns have shown that it is possible to prepare two highly specific antibody markers for the localization of A1 and A2 myosin in muscle.*

**For simplicity, we have assumed in the experiments that follow that only A1 and A2 myosin occur in these muscles; if subsequent experiments show myosin heterodimers to be present in significant amounts, the interpretation of these staining patterns will have to be modified accordingly.*

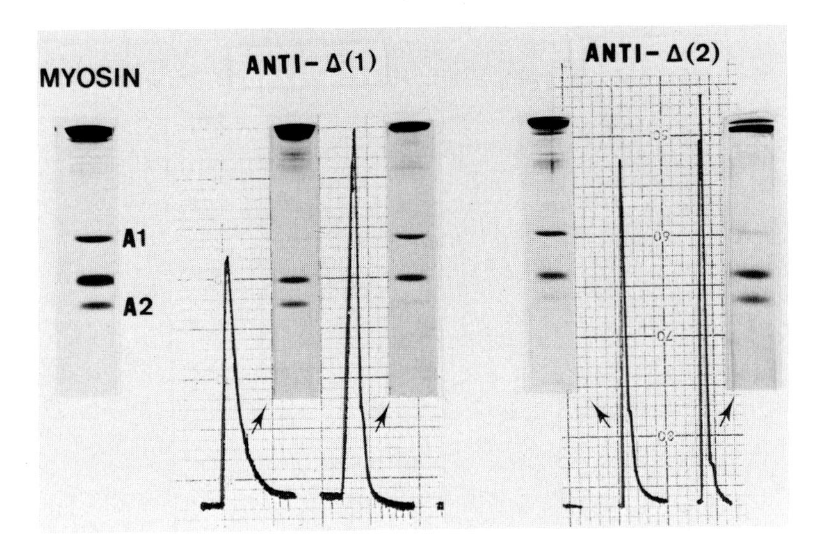

Fig. 9. Separation of myosin isoenzymes by affinity columns specific for A1 or A2. From left to right: the starting material of chicken myosin; the unretained alkali 2 fraction and the 4 M *guanadine·HCl eluted alkali 1 fraction from an anti-Δ1 affinity column. When the same material was applied to an anti-Δ2 column, alkali 1 was not retained, and alkali 2 was bound until eluted with 4* M *guanidine·HCl.*

The following experiments show the distribution of myosin isoenzymes among fibers, myofibrils, and finally myofilaments of chicken pectoralis muscle. Figure 10 shows a cryostat section of chicken pectoralis reacted with antibodies specific for either A1 myosin (Anti-Δ1) or A2 myosin (anti-Δ2). Indirect immunofluorescence was used; meaning, the sections were stained with primary antibody (anti-Δ) and then reacted with goat antibody to rabbit IgG coupled with fluorescein isothiocyanate. It is readily apparent that both isoenzymes are located within the same fibers, a result consistent with earlier electrophoretic experiments demonstrating the presence of both A1 and A2 light chains in a single fiber from rabbit psoas (Weeds *et al.*, 1975). This region of the pectoralis is atypical in that it has a number of fibers which do not react with antibodies to white myosin. It was chosen to demonstrate the specificity of the reaction and the absence of nonspecific staining, but it must be emphasized that this so-called "red region" of the pectoralis accounts for only about 1% of the muscle mass (Gauthier and Lowey, 1977). The experiments with fibrils and filaments were made on randomly chosen specimens of pectoralis muscle, which consist primarily of fast, white fibers and,

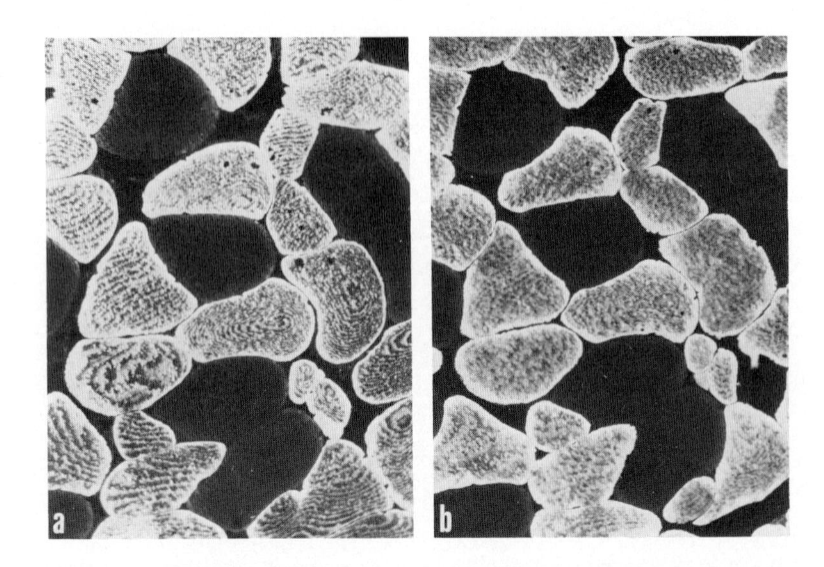

Fig. 10. Chicken pectoralis, "red region". Serial transverse sections. (a) Anti-Δ1. (b) Anti-Δ2. This region of the muscle consists of two different types of fibers. Those which react with fluorescein-labeled antibodies appear bright. The same population of fibers have reacted strongly with both anti-Δ1 (a) and anti-Δ2 (b), which indicates that A1 and A2 myosin are present within the same fibers. The remaining fibers are unreactive, and this illustrates the specificity of these antibodies. ×200.

therefore, all fibers would be expected to contain both A1 and A2 myosin. Figure 11 shows that all the myofibrils fluoresce when reacted with labeled anti-Δ1. Similar results were obtained with anti-Δ2. (Appropriate controls were carried out with antibodies absorbed with the homologous antigens.) Thus both light chain isoenzymes are present within a single myofibril.

The most intriguing question now remaining is whether A1 and A2 myosin occur within a single filament. A suspension of myofilaments was reacted with antibody to Δ1 or to Δ2 and examined in the electron microscope. The control consisted of myofilaments washed with nonimmune IgG, or antibody absorbed with the homologous light chains. A slight accentuation of the cross-bridges was observed on filaments reacted with antibody, and the bare zone was better defined in this preparation; however, the contrast with filaments subjected to nonimmune IgG was not striking. A much more impressive image could be obtained by treating those filaments stained with anti-Δ with a second antibody, namely goat antibody against rabbit IgG.

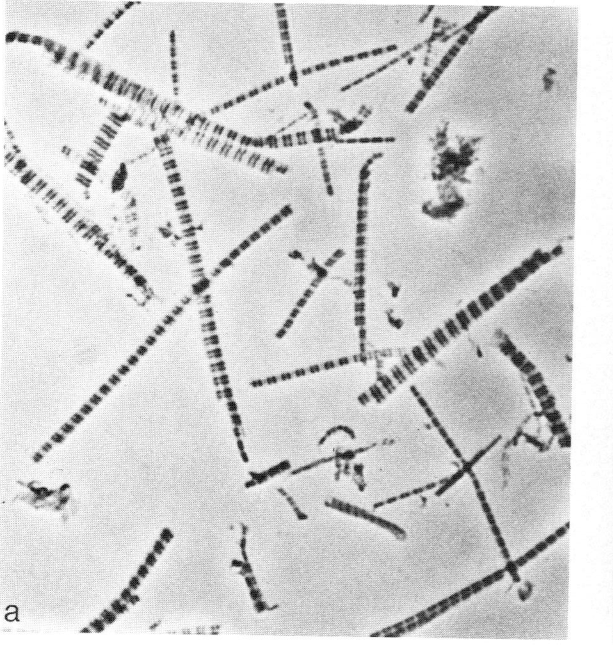

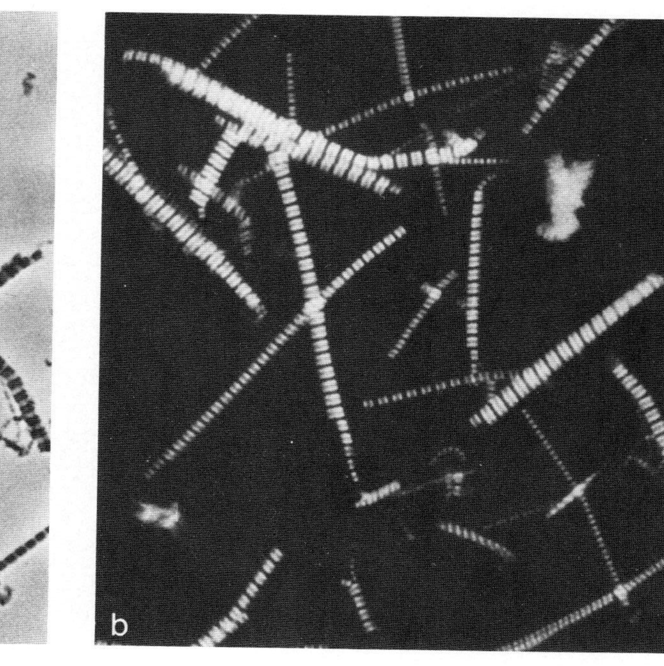

Fig. 11. *Light microscopy of myofibrils reacted with fluorescein-labeled antibody specific for A1. (a) Phase contrast. (b) Dark field fluorescence image of the same field. ×830.*

This procedure resulted in sufficient buildup of antibody on the filaments to make the identification of reacted filaments unmistakeable, as in Fig. 12. The control was subjected to the same treatment except that rabbit IgG was substituted for specific antibody. Remarkably little nonspecific binding of IgG to either the actin or the myosin filaments occurred. These experiments suggest that both A1 and A2 myosin can coexist within a single filament. To determine whether there is any preferential distribution of these isoenzymes along the length of the filament is beyond the resolution of this technique. Staining of whole glycerinated muscle with antibody, by the procedures developed for the localization of C-protein, may provide a method for detecting any differences in packing which may exist *in vivo* (Craig and Offer, 1976).

One interesting observation which suggests that the isoenzymes are not distributed at random is the finding that a slow muscle such as the anterior latissimus dorsi (ALD) does not contain equal amounts of A1 and A2 myosin. Although the ALD consists primarily of slow red fibers, a small percentage of the fiber population reacts with antibodies to fast pectoralis myosin (Arndt and Pepe, 1975; Gauthier and Lowey, 1977). These reactive fibers, however, show far weaker fluorescence with anti-Δ2 than with anti-Δ1, Fig. 13. Unequal distribution of A1 and A2 myosins has also been found in mammalian muscles such as the rat soleus and rat diaphragm (Gauthier and Lowey, manuscript in press, J. Cell Biol.)

What can be concluded about the function of these isoenzymes? Do they modulate the speed of the muscle? Or do they perhaps constitute a vernier type of mechanism for limiting the growth of the filament? If the myosin isoenzymes also show the differences in actin-activated ATPase reported for S1 and HMM (Wagner and Weeds, 1977; Wagner, 1977), they may provide a gradation in the strength of cross-bridge attachment to actin along the length of the filament. Despite a considerable effort by many laboratories over the past years, surprisingly little is still known about the functional role of the "alkali" light chains in myosin. It is hoped that by the isolation of individual isoenzymes, and by the ability to detect these isoenzymes in muscles with diverse physiological properties, we may soon gain fresh insights into why nature created these diverse forms.

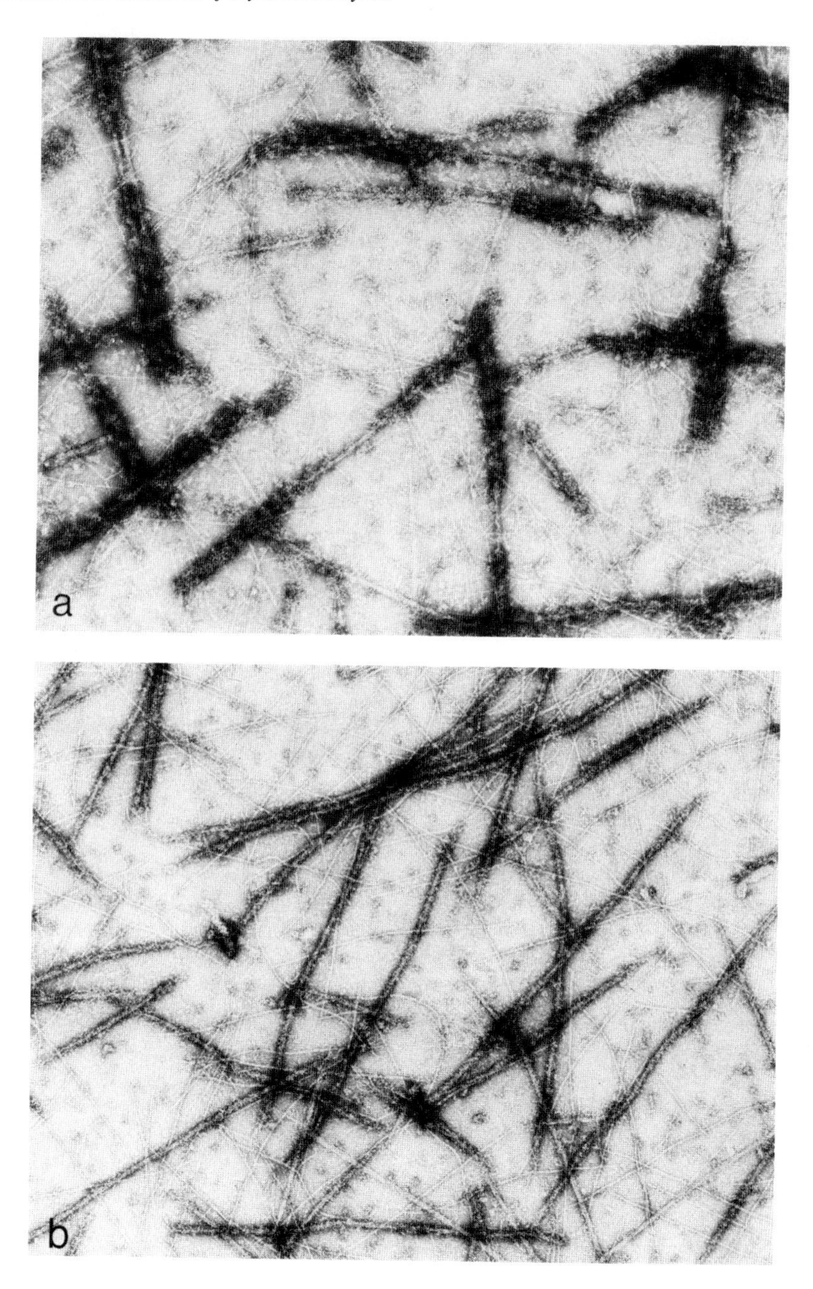

Fig. 12. *Electron microscopy of myofilaments reacted with antibody specific for A1. (a) Filaments were reacted with anti-Δ1, followed by goat anti-rabbit antibody. (b) Filaments were reacted with rabbit nonimmune IgG, followed by goat anti-rabbit antibody, as above. The image in (a) is typical of many fields which showed that essentially all myosin filaments react with either anti-Δ1 or anti-Δ2. ×44,000.*

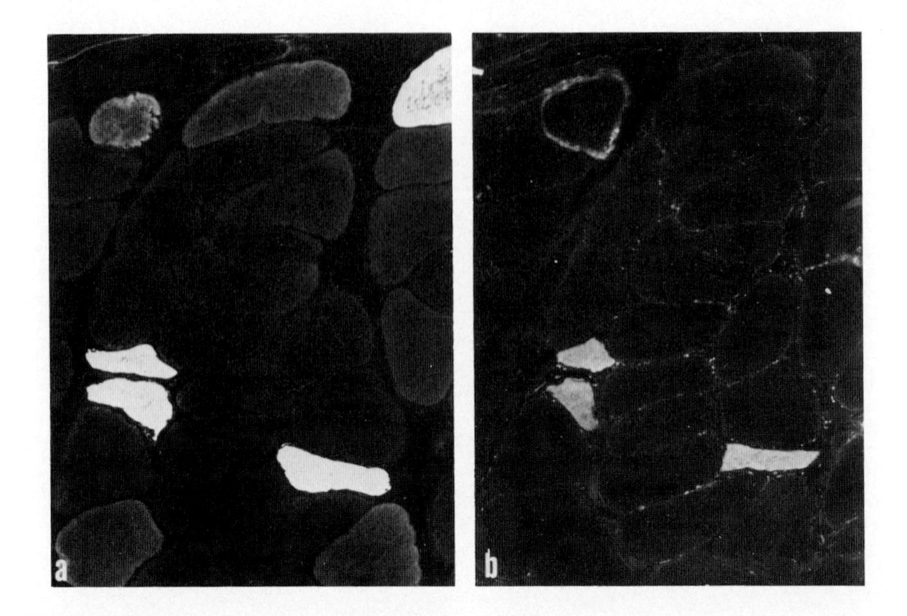

Fig. 13. Chicken anterior latissimus dorsi (ALD), serial transverse sections. (a) Anti-Δ1. (b) Anti-Δ2. Although most fibers in this "slow red" muscle fail to react with either antibody, a minority are reactive. However, the response to Anti-Δ1 (a) is more intense than the response to Anti-Δ2 (b). This suggests that the reactive fibers contain different amounts of A1 and A2 myosins. Sections of posterior latissimus dorsi (PLD) exposed to identical conditions responded to these two antibodies with equal intensity, indicating that the differences observed were not a result of varying levels of antibody concentration. ×200.

ACKNOWLEDGMENTS

This study was supported by Grants AM-17964 (G.F.G) and AM-17350 (S.L.) from the United States Public Health Service. GB-38203 (S.L.) from the National Science Foundation, and by grants from the Muscular Dystrophy Association, Inc., to G.F.G. and to S.L.

REFERENCES

Adelstein, R. S., Chacko, S., Barylko, B., and Scordilis, S.P. (1976). *In* "Contracile Systems in Non-Muscle Tissues" (S.V. Perry *et al.*, eds.), pp. 153-163. Elsevier/North Holland Biochemical Press.

Arndt, I., and Pepe, F. A. (1975). *J. Histochem. Cytochem. 23,* 159-168.

Clarke, M., and Spudich, J. A. (1977). *Ann. Rev. Biochem. 46,* 797-822.

Cohen, C., Lowey, S., Harrison, R. G., Kendrick-Jones, J., and Szent-Györgyi, A. G. (1970). *J. Mol. Biol. 47,* 605-609.

Craig, R. C., and Offer, G. W. (1976). *Proc. R. Soc. London, Ser. B. 192,* 451-461.

Elliott, A., Offer, G., and Burridge, K. (1976). *Proc. R. Soc. London, Ser. B. 193,* 45-53.

Frank, G., and Weeds, A. G. (1974). *Eur. J. Biochem. 44,* 317-334.

Gauthier, G. F., and Lowey, S. (1977). *J. Cell Biol. 74,* 760-779.

Holt, J. C., and Lowey, S. (1975). *Biochem. 14,* 4600-4609.

Holt, J. C., and Lowey, S. (1977). *Biochemistry 16,* 4398-4402.

Lowey, S., and Risby, D. (1971). *Nature 234,* 81-85.

Lowey, S., Slayter, H. S., Weeds, A. G., and Baker, H. (1969). *J. Mol. Biol. 42,* 1-29.

Mendelson, R. A., Morales, M. F., and Botts, J. (1973). *Biochemistry 12,* 2250-2255.

Moore, P. B., Huxley, H. E., and DeRosier, D. J. (1970). *J. Mol. Biol. 50,* 279-295.

Pollard, T. D., and Weihing, R. R. (1974). *CRC Crit. Rev. Biochem. 2,* 1-65.

Sarkar, S. (1972). *Cold Spring Harbor Symp. Quant. Biol. 37,* 14-17.

Slayter, H. S., and Lowey, S. (1967). *Proc. Nat. Acad. Sci. U.S.A. 58,* 1611-1618.

Wagner, P. D. (1977). *FEBS Lett. 81,* 81-85.

Wagner, P. D., and Weeds, A. G. (1977). *J. Mol. Biol. 109,* 455-473.

Weeds, A. G., Hall, R., and Spurway, N.C.S. (1975). *FEBS Lett. 49,* 320-324.

Weeds, A. G., and Lowey, S. (1971). *J. Mol. Biol. 61,* 701-725.

Motility in Cell Function
Proceedings of the First John M. Marshall Symposium in Cell Biology

MYOSIN FILAMENTS AND CROSS BRIDGE MOVEMENT*

William F. Harrington,[*] Kazuo Sutoh,[†]
and
Yu-Chih Chen Chiao[*]

[*]Department of Biology and McCollum-Pratt Institute
Johns Hopkins University
Baltimore, Maryland

[†]Department of Biophysics and Biochemistry,
Faculty of Science
University of Tokyo
Hongo, Tokyo, Japan

INTRODUCTION

The question of how the myosin cross bridges move to interact with neighboring thin filaments following activation is one of the most fascinating problems in the muscle field at the present time. According to current ideas, which are based mainly on low-angle X-ray diffraction studies (Huxley and Brown, 1967; Huxley, 1968; Haselgrove and Huxley, 1973; Haselgrove, 1975), the bridges are held rigidly in close proximity to the thick filament backbone in a highly ordered helical arrangement in the resting state of muscle. Changes in the equatorial reflections in the X-ray diagram have been interpreted as indicating that the cross bridges undergo both a radial and azimuthal movement to make contact with their appropriate thin filaments following stimulation (Huxley, 1968; Haselgrove and

**This work was supported by Research Grant AM 04349 from the National Institutes of Health. Contribution No. 976 from the McCollum-Pratt Institute, The Johns Hopkins University.*

ISBN 0-12-551750-5

Huxley, 1973). In the two-stranded model of the vertebrate thick filament originally proposed by Huxley and Brown (1967), the bridges extend out to a radius of about 130 Å from the center of the filament and there is a gap of 50--60 Å between the outer tips of the bridges and the surface of the thin filaments which lie on a radius of 180--190 Å. Thus a very substantial radial movement would be required to allow attachment to the thin filament surface. In the recently proposed three- and four-stranded models of Squire (1972) the myosin projections reach out to a radius of 180--200 Å in their resting state and would therefore require little or no radial movement to make contact with neighboring thin filaments. When muscle contracts, however, some type of radial movement of the cross bridges would seem to be necessary to allow the bridges to accommodate to the lateral expansion of the double hexagonal lattice formed by the thick and thin filaments in the overlap regions as the thin filaments are drawn into the A-bands. Present evidence (Huxley, 1972; 1976; Taylor, 1975) suggests that the rod segment which connects the LMM and head portions of the molecule, the S-2 region, may swing away from the surface of the thick filament backbone to permit the S-1 subunits to search out and attach to the thin filaments at all sarcomere lengths during each contractile cycle of a cross bridge. As is now well established, the isolated S-2 fragment, unlike the LMM segment of the myosin tail, shows little tendency to associate either with itself or with light meromyosin over a broad range of pH and ionic strengths (Lowey *et al.*, 1967; 1969). Thus, it has been argued (Huxley, 1969), this segment could disengage and swing freely away from the filament backbone.

In pondering how the cross bridge might move to attach to its appropriate actin filament, two questions come to mind. First, what is the process which unlocks the bridges from their ordered, resting state orientation, where they are presumably held close to the thick filament surface, and allows them to enter the cross bridge cycle? Second, does movement of the S-1 subunit really involve release of the S-2 segment from the filament backbone? Here I would like to summarize recent work in our laboratory which bears on these questions. In the first part I will review briefly our crosslinking studies which provide information on the freedom of movement of the cross bridges in synthetic filaments as well as in glycerinated myofibrils obtained from rabbit psoas muscle in rigor under various environmental conditions. Some of these results have already been published (Sutoh and Harrington, 1977). Then, I will present some of our recent experiments on the solution properties of the long S-2 fragment recently isolated by Weeds and Pope (1977) which may help to clarify the role of this region of the myosin molecule in the contractile cycle.

CROSSLINKING OF SYNTHETIC MYOSIN FILAMENTS

Changes in the orientation and radial disposition of the S-1 subunits in myofilaments under various environmental conditions can in principle be investigated through the use of bifunctional crosslinking reagents. The kinetics of crosslinking should be a sensitive indicator of the freedom of movement of the bridges. In our laboratory we have employed two imido ester crosslinkers for this purpose (dimethyl 3,3'-dithiobispropionimidate (DTBP) and methyl 4-mercaptobutyrimidate (IM-SH)). These reagents react very specifically with lysine side chains without altering the net charge on the protein (Wang and Richards, 1974; Traut *et al.*, 1973). In a typical experiment (Sutoh and Harrington, 1977) synthetic filaments were prepared at pH 8 (Josephs and Harrington, 1966), then dialyzed against neutral pH solvent (pH 7.4) prior to addition of the crosslinker. The crosslinking reaction was allowed to proceed for various periods of time at a constant temperature (4°C) in the presence of a fixed amount of the crosslinking reagent or for a fixed time with differing concentrations of crosslinker. To follow the time course of crosslinking the head and rod segments of the myosin molecules, aliquots of partially crosslinked filaments were removed at various times of reaction and digested with papain or chymotrypsin in the presence of EDTA to cleave the head-rod junction (Margossian and Lowey, 1973; Weeds and Taylor, 1975; Bagshaw, 1977). The proteolytic digest was analyzed by densitometry following electrophoresis on polyacrylamide gels in the presence of SDS.

When synthetic myosin filaments are crosslinked with DTBP and samples of this reacting system digested with papain or with chymotrypsin (in the presence of EDTA) at various stages of the reaction, three bands corresponding to the myosin heavy chain, myosin rod, and S-1 subunit are observed on SDS gels. During the course of the reaction these bands decrease in intensity indicating that both the rod and head segments of myosin are immobilized. No new bands are detected on the SDS gels during the crosslinking reaction, but we do see a rapid increase in high molecular weight crosslinked material at the tops of the gels. The myosin heads are clearly crosslinked to the thick filament surface much more rapidly than to adjoining heads of the same molecule since there is no indication of a band corresponding to the S-1 dimer. During the crosslinking reaction the time course of the intensity change of bands corresponding to the heavy chain and rod segment of myosin decrease virtually at the same rate consistent with a low level of intramolecular head-head crosslinking (Fig. 1). The rate of the intensity change in the S-1 band is about one third of that observed for the heavy chain and rod suggesting that the rate of immobilizing

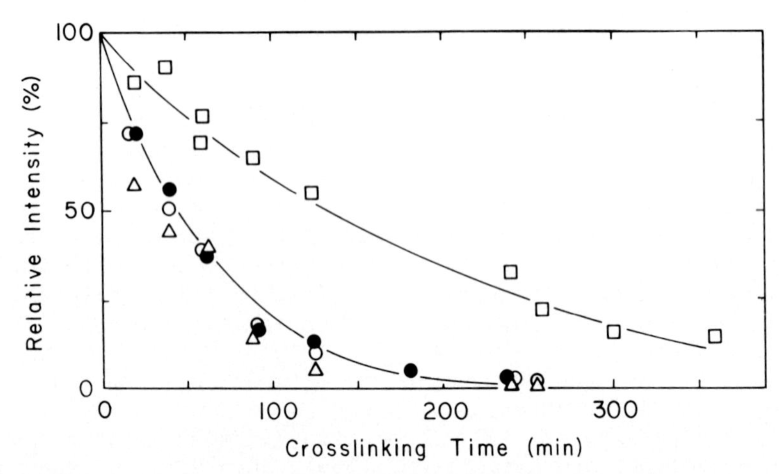

Fig. 1. Time course of crosslinking various segments of myosin. The relative intensities of the heavy chain band (O). the rod band (Δ), and the subfragment-1 (S-1) band (□) was determined by densitometry of SDS gels (5% acrylamide), following electrophoresis of crosslinked synthetic thick filaments digested with papain. The relative intensity of the heavy chain band of crosslinked synthetic thick filaments in the absence of proteolytic digestion is also shown (●). Solvent conditions for the crosslinking reaction: 80mM KCl, 40 mM imidazole, and 0.1 mM MgPPi (pH 7.4). Crosslinker: DTBP (0.6 mg/ml). Temperature of crosslinking, 4°. (From Sutoh and Harrington, 1977.)

either the intact heavy chain or the rod segment is about three times as fast as the rate of immobilizing the S-1 subunits at neutral pH.

The time course of crosslinking the S-1 subunits to the filament core is unchanged in the presence of the low molecular weight ligands responsible for activation of muscle. Figure 2 shows that within the limits of experimental error, all the points fit a single decay curve regardless of the presence or absence of MgATP (5 m*M*), or Ca^{2+} ions (1 m*M*). The same result was observed when IM-SH was used as crosslinker. It appears from these studies that the S-1 subunits are held close to the thick filament surface under solvent conditions where the enzymatically active myosin heads are maintained in the high energy ($Mg^{**}ADP \cdot P_i$ state (Lymn and Taylor, 1970) and where the divalent-metal binding sites of myosin (the DTNB light chains located on the S-1 subunits (Morimoto and Harrington, 1974)) are fully saturated with calcium. This result is consistent with recent depolarization measurements (Mendelson and Cheung, 1976) which have revealed that the rotational Brownian motion of fluorophore-labelled myosin heads is strongly hindered in synthetic thick filaments and relaxed glycerinated myofibrils

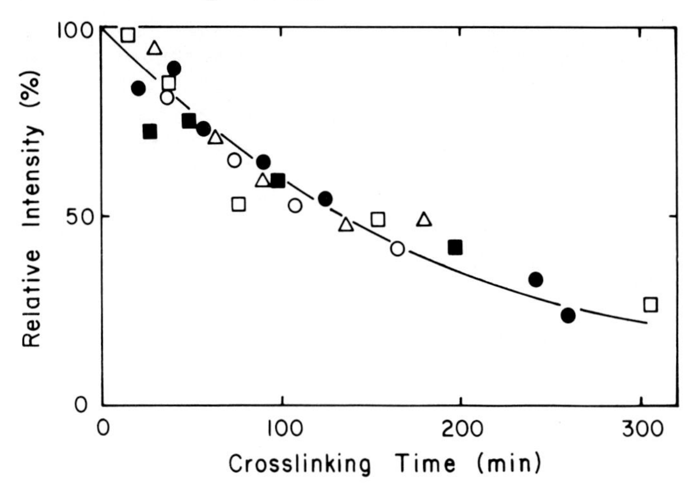

Fig. 2. Effect of various ligands on the time course of crosslinking myosin heads to the thick filament surface with DTBP (0.6 mg/ml). (Δ) No ligand; (●) 0.1 mM MgPPi; (O) 1 mM MgADP; (□) 5 mM MgATP, 1 mM $CaCl_2$; (■) 5 mM MgATP, 1 mM EGTA. The buffer system consists of 80 mM KCl and 40 mM imidazole (pH 7.4). Temperature of crosslinking, 4°. (From Sutoh and Harrington, 1977.)

at neutral pH as compared to their rotational mobility in dispersed myosin molecules. The mobility of the heads is unchanged in synthetic thick filaments in the presence of Ca^{2+} ions, contrary to the result expected if the S-1 subunits were detached from the filament surface and allowed to oscillate about the head-rod hinge.

CROSSLINKING OF GLYCERINATED MYOFIBRILS IN RIGOR

Since the radial and angular distribution of S-1 subunits in synthetic thick filaments could be quite different from their arrangement in native thick filaments we have also investigated the crosslinking kinetics of glycerinated myofibrils (at rest length) where the contractile apparatus is virtually intact. It is now generally believed that under rigor conditions most if not all of the cross bridges are bound along the thin filament surface with the long axis of the S-1 subunits lying at an angle of about 45° to the thin filament axis (Huxley, 1972). Changes in the equatorial X-ray reflections when a living resting muscle goes into rigor have been interpreted as indicating that the cross bridges swing away from the surface of the filament core and lie with their center of mass nearer to the axis of the actin filaments (Huxley, 1968; Haselgrove and Huxley, 1973). We therefore expected to find a dramatic change in the time course of crosslinking the S-1 subunits in glycerinated myofibrillar preparations in rigor as compared to synthetic filaments. In these studies, where the conditions of the crosslinking reaction were identical to those employed for the synthetic filaments, we found that the rate of crosslinking the constituent components of the thin filament complex (actin, tropomyosin and troponin) by DTBP is much lower than that of the myosin heavy chains. The relative intensities of the subunit bands of these species remain nearly unchanged with time on SDS gels. When the rates of crosslinking the various regions of the myosin molecule are followed by digesting either with papain or with chymotrypsin in the presence of EDTA, the digested myofibrils show very similar banding patterns following electrophoresis on SDS gels. During the crosslinking reaction the relative intensities of bands corresponding to the S-1 subunit and the rod segment of myosin decrease with a first-order decay just as they do in the case of the synthetic thick filaments. Moreover, no new bands were detected on the gels during the crosslinking reaction which would correspond to the S-1-actin complex. Indeed, in separate crosslinking experiments of the Acto-S-1 complex using DTBP as crosslinker we have been unable to detect crosslinking of the myosin head to actin under these solvent conditions (see legend to Fig. 1) although a small amount of S-1 dimer is formed.

To compare the kinetics of crosslinking of myosin heads in the glycerinated myofibrils with that of the synthetic filaments, the time-dependent changes in the relative intensity of the heavy chain bands in the earlier synthetic filament experiment and the rigor myofibril system have been normalized by adjusting the abscissa scales so that the two curves corresponding to the relative intensities of the rod band superimpose. This procedure compensates for differences in the overall rates of the crosslinking reaction due to differences in the concentrations of the myosin filaments as well as differences in the crosslinking environment in the two systems. When this is done it can be seen (Fig. 3) that there is no observable difference between the two S-1 kinetic curves suggesting that the proximal surface of the S-1 subunits remain attached or at least in close physical proximity to the thick filament surface in the rigor myofibril just as in the crosslinking studies of the synthetic thick filament system.

It is unlikely that the myosin heads are dissociated from the thin filament surface as a result of chemical modification during the crosslinking of glycerinated myofibrils with DTBP since we find that after extensive reaction of glycerinated myofibrils with the homologous crosslinker dimethyl malonimidate, followed by digestion with chymotrypsin, the S-1 subunits are

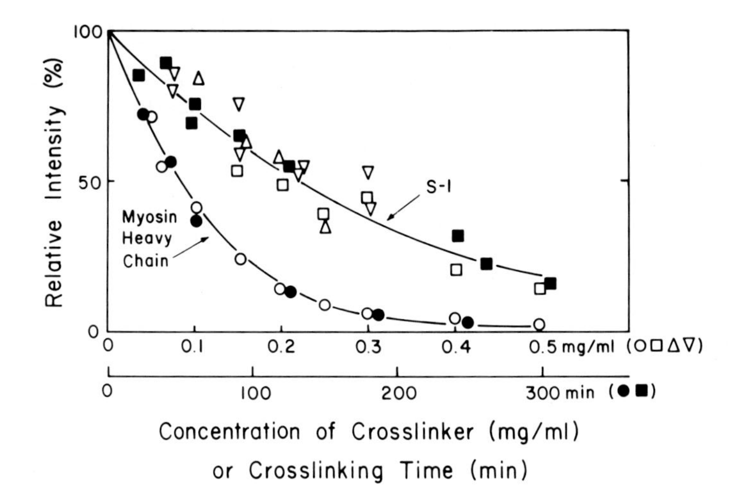

Fig. 3. Kinetics of crosslinking myosin heavy chains and myosin heads in the glycerinated rigor myofibril (open symbols) and in the synthetic thick filament (closed symbols). The relative intensities of the myosin heavy chain band (○,●), the S-1 band after papain digestion (□,■), and the S-1 band after chymotrypsin digestion ((△); 0.2 mg/ml for 10 min; (▽) 0.3 mg/ml for 15 min are shown. Note that the abscissa scales are normalized so that the intensity change in the myosin heavy chain band follows a single decay curve in both cases. Conditions for crosslinking myofibrils: solvent, 80 mM KCl, 40 mM imidazole, and 1 mM EGTA (pH 7.4); crosslinker, DTBP; crosslinking time, 16 hr at 4°C. See Fig. 1 for conditions of crosslinking synthetic thick filaments. (From Sutoh and Harrington, 1977.)

still able to bind to the thin filaments to form the rigor complex in the absence of MgATP and can be centrifuged down with the modified myofibrils. In the presence of MgATP (5 m*M*) virtually all of the heads are released from the thin filaments and remain in the supernatant after centrifugation. Although dimethyl malonimidate reacts with lysine side chains in an identical manner to DTBP this bifunctional reagent has only one methylene group linking the two reactive imidate groups and, because of its shorter length, is unable to bridge between the myosin heavy chains in the myofibrils even after extensive amidination. It is also unlikely that the myosin cross bridges are forced down onto the thick filament surface during crosslinking of the myofibrils as a result of lateral contraction of the filament lattice since X-ray diffraction measurements (J. C. Haselgrove and J. Murray, personal communication) show no measurable difference in lattice spacing between noncrosslinked and crosslinked samples where 80% of the myosin heavy chains were immobilized by the bifunctional reagent.

The simplest interpretation of these results is that in all stages of the crosslinking reaction of rigor myofibrils at rest length the lower surface of the S-1 subunit is held close to the thick filament core.

EVIDENCE FOR CROSS BRIDGE RELEASE AT HIGHER pH

Although the S-1 subunits are rapidly crosslinked to the filament surface at neutral pH, their rate of crosslinking falls off rapidly at higher pH suggesting that the myosin heads may be released from the filament surface under these conditions. In these studies dimethyl suberimidate was used in place of DTBP since the disulfide bond in this cleavable crosslinker is expected to be unstable at alkaline pH.

Synthetic filaments were crosslinked for various periods of time at different pH values in the range 6.8-8.3, then digested with chymotrypsin in the presence of EDTA to cleave the head-rod linkage and the time course of crosslinking monitored by densitometry of SDS gels following electrophoresis. At each pH the time course of crosslinking the S-1 subunits as well as the rod segments of myosin followed first order kinetics. The rate of crosslinking each segment is expected to increase about twofold over the pH range examined due to the pH dependence of the amidination reaction (Hunter and Ludwig, 1962) and to eliminate this factor, which is unrelated to structural changes within the thick filament, we have compared the ratio of crosslinking rates of the two fragments (k_{S-1}/k_{rod}) at each pH. Figure 4 shows that the normalized rate of crosslinking the S-1 subunit falls rapidly over the pH range 7.4-8.0 and levels off at about 10% of its neutral pH value above pH 8.0. The presence of 1 m*M* $CaCl_2$ or 1 m*M* EGTA in the crosslinking solvent had no detectable effect on the crosslinking rate of the S-1 subunit at pH 8.3.

Similar experiments with glycerinated muscle fibers in rigor show that crosslinking of the S-1 subunits in these preparations is also markedly depressed at high pH (8.0-8.3) suggesting an outward radial movement of the myosin heads as compared with their immobilized state near the filament surface at neutral pH. This result is consistent with Rome's 1968 observation that the lattice spacing of glycerinated muscles expands on raising the pH. The depolarization of fluorescence measurements of Mendelson and Cheung (1976) also suggest that the heads are released at high pH. These workers observed a marked increase in the rotational mobility of the S-1 subunits of synthetic filaments at pH 8.3 as compared to its value at neutral pH.

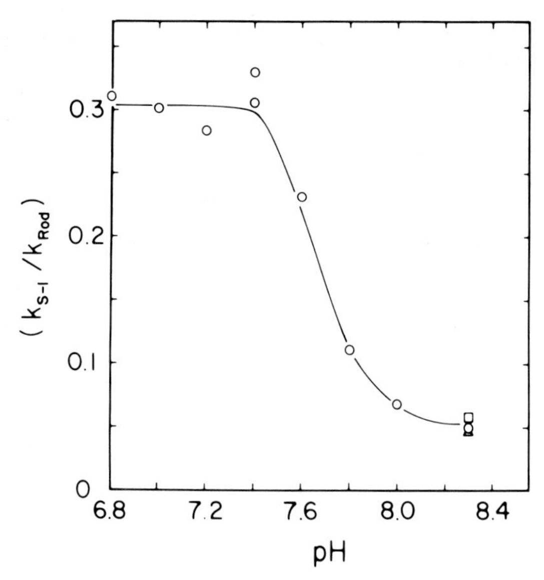

Fig. 4. Titration curve of the "normalized" rate of crosslinking myosin heads (k_{S-1}/k_{rod}). Crosslinking was carried out in (a) 80 mM KCl, 40 mM imidazole, and 0.1 mM $MgPP_i$ (pH 6.8-7.4) or (b) 80 mM KCl and 40 mM triethanolimine (pH 7.4-8.3). Crosslinking was also carried out in solvent (b) in the presence of 1mM $CaCl_2$ (□) and 1 mM EGTA (Δ) at pH 8.3. (From Sutoh et al., *1978a).*

Although the myosin heads appear to be released from the thick filament surface above pH 8 in both the synthetic filament and glycerinated fiber preparations, our present evidence suggests that a major fraction of the S-2 segment of HMM remains associated with the surface under these conditions. For one thing, densitometry of the SDS gels shows a gradual monotonic increase in the rate of crosslinking the rod segment of myosin in synthetic filaments over the pH range 6.8-8.3. This result is consistent with the pH dependence expected for the amidination reaction and suggests that the effective surface area for crosslinking the rod segment remains unchanged over this pH range. Additionally, we find that the time course of crosslinking HMM at pH 8.3 is comparable to that of the LMM segments of myosin, which are rigidly locked into the filament core, and very much faster than the rate of immobilization of the S-1 subunits. Figure 5 shows the decay in the relative intensities of the LMM and HMM bands compared to that of the S-1 band on SDS gels as these regions of the myosin molecule were gradually immobilized in synthetic thick filaments at pH 8.3 in the presence of dimethyl suberimidate. In this experiment the partially crosslinked filaments were digested at each stage of the reaction with chymotrypsin either in the presence

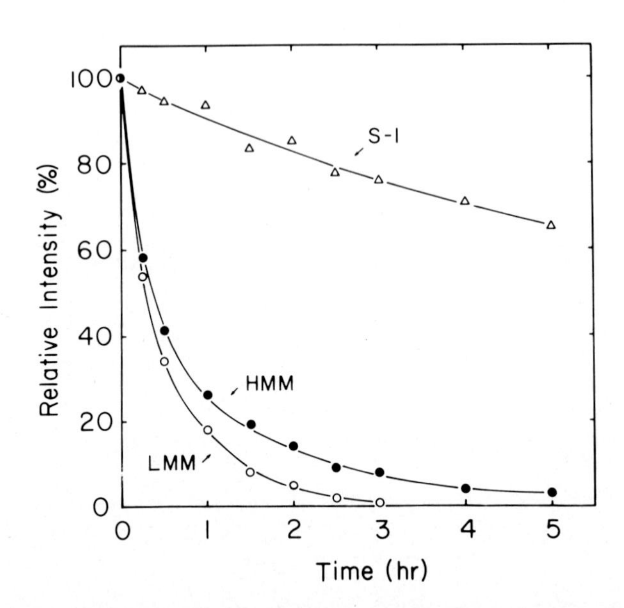

Fig. 5. Time course of crosslinking S-1, HMM and LMM segments in synthetic thick filaments at pH 8.3. The relative intensities of the S-1 (Δ), HMM heavy chain band (●), and the LMM band (O) were determined by densitometry of SDS gels of crosslinked filaments digested by chymotrypsin in the absence (S-1) and in the presence (LMM and HMM) of 1 mM $CaCl_2$. Crosslinking conditions: solvent, 80 mM KCl and 40 mM triethanolamine (pH 8.3); myosin filament 1 mg/ml; crosslinker: dimethyl suberimidate, 0.6 mg/ml; temperature, 4°C. (From Sutoh et al., *1978a.)*

of EDTA (1 m*M*) to release the S-1 subunits or in the presence of Ca^{2+} ions (1 m*M*) to release the noncrosslinked LMM and HMM fragments.

ASSOCIATION BEHAVIOR OF LONG S-2

There is good evidence that the S-2 fragment, isolated from papain digests of insoluble myosin, or by tryptic cleavage of dispersed myosin rod, has little tendency to associate either with itself or with LMM over a wide range of ionic strengths and pH conditions. This S-2 fragment has a weight in the range of 60-80,000 daltons (Lowey *et al.*, 1967; Biro *et al.*, 1972; Goodno *et al.*, 1976). Weeds and Pope (1977) have recently demonstrated that the S-2 tail segment of HMM, released by chymotryptic digestion of myosin in the presence of millimolar concentrations of Ca^{2+} or Mg^{2+} ions, is considerably larger than this, with weight of the isolated S-2 rod segment near 120,000 daltons as determined by its electrophoretic mobility on SDS gels. In view of our crosslinking experiments showing that this portion of the myosin molecule is held close to the thick filament surface in all of the solvent systems so far examined, it was of interest to investigate the association properties of this long S-2 fragment.

Figure 6 shows velocity sedimentation patterns of synthetic filaments which had been digested by trypsin in the presence of Ca^{2+} (1 m*M*) at pH 8.3 and the reaction stopped by addition of trypsin inhibitor. SDS gel electrophoresis patterns revealed that under the digestion conditions used virtually all of the myosin molecules have been cleaved into LMM and HMM containing the long S-2 tail. The faster sedimenting peak is the residual

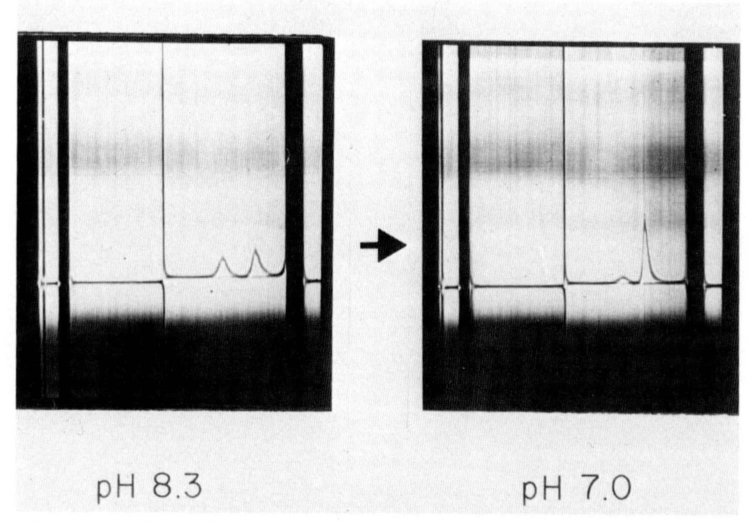

Fig. 6. Velocity sedimentation patterns of synthetic thick filaments which had been digested by trypsin in the presence of Ca^{2+} *(1 mM) at pH 8.3. Concentration of filaments, 1 mg/ml; proteolysis was carried out for 8 min (5°C) at a trypsin: myosin ratio (w:w) of 1/50 in pH 8.3 buffer consisting of 0.08* M *NaCl, 0.04* M *triethanolamine, 1* mM Ca^{2+}. *Solution dialyzed overnight (5°) versus 0.08* M *NaCl, 0.04* M *imidazole, 1* mM Ca^{2+}, *0.1* mM *MgPPi, pH 7. Rotor speed, 15,000 rpm; 30* mm *synthetic-boundary cell. Tryptic digestion terminated by addition of trypsin inhibitor (w:w = 1/25). SDS gel electrophoresis patterns revealed that virtually all of the myosin had been transformed into LMM and HMM (long S-2).*

LMM core of the thick filaments and its associated HMM. The slower peak is HMM. Judging from the areas of the two peaks, about 50% of the HMM is released at pH 8.3 and 50% is associated with the LMM core. When the pH of this system is lowered by dialysis against a pH 7 buffer at the same ionic strength (0.12 *M*) nearly all of the HMM is found to be associated with the LMM filament core. Thus this experiment indicates that the long S-2 segment has a measurable tendency to associate with the LMM backbone over this pH range.

In contrast to the low molecular weight (60-80,000 daltons) S-2 fragment, the long S-2 particle isolated from chymotryptic (or tryptic) HMM through further chymotryptic proteolysis of this myosin fragment in the absence of divalent metal ions (1 m*M* EDTA) to remove the globular heads, shows a marked tendency to self-associate. The reduced viscosity ($\eta_{sp/c}$) of this water-soluble, α-helical fragment increases rapidly with concentration (Fig. 7) at neutral pH giving a concave-down

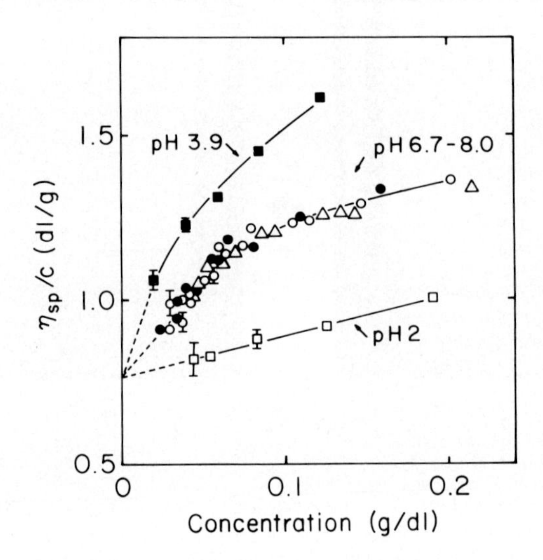

Fig. 7. Reduced viscosity ($\eta_{sp/c}$) versus protein concentration plots of long S-2 fragment of myosin in various solvent systems. (△) 0.1 M *NaCl, 20 mM phosphate (pH 8.0); (○), 0.1* M *NaCl, 20 mM phosphate (pH 7.1); (●) 0.1* M *NaCl, 20 mM phosphate (pH 6.2); (■) 0.1* M *NaCl, 20 mM Na acetate-acetic acid (pH 3.9); (□) 0.1* M *NaCl, 10 mM HCl (pH 2). Temperature = 10°C. The long S-2 fragment was isolated either from tryptic HMM or chymotryptic HMM prepared according to the procedure of Weeds and Taylor (1975) HMM (3 mg/ml in NaCl, 20 mM phosphate, 2 mM EDTA was further digested by chymotrypsin (0.01 mg/ml) for 10 min at 25°. The long fragment was purified by ethanol fractionation (3:1, v:v) followed by dialysis versus 20 mM phosphate (pH 4.5). (From Sutoh* et *al., 1978b.)*

profile indicative of association of the S-2 particle to form higher molecular weight polymeric species with axial asymmetry greater than that of the monomer. This concentration-dependent profile is unchanged over the range pH 6.8-8.3. At pH 2, self-association is depressed and the reduced viscosity versus concentration plot is linear with slope and intrinsic viscosity characteristic of a monomeric species of high axial ratio.

The tendency for the long S-2 fragment to undergo self-association has also been confirmed by sedimentation equilibrium experiments. In these studies, we have used a column purified (G-200 Sephadex) sample of long S-2 which gives a single band of subunit molecular weight 60,000 daltons on SDS gels. Reciprocal molecular weight versus concentration plots derived from high-speed meniscus depletion studies at several loading concentrations and rotor velocities show a relatively small concentration dependence of the apparent number average, weight average, and Z-average molecular weights at pH 7 (Fig. 8). Although the modest concentration dependence observed at neutral pH would normally be regarded as a good test for the absence of significant self-association, in the case of the highly asymmetric S-2 fragment it is clear that a large nonideal term (2nd virial coefficient) is acting to mask the expected downward-curvature in the reciprocal plots. This interpretation is supported by a set of sedimentation equilibrium studies at pH 2 where, according to the viscosity measurements, self-association of the S-2 particles should be eliminated. In this case, the reciprocal plots are linear with a large positive concentration dependence corresponding to the 2nd virial coefficient. The ratio of slopes for the three molecular weight averages is that expected for a homogeneous, monomeric system ($B_N = 1/2B_W$ and $B_W = 1/2B_Z$). It can be shown (Roark and Yphantis, 1972) that the net positive charge on the S-2 particles at pH 2 accounts for only a fraction of the 2nd virial term under these ionic conditions ($I = 0.12\ M$).

We have employed the reciprocal molecular weight versus concentration plots shown in Fig. 8 to search for the assembly mode of long S-2 using a variety of procedures currently in use to analyze nonideal self-associating systems. Neither a simulated monomer-dimer or monomer-nmer type association appears to fit the experimental data nearly as well as an indefinite (isodesmic) association. For this mode of association 2 $M_N\ M_W$ should be invariant with protein concentration in an ideal system and we find that this quantity, corrected for the virial coefficient, remains constant within a few percent over the concentration range investigated in the present study (0.01-0.3 g/dl).

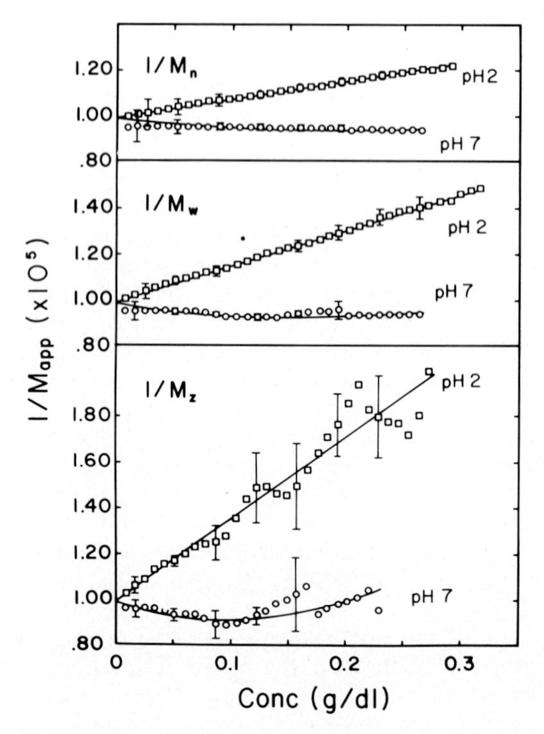

Fig. 8. Reciprocal molecular weight versus concentration plots of purified long S-2 fragment. Solvent used in pH 7 studies was 0.1 M *NaCl, 20* mM *phosphate; solvent used in pH 2 studies was 0.1* M *NaCl, 20* mM *HCl. Data was collected from three cell-loading concentrations (0.5, 1, and 2 mg/ml) and three rotor speeds (17,000, 18,000, and 22,000 rpm) at each pH and the nine experiments averaged to give the results shown. Upper frame,* $1/M_{N,app}$, *center frame,* $1/M_{W,app}$, *lower frame,* $1/M_{Z,app}$. *Error bars are standard deviations of the mean. Temperature, 5°C. (From Sutoh* at al., *1978b)*

Figure 9 shows the fit of the smoothed curves (points) derived from the reciprocal plots, corrected for the virial coefficient, with the expected profiles for the three molecular weight averages in an indefinite association system where $K = 1.2$ dl/g and the 2nd virial coefficient $B = 10 \times 10^{-6}$ mole-dl/g^2. The simulated data is derived by combining various equations for the three molecular weight averages which are given in terms of the 2nd virial coefficient and equilibrium constant (see Teller, 1973).

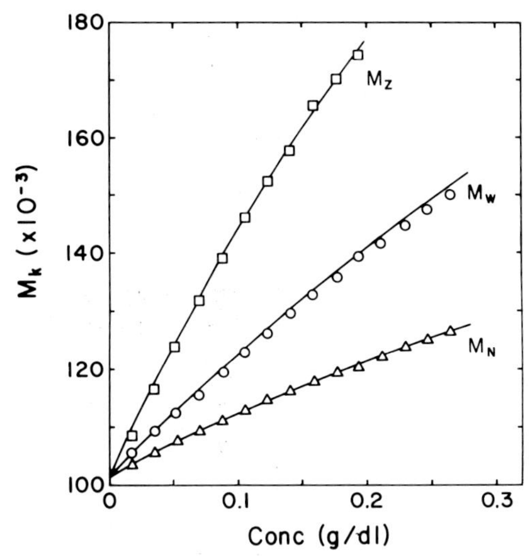

Fig. 9. Association of long S-2 fragment at pH 7. Points: taken from smoothed curves of Fig. 8, corrected for virial coefficient. (B = *10* × *10*$^{-6}$ *mole-dl/g*^{2}*); solid lines: theoretical profiles for indefinite (isodesmic) association assuming the equilibrium constant,* K = *1.2 dl/g. (From Sutoh* et al., *1978b.)*

The indefinite self-association observed for the long S-2 fragment suggests that this segment of the myosin molecule has a sticky end near the LMM-HMM junction which is missing from the shorter S-2 fragment investigated by earlier workers.

It has been known for many years that proteolytic digestion of dispersed myosin in high salt rapidly digests the junction between LMM and HMM yielding an HMM particle with an S-2 tail segment which varies somewhat in size but generally has a length of 400-500 Å (Lowey, 1971). Brief tryptic digestion of this HMM fragment (Balint *et al.*, 1975) in the absence of Ca^{2+} removes the heads and yields predominantly a 37,000 dalton subunit S-2 band following electrophoresis on polyacrylamide-SDS gels rather than the high molecular weight species (60,000 daltons on SDS gels) prepared by the Weeds and Pope procedure. Since this S-2 fragment does not undergo self-association at neutral pH, it seems likely that the sticky segment of long S-2 is the region near the LMM-HMM junction which is digested away during formation of the S-1 and S-2 fragments from HMM. We find (Sutoh *et al.*, 1978b) that the long S-2 fragment is particularly vulnerable to further proteolysis and is rapidly cleaved by trypsin to form a more resistant core segment of molecular weight ∿65,000 daltons. The rate of conversion of the long S-2 fragment into the short S-2 particle is about 20%

of the rate of conversion of rod into long S-2 and LMM suggesting again that the fragment of long S-2 digested away by trypsin resides near the LMM-HMM junction. The residual core fragment is similar in size to the S-2 fragment prepared by Lowey *et al.* (1967, 1969) and like the particle investigated by these workers shows no apparent tendency to self-associate over the range pH 2.0-7.1 in 0.12 *M* salt (see Fig. 10).

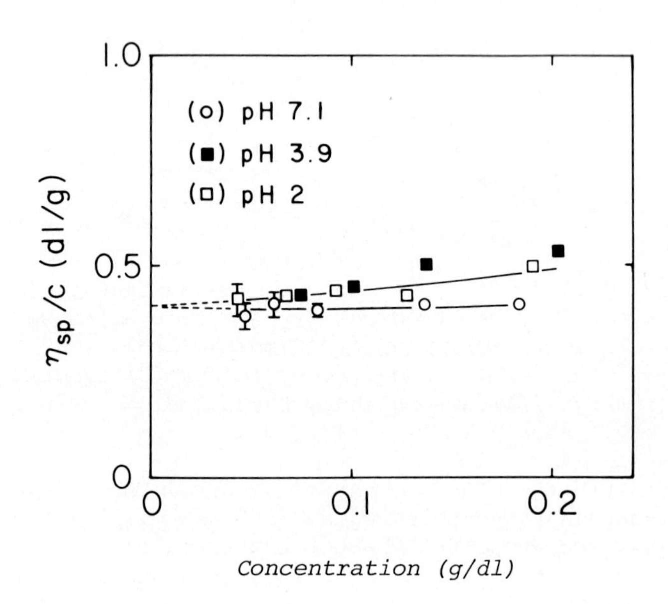

Fig. 10. Reduced viscosity versus concentration plots of the proteolytic fragments *of long S-2 in various solvent systems. (0) 0.1* M *NaCl, 20* mM *phosphate (pH 7.1); (●)* M *NcCl, 20* mM *Na acetate-acetic acid (pH 3.9); (□) 0.1* M *NaCl, 10* mM *HCl (pH 2). Temperature, 10°C. This proteolytic fragment was prepared by tryptic (0.5 mg/ml) digestion of long S-2 (1 mg/ml) in 0.5* M *NaCl, 20* mM *phosphate (pH 7.0) for 1 min at 25°. Following addition of trypsin inhibitor (1 mg/ml), the digest was dialyzed versus 20* mM *phosphate (pH 4.5) and the resulting precipitate dissolved in 20* mM *phosphate (pH 6.2). Electrophoresis on polyacrylamide-SDS gels showed one major band (80% of the total mass detected) corresponding to a molecular weight of 40,000 and several minor bands. (From Sutoh* et al., *1978b.)*

The molecular weight of long S-2 is near 100,000 based on the equilibrium sedimentation studies. This value is about 15% lower than the weight deduced from its electrophoretic mobility on SDS gels. The length of this segment of myosin, assuming that it is a two-chain coiled-coil of α-helices, would be 650 Å suggesting that the length of the sticky end segment is 200-250 Å. An indefinite association process is consistent with parallel association of the long S-2 fragments since anti-parallel assembly, which would involve mutual interaction of the two sticky ends, would be expected to stop association at the dimer state.

THE ROLE OF THE S-2 SEGMENT IN CROSS BRIDGE MOVEMENT

Parallel association of the long S-2 segments (i.e., with trypsin-sensitive regions pointing in the same direction) would also be consistent with their arrangement in each half of the bipolar thick filament. Along a row of cross bridges the myosin molecules are arranged systematically with a 430 Å displacement between heads (Huxley and Brown, 1967; Squire, 1972) and this displacement has also been deduced (Harrington and Burke, 1972) for the myosin dimers detected in solution (Godfrey and Harrington, 1970). It will be clear from Fig. 11, which depicts the relative arrangement of two myosin molecules in the myofilament along a row of cross bridges, that the sticky, protease-sensitive end-segment of the S-2 moiety of one molecule (hatched region) can interact over a length of ∿200 Å with the protease-resistant S-2 region of a neighboring molecule near its head/rod junction. It seems reasonable to assume that the interactions responsible for self-association of the long S-2 fragment at neutral pH and physiological ionic strength are conserved in the native thick filament and serve to hold the S-1 subunits close to the filament surface under these conditions. This interpretation would be consistent with our finding that the HMM fragment with the long S-2 tail produced by proteolytic-digestion of thick filaments in the absence of Ca^{2+} binds to the residual filament core under these ionic conditions. It is also supported by the crosslinking studies summarized earlier which reveal that the myosin heads are rapidly cross-linked to the thick filament surface at neutral pH. The depolarization of fluorescence measurements of Mendelson and Cheung (1976) provide additional evidence showing that the rotational Brownian motion of the myosin head is highly restricted in synthetic thick filaments and relaxed glycerinated myofibrils at neutral pH.

Schematic Representation of Two Myosin Molecules in the Thick Filament

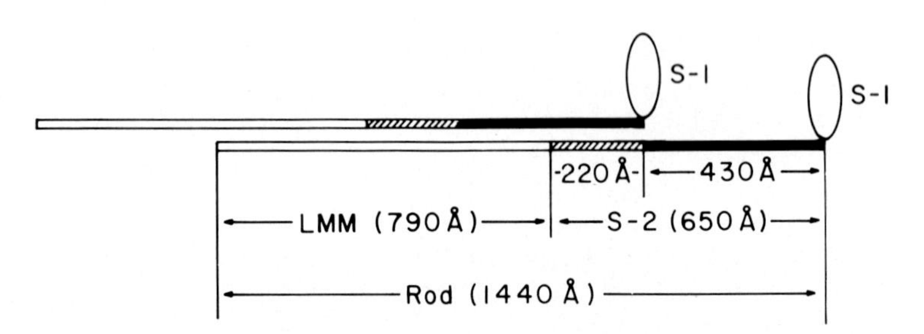

Fig. 11. Schematic representation of the arrangement of two myosin molecules (myosin dimer) along a row of cross bridges in the vertebrate thick filament. Lengths of various segments LMM, S-2, and rod were calculated from their molecular weights determined by sedimentation equilibrium. Hatched region corresponds to the segment of long S-2 digested away by trypsin. The shaded region corresponds to the small S-2 fragment. (From Sutoh et al., *1978b.)*

If, as suggested by the crosslinking studies, the lower surface of the S-1 subunit is always held close to the thick filament core in a cross bridge cycle, then some type of azimuthal or swivelling motion may allow the bridge to interact with its appropriate *actin filament*. As we have noted earlier, the myosin projections from the three- and four-stranded thick filament models of Squire (1972) extend to a radius of 180-200 Å in a relaxed muscle. Thus the outer tip of the cross bridge may not require significant radial displacement to make contact and may simply undergo a swivelling movement as has been recently proposed by Lymn and Cohen (1975) based on an analysis of the higher order equatorial reflections observed when vertebrate striated muscle is activated or goes into rigor. It is difficult to see how this swivelling process could accommodate the lateral expansion of the filament lattice during contraction, however, particularly if the bridge is both slewed and angled at 45° to the thin filament axis as it is assumed to be (Huxley, 1972) at the completion of its drive stroke.

If, on the other hand, the S-2 segment does actually swing away from the thick filament surface during each cross bridge cycle to accommodate the variable lattice expansion, then some type of transient modulation of the bonding interaction between the trypsin-sensitive end-region of the S-2 moiety of one myosin molecule and the trypsin-resistant S-2 region of a neighboring molecule would seem to be required. Although it is not possible to decide convincingly between these two alternatives, our recent optical rotation measurements of the purified S-2 fragment suggests that the thermal stability of this structure could be highly relevant to the swinging cross bridge mechanism. We find that the long S-2 fragment exhibits a cooperative helix-coil transition with midpoint (T_m) of the transition at 45°C, independent of pH over the range pH 6.8-8.3. No significant difference was observed in the melting profiles of the long S-2 fragment and its short proteolytic fragment at neutral pH. The protein concentration employed in these studies was ∿0.07 mg/ml. At this concentration over 95% of the long S-2 particles are expected to be present in the monomeric state at 5°C based on the equilibrium constant (1.2 dl/g) for indefinite association derived from the sedimentation equilibrium studies.

The striking feature of the melting profile shown in Fig. 12a is that 15-25% of the double α-helical structure is melted at 37°. Assuming that the S-2 segment is held rigidly onto the thick filament surface its melting temperature would be expected to be raised well above the value (45°C) estimated in free solution as a result of the stabilizing effect of the secondary bonding forces between the neighboring S-2 segments within the thick filament core. If the S-2 segment were to swing away from the surface, 15-25% of the double α-helical structure should undergo immediate melting at physiological temperature (Fig. 12b). This partially melted, flexible structure would be a good candidate for the series elastic element in the Huxley/Simmons (1971) model of contraction. It is well-established that such partially crystalline systems give length-tension behavior obeying Hooke's law (see, e.g., Flory, 1956) if the length change under tension is less than ∿50% of the maximum extension (β-configuration) of the random coil polypeptide chains.

If release of the S-2 element from the thick filament surface were to occur *after* its S-1 subunits were bound to the thin filament in the cross bridge cycle, the rapid partial melting of the S-2 segment would provide a large retractive force on the thin filament in the direction of the *M*-line (Harrington, 1971). It will be clear that this latter type of process could serve to generate tension in the absence of any angular movement of the subunit while it is attached to the actin filament.

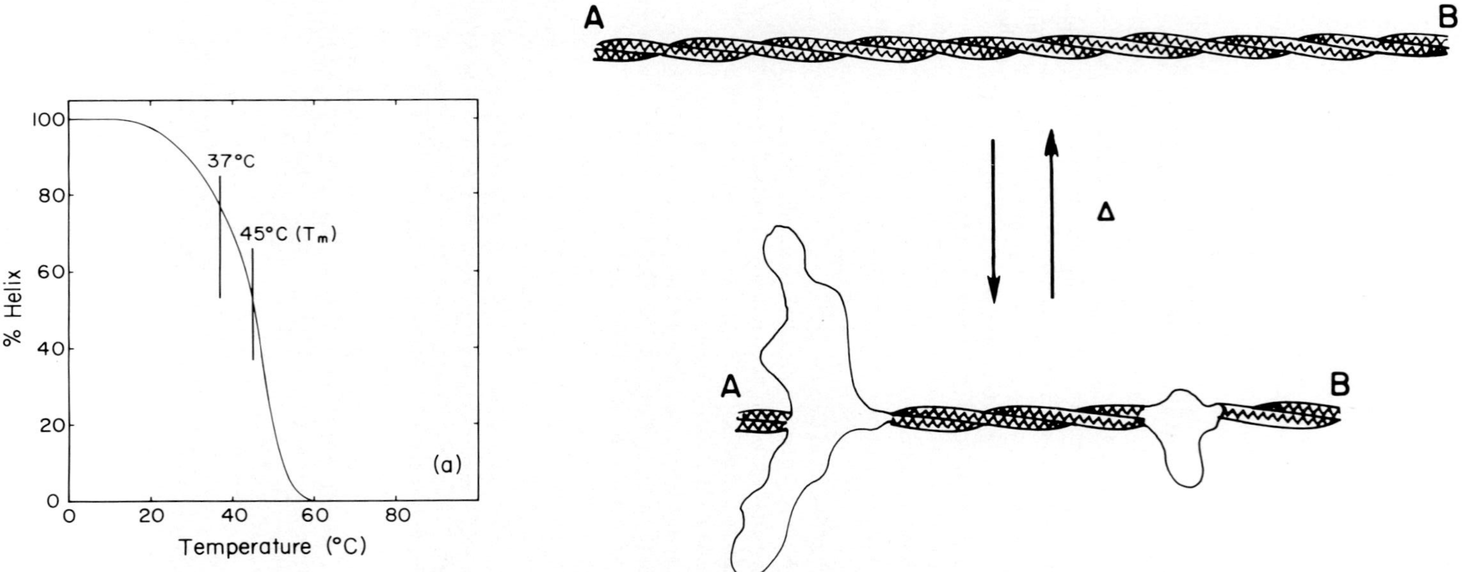

Fig. 12. (a) *Melting profile of long S-2 segment of myosin derived from optical rotation* [[M]$_{233}$] *versus temperature measurements of this fragment at pH 7.0. Solvent is 0.1* M *NaCl, 20 mM phosphate; protein concentration, 0.07 mg/ml. The midpoint of the double-helix→coil transition* (T_m) *occurs at 45°C. Note that at 37°C about 15–25% of the S-2 structure has undergone the sharp cooperative transition between the α-double helical and random coil states.* (b) *Schematic representation of the melting process in the S-2 segment where the crystalline (α-helical) order has disappeared completely in part of the structure and the constituent polypeptide chains have assumed a random chain configuration. This region should be rubberlike and readily extensible under tension. Note that if* A *and* B *are held fixed, melting of S-2 will generate a large contractile force due to the entropic change as the polypeptide chains gain configurational freedom.*

REFERENCES

Bagshaw, C. R. (1977). *Biochemistry 16,* 59.

Balint, M., Sreter, F. A., Wolf, I., Nagy, B., and Gergely, J. (1975). *J. Biol. Chem. 250,* 6168.

Biro, N. A., Szilagyi, L., and Balint, M. (1972). *Cold Spring Harbor Symposium Quant. Biol. 37,* 55.

Flory, P. J. (1956). *Science 124,* 53.

Godfrey, J. E., and Harrington, W. F. (1970). *Biochemistry 9,* 894.

Goodno, C. G., Harris, T. A., and Swenson, C. A. (1976). *Biochemistry 15,* 5157.

Harrington, W. F. (1971). *Proc. Nat. Acad. Sci. U.S. 68,* 685.

Harrington, W. F., and Burke, M. (1972). *Biochemistry 11,* 1448.

Haselgrove, J. C. (1975). *J. Mol. Biol. 92,* 113.

Haselgrove, J. C., and Huxley, H. E. (1973). *J. Mol. Biol. 77,* 549.

Hunter, M. J., and Ludwig, M. L. (1962). *J. Am. Chem. Soc. 84,* 3491.

Huxley, A. F., and Simmons, R. M. (1971) *Nature 233,* 533.

Huxley, H. E. (1968). *J. Mol. Biol. 37,* 507.

Huxley, H. E. (1969). *Science 164,* 1356.

Huxley, H. E. (1972). *In* "The Structure and Function of Muscle" (G. H. Bourne, ed.), Vol. I, pp. 301-387. Academic Press, New York.

Huxley, H. E. (1976). *In* "Molecular Basis of Motility" (L. M. G. Heilmeyer, Jr., J. C. Ruegg, and Th. Wieland, eds.), pp. 9-23. Springer-Verlag, Berlin.

Huxley, H. E., and Brown, W. (1967). *J. Mol. Biol. 30,* 383.

Josephs, R., and Harrington, W. F. (1966). *Biochemistry 5,* 3437.

Lowey, S. (1971). *In* "Subunits in Biological Systems" (S. N. Timasheff and G. D. Fasman, eds.) pp. 201-255. Marcel Dekker, New York.

Lowey, S., Goldstein, L., Cohen, C., and Luck, S. M. (1967). *J. Mol. Biol. 23,* 287.

Lowey, S., Slayter, H. S., Weeds, A. G., and Baker, H. (1969). *J. Mol. Biol. 42,* 1.

Lymn, R. W., and Cohen, G. H. (1975). *Nature 258,* 770.

Lymn, R. W., and Taylor, E. W. (1970). *Biochemistry 9,* 2975.

Margossian, S. S., and Lowey, S. (1973). *J. Mol. Biol. 74,* 301.

Mendelson, R. A., and Cheung, P. (1976). *Science 194,* 190.

Morimoto, K., and Harrington, W. F. (1974). *J. Mol. Biol. 88,* 693.

Roark, D. E., and Yphantis, D. A. (1971). *Biochemistry 10,* 3241.

Rome, E. (1968). *J. Mol. Biol. 37,* 331.

Squire, J. M. (1972). *J. Mol. Biol. 72,* 125.

Sutoh, K., and Harrington, W. F. (1977). *Biochemistry 16,* 2441.

Sutoh, K., Chiao, Y-C. C., and Harrington, W. F. (1978a). *Biochemistry 17,* 1234.

Sutoh, K., Sutoh, K., Karr, G., and Harrington, W. F. (1978b). *J. Mol. Biol.,* (in press)

Taylor, D. L. (1975). *J. Supramolecular Structure 3,* 181.

Teller, D. C. (1973). *In* "Methods in Enzymology" (W. Hirs and S. N. Timasheff, eds.), Vol. 27, p. 346. Academic Press, New York.

Traut, R. R., *et al.* (1973). *Biochemistry 12,* 3266.

Wang, K., and Richards, F. (1974). *J. Biol. Chem. 249,* 8005.

Weeds, A. G., and Pope, B. (1977). *J. Mol. Biol. 111,* 129.

Weeds, A. G., and Taylor, R. S. (1975). *Nature 257,* 54.

Motility in Cell Function
Proceedings of the First John M. Marshall Symposium in Cell Biology

ELECTRON MICROSCOPE STUDIES ON MUSCLE THICK FILAMENTS

Roger Craig[†] *and Joseph Megerman*[*]

Rosenstiel Basic Medical Sciences Research Center
Brandeis University
Waltham, Massachusetts

Muscle thick filaments are built with a polarity that enables them to draw appropriately oriented thin filaments from opposite ends of a sarcomere towards each other, producing shortening. They also contain proteins in addition to myosin, which may act to modify myosin assembly or function. It is these two aspects of thick filament structure--polarity and nonmyosin proteins--that we shall discuss here.

NONMYOSIN COMPONENTS OF SKELETAL THICK FILAMENTS

The possible existence of proteins other than myosin in the cross-bridge region of vertebrate skeletal thick filaments was suggested by Huxley (1967). In longitudinal sections of the frog sartorius A-band, he observed an array of transverse stripes (Fig. 1c), and from their spacing, suggested that these gave rise to the 442 Å meridional reflection in the X-ray pattern of muscle. Since this was different from the 430 Å repeat of the myosin molecules, he suggested that the stripes were due to some component other than myosin.

The first biochemical suggestion of nonmyosin components came from the work of Starr and Offer (1971) on impurities in myosin preparations. (These were named alphabetically according to their positions on SDS gels.) The persistence of these

[†]*Present address: A. R. C. Unit, Department of Zoology, South Parks Road, Oxford, England.*

[*]*Present address: Vascular Research Laboratory, Massachusetts General Hospital, Boston, Massachusetts 02114.*

ISBN 0-12-551750-5

impurities in myosin purified by repeated precipitation at low ionic strength suggested that they had a high affinity for myosin and therefore that they might derive from the thick filament. The most abundant impurity, C-protein, has been purified and characterized (Offer *et al.*, 1973). It is a monomer with a molecular weight of 140,000; it is an elongated molecule but contains no α-helix.

The location of C-protein was discovered by antibody labelling. Glycerinated rabbit psoas muscle was incubated with purified antibodies to C-protein (Offer, 1976) and the distribution of label in sections of the muscle was observed in the electron microscope. Antibodies to C-protein label seven stripes approximately 430 Å apart in each half of the A-band (Fig. 1d; Craig and Offer, 1976a). Pepe and Drucker (1975) have come to similar conclusions regarding the location of C-protein in chicken breast muscle. Our original conclusion that C-protein occurs on nine stripes in each half of the A-band (Offer, 1972) was arrived at using unpurified antiserum. With antiserum purified on a C-protein affinity column, only the outer seven of the nine stripes are labelled, the inner two being due to impurities (Craig and Offer, 1976a).

Craig *et al.* (in preparation) have isolated a contaminant of C-protein named H-protein (according to the same alphabetical nomenclature), and have used an H-protein affinity column to isolate antibodies to H-protein from the unpurified anti-C serum. Anti-H protein labels a single stripe in each half of the A-band, the innermost of the original nine (Fig. 1e). Starr and Offer (in preparation) are currently attempting to isolate the impurity responsible for the other stripe labelled with unpurified anti-C serum.

These nine stripes correspond in position to nine prominent stain-excluding stripes seen in negatively stained A-segments (isolated thick filament assemblies held together at the M-line: Fig. 1b; Hanson *et al.*, 1971; Craig, 1977), and also to the stained stripes observed by Huxley (1967) in sectioned A-bands (Fig. 1c). Both these sections and the A-segments show two further prominent stripes proximal to the nine, and these undoubtedly correspond to two further, probably different, thick filament components. Including the two proteins currently thought to be present in the M-line (Turner *et al.*, 1973; Trinick and Lowey, 1977), it is clear that the thick filament contains at least seven different components and possibly several more. Their arrangement, together with the distribution of myosin heads deduced from A-segments (Craig, 1977), frozen sections (Sjöström and Squire, 1977) and anti-S1 labelling (Craig and Offer, 1976b) is summarized in Fig. 1.

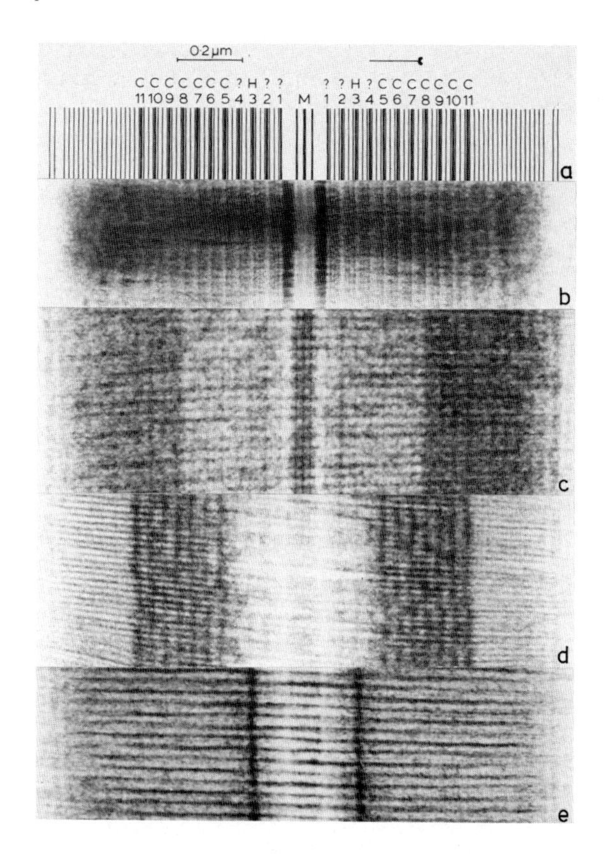

FIGURE 1. Nonmyosin proteins in the thick filament. (a) Model derived from our current knowledge of the distribution of cross bridges and nonmyosin components deduced from A-segments, frozen sections, and antibody-labelled sections. (b) Negatively stained A-segment showing nonmyosin components as prominent white stripes. (c) Section of frog sartorius A-band showing nonmyosin components as narrow, stained stripes. (d) and (e) Rabbit psoas A-bands labelled with antibodies to (d) C-protein and (e) H-protein. Magnifications: ×45,000.

THE PERIODICITY OF C-PROTEIN

The 430-440 Å periodicity of the anti-C stripes measured by electron microscopy, taken together with the 442 Å meridional reflection in the X-ray pattern of muscle, would suggest that C-protein has a 442 Å repeat--different from that of myosin. Rome *et al.* (1973) have confirmed that C-protein does indeed contribute to the 442 Å reflection by showing that in muscles labelled with antibodies to C-protein, this reflection is intensified. However, they have observed that a reflection at 418 Å is also enhanced in labelled specimens, and have tried to reconcile these observations with the conclusion from other work (see Craig and Offer, 1976a) that myosin and C-protein share a common 430 Å periodicity. They have shown that reflections at 442 and 418 Å could arise from a 430 Å repeat if interference between the arrays of stripes in the two halves of the A-band caused the 430 Å reflection to be split. Measurement of electron micrographs reveals that the two arrays are approximately half a repeat out of phase, so that if the repeat were 430 Å, interference would be expected to produce a minimum in the diffraction pattern at 430 Å and maxima at 442 Å and 418 Å.

By optical diffraction of antibody-labelled electron micrographs it has been demonstrated that such interference effects do in fact occur (Craig and Offer, in preparation). Figure 2a is the diffraction pattern of *half* an A-band labelled with anti-C (e.g., the left half of Fig. 1d). On the meridian is a strong reflection coming from the antibody repeat, and further out a weak reflection, at about one-third of the antibody spacing, thought to arise from the 143 Å myosin repeat. When the optical diffractometer mask is opened up to include both halves of the A-band together (e.g., the whole of Fig. 1d), the antibody reflection is split (Fig. 2b) producing reflections at about 420 and 450 Å (assuming that the unsplit reflection is at 430 Å). This confirms optically that interference between the two halves of the A-band can significantly modify the meridian of the diffraction pattern of muscle. It supports Rome's suggestion that interference is responsible for the contribution of C-protein to X-ray reflections at two different spacings (442 and 418 Å).

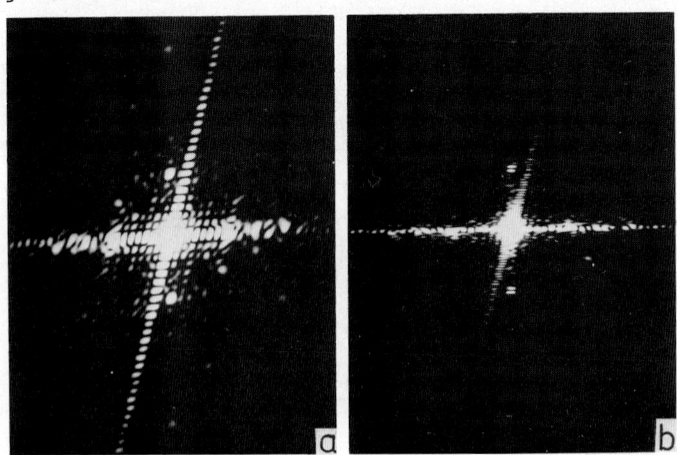

FIGURE 2. Diffraction patterns of (a) half A-band and (b) whole A-band labelled with antibodies to C-protein to demonstrate the effects of interference between the two halves of the A-band. The meridian of each pattern is vertical, as is the long axis of the A-band giving rise to them.

Clearly the period of C-protein cannot be measured accurately from diffraction patterns (either X-ray or optical) where reflections are altered by interference. However, in optical diffraction patterns interference can be avoided by diffracting only half an A-band at once (Fig. 2a), and such patterns have been used to estimate the C-protein periodicity. The ratio between the spacing of the antibody reflection and that of the reflection thought to come from the 143 Å myosin repeat is integral (3.014 ± .015), and the repeat of C-protein thus calculated is 430 Å, the same as the helical repeat of myosin. (The 143 Å reflection probably receives little or no contribution from the antibody stripes considering their width and disorder.) Several other lines of reasoning also lead to the conclusion that myosin and C-protein have the same repeat (see Craig and Offer, 1976a); it therefore appears that the location of C-protein is determined by the packing of the underlying myosin molecules. Assuming that C-protein binds to equivalent sites in the thick filament, we conclude that the seven stripes of C-protein represent a region of constant myosin packing and that this packing is different where the filament tapers towards its end (no C-protein binding). The packing also appears to change continuously near the middle of the filament, judging by the fact that three or possibly four different proteins bind on the proximal four of the eleven stripes (for example, H-protein recognizes only a single site in each half of the filament). Changes in this region are not surprising, considering that it represents a transition from myosin molecules with antiparallel overlaps (stripe 1) to those with purely parallel interactions (stripe 4).

POLARITY OF SMOOTH MUSCLE THICK FILAMENTS

In striated muscle thick filaments, the myosin molecules in one half are oriented with the same polarity and those in the other half with the opposite polarity (Fig. 3a). The reversal occurs half way along the filament at the "bare zone" (a region free of myosin heads), where antiparallel overlaps between myosin tails occur. As pointed out by Huxley (1963), this bipolarity is an essential feature of the sliding filament model of contraction for striated muscle, where thin filaments of opposite polarity are pulled towards each other from opposite ends of a sarcomere.

In vertebrate smooth muscle it is generally assumed that similar bipolar thick filaments also occur, although no clear demonstration of a central bare zone has been made. One suggestion, however, that the polarity might be different comes

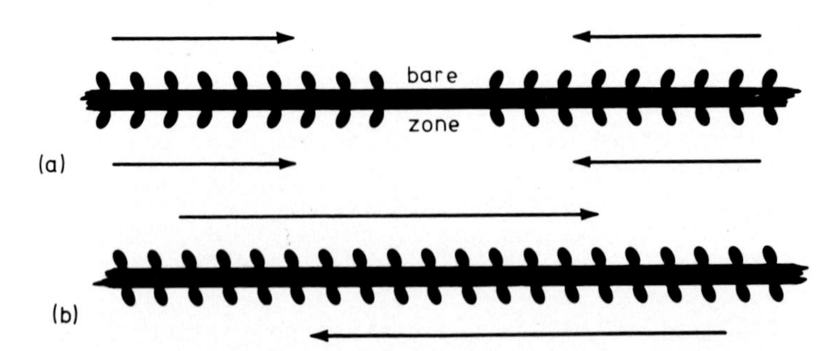

FIGURE 3. Diagram to illustrate the functional similarity between (a) bipolar filaments and (b) face-polar ribbons (ribbon drawn in edge view with one face on each side). The arrows indicate that both structures have two separate cross bridge domains with opposite polarity.

from the electron microscope studies of Small and Squire (1972) on sectioned *taenia coli* muscle. These authors concluded that myosin in this smooth muscle occurred as broad ribbons where the polarity along the entire length of one face of the ribbon was the same and opposite from the polarity on the other face. In a functional sense ribbons were still bipolar (opposite faces of a ribbon could draw thin filaments in opposite directions), but different polarities occurred on different faces, not at different ends (Fig. 3b).

It is now known that ribbons are artifacts that form by the lateral aggregation of narrower filaments (Jones *et al.*, 1973; Shoenberg and Haselgrove, 1974; Somlyo *et al.*, 1971, 1973). Yet there has been no published account of how aggregation of conventional bipolar filaments could lead to ribbons with the face-polarity described by Small and Squire. Based on our *in vitro* studies of the formation of filaments from calf aorta myosin, we reconsider here the possibility that in some vertebrate smooth muscles, myosin filaments may possess a polarity similar to that described by Small and Squire for their ribbons. A fuller account of this work has been published by Craig and Megerman (1977).

Myosin was prepared as described by Craig and Megerman (1977). It contained less than 10% impurities (occurring as high molecular weight proteins), was monodisperse in the analytical ultracentrifuge, and had ATPase activity typical of published values for smooth muscle myosin.

When dialysed from the monomer state (0.6 *M* KCl) to lower ionic strength, two types of filament having distinctly different structures are formed. Depending on the dialysis conditions the two types may coexist, or one or the other may occur exclusively. The first type is a short bipolar filament typical of those formed in the early stages of assembly of striated muscle myosin (Fig. 4a). The myosin molecules interact with antiparallel overlaps to form a filament having cross bridges at each end and a bare zone 1800 Å long at the center. These bipolar filaments never grow longer than about 0.5 μm, so there is never a region containing myosin molecules with purely parallel interactions as there are in the lateral portions of skeletal thick filaments.

The other type of filament looks like a bipolar filament that has been skewed into the shape of an oblique parallelogram (Fig. 4b). Two sides of the parallelogram bear cross bridges while the other two sides are smooth. These filaments grow (up to 6 μm in length) by elongation of the cross bridge bearing sides, while the bare sides at each end remain the same length (Fig. 5a). Our interpretation of this structure is shown in Fig. 5b. The myosin tails have antiparallel overlaps along the entire length of the filament and run at a small angle to the filament axis. The cross bridges along one side would clearly all have the same polarity while those along the other side would have the opposite polarity.

This "side-polarity" we propose is clearly evident from micrographs where the cross bridges are especially well preserved (Fig. 6a). Optical diffraction patterns of such micrographs (Fig. 6b) show a layer line, at a spacing of about 150 Å, which is asymmetric, consisting of a meridional reflection and an additional reflection on one side of the meridian only.

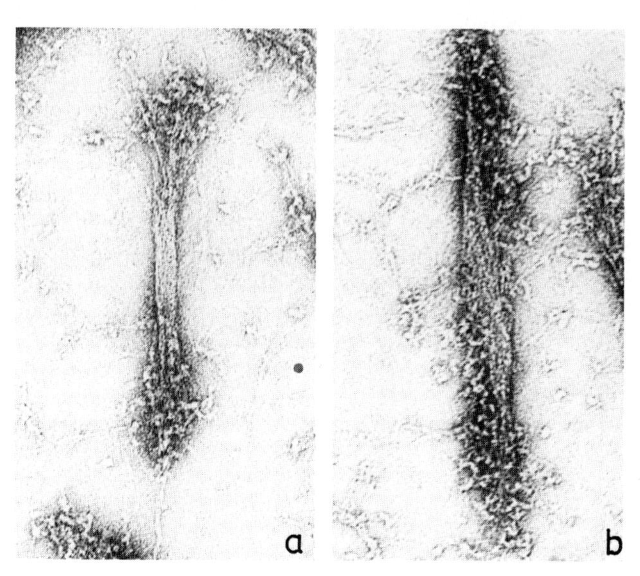

FIGURE 4. Filaments formed from calf aorta myosin negatively stained with 1% uranyl acetate. (a) Short bipolar filament and (b) short side-polar filament, both formed at 0.2 M *KCl, pH 7.* ×230,000. *(From Craig and Megerman, 1977.)*

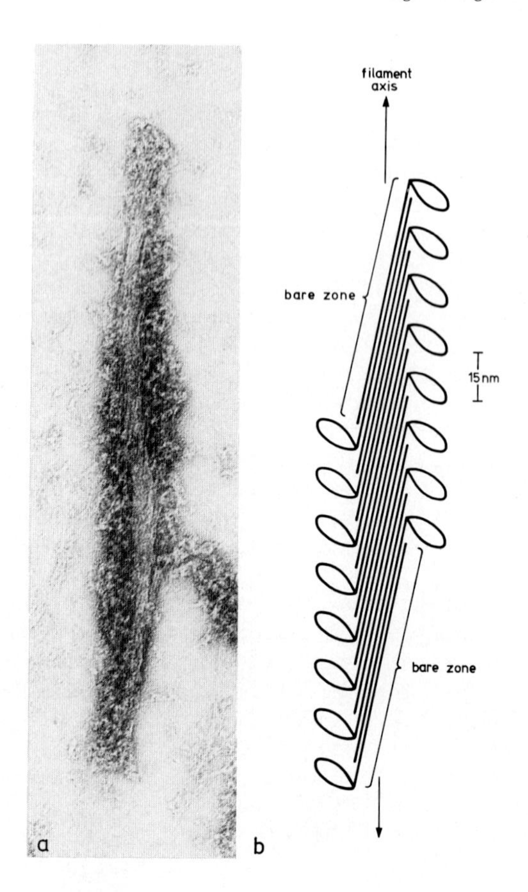

FIGURE 5. (a) Longer side-polar filament formed at 0.3 M *KCl, pH 6. ×205,000. (b) Schematic diagram showing essential features of side-polar structure: opposite polarity on opposite sides, antiparallel overlaps along the entire length of filament, absence of central bare zone but bare regions at each end. (From Craig and Megerman, 1977.)*

This asymmetry reflects a corresponding asymmetry in the cross bridge arrangement. Optical filtering of the image in Fig. 6a (allowing only the strong reflections in Fig. 6b to recombine to form the image) reveals the opposite polarity of the cross bridges on opposite sides of the filament with particular clarity.

The structure we observe by negative staining has cross bridges along two sides only. To check whether this noncylindrical distribution might be an artifact produced during drying

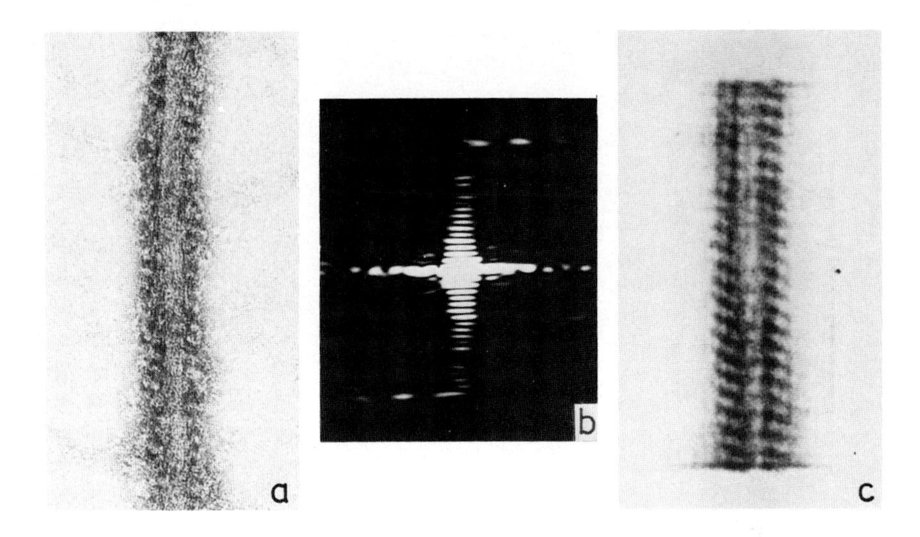

FIGURE 6. (a) Portion of side-polar filament, ×160,000. *(b) Optical diffraction pattern of (a). (c) Filtered image of (a) allowing only the strong reflections in (b) to recombine to form the reconstructed image.* ×160,000. *((a) and (b) from Craig and Megerman, 1977.)*

of the negative stain, we have also sectioned a pellet of side-polar filaments to observe the cross bridge arrangement in transverse section. Figure 7 shows that the filament backbone is square in cross section and possesses cross bridges along only two sides of the square, the other sides being bare. The noncylindrical distribution of cross bridges thus appears to be real, and we conclude that side-polar filaments do not have helical symmetry.

The assembly of side-polar filaments differs from other modes of myosin assembly. The initial growth of striated muscle thick filaments depends on the antiparallel interaction of myosin molecules. Beyond a length of about 0.5 μm, myosin molecules are added with exclusively parallel interactions (Huxley, 1963). In side-polar filaments, antiparallel interactions appear to be essential for growth. They occur along the entire length of the filament, and the filament can be thought of as growing by the addition of bipolar myosin units, producing a structure that appears to have uniform packing throughout its length.

It is worth reconsidering the possibility that the *in vivo* state of myosin in some vertebrate smooth muscles is a side-polar structure comparable to that described here. Several arguments are presented by Craig and Megerman (1977), but the most striking is the remarkable similarity between negatively stained side-polar filaments (especially their diffraction patterns and their filtered images (Fig. 6)) and the edge-views

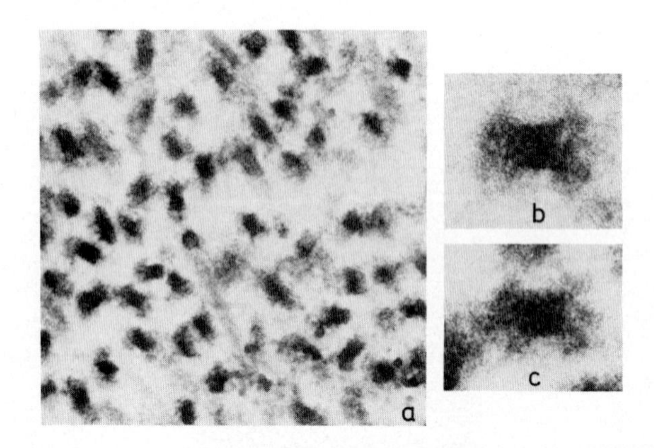

FIGURE 7. Sectioned pellet of side-polar filaments, some of which are seen in approximately transverse section (a). The backbone is roughly square, and cross bridges project from only two sides of the square ((b) and (c)). (a): ×65,000. (b) and (c): ×200,000. (From Craig and Megerman, 1977.)

and diffraction patterns of myosin ribbons observed by Small and Squire (e.g., Plate XII). These face-polar ribbons are now thought to be aggregates of *in vivo* filaments. Lateral aggregation of side-polar filaments through their bridge-free sides; i.e., aggregation perpendicular to the plane of the micrographs (e.g., Fig. 6a) could produce ribbons having cross bridges with uniform polarity on one face and opposite polarity on the other. Side-polar filaments would thus provide a simple basis for the ribbon observations, while it is not clear how aggregates of bipolar filaments could give such an appearance.

While a case is made here for a close relationship between the polarity of our filaments and that of the ribbons, we should emphasize that our interpretation of the packing in side-polar filaments differs in one major way from the ribbon model of Small and Squire. The ribbon model has a core of nonmyosin protein which interposes between the myosin molecules on the two faces and which determines their polarity. In side-polar filaments there is no core and antiparallel interactions between myosin tails are responsible for the opposite polarity on opposite sides.

Side-polar filaments could readily explain the ability of smooth muscle to shorten and maintain tension over a wide range of lengths (in a manner similar to that suggested by Small and Squire for their ribbons). During shortening a thin filament could slide all the way along a thick filament "side" until the end of the thin filament was reached. If the thick filament were bipolar, sliding would be impeded halfway along, when the thin filament encountered cross bridges of opposing polarity. The side-polar mode of assembly could thus confer a distinct structural advantage on a sliding filament system required to shorten by large amounts.

ACKNOWLEDGMENTS

We thank Drs. G. Offer, C. Cohen, P. Vibert, and S. Lowey for advice. Part of this work was done during the tenure of a Research Fellowship of Muscular Dystrophy Associations of America (R.C.) and of a National Institutes of Health postdoctoral fellowship (J.M.), and was supported by National Institutes of Health Grants AM17346-05 to C. Cohen, and AM17350 to S. Lowey; National Science Foundation Grants PCM76-10558 to C. Cohen and GB-38203 to S. Lowey; and Muscular Dystrophy Association Grants to C. Cohen and S. Lowey.

REFERENCES

Craig, R. (1977). *J. Mol. Biol. 109,* 69-81.

Craig, R., and Megerman, J. (1977). *J. Cell Biol. 75,* 990-996.

Craig, R., and Offer, G. (1976a). *Proc. Roy. Soc., Ser. B 192,* 451-461.

Craig, R., and Offer, G. (1976b). *J. Mol. Biol. 102,*325-332.

Hanson, J., O'Brien, E. J., and Bennett, P. M. (1971). *J. Mol. Biol. 58,* 865-871.

Huxley, H. E. (1963). *J. Mol. Biol. 7,* 281-308.

Huxley, H. E. (1967). *J. Gen. Physiol. 50,* (suppl.), 71-83.

Jones, A. W., Somlyo, A. P., and Somlyo, A. V. (1973). *J. Physiol. (Lond.) 232,* 247-273.

Offer, G. (1972). *Cold Spring Harbor Symp. Quant. Biol. 37,* 87-93.

Offer, G. (1976). *Proc. Roy. Soc., Ser. B 192,* 439-449.

Offer, G., Moos, C., and Starr, R. (1973). *J. Mol. Biol. 74,* 653-676.

Pepe, F. A., and Drucker, B. (1975). *J. Mol. Biol. 99,* 609-617.

Rome, E., Offer, G., and Pepe, F. A. (1973). *Nature New Biol. 244,* 152-154.

Shoenberg, C. F., and Haselgrove, J. C. (1974). *Nature 249,* 152-154.

Sjöström, M., and Squire, J. M. (1977). *J. Mol. Biol. 109,* 49-68.

Small, J. V., and Squire, J. M. (1972). *J. Mol. Biol. 67,* 117-149.

Somlyo, A. P., Somlyo, A. V., Devine, C. E., and Rice, R. V. (1971). *Nature New Biol.* *231*, 243-246.

Somlyo, A. P., Devine, C. E., Somlyo, A. V., and Rice, R. V. (1973). *Phil. Trans. Roy. Soc., Ser. B* *265*, 223-229.

Starr, R., and Offer, G. (1971). *FEBS Lett.* *15*, 40-44.

Trinick, J., and Lowey, S. (1977). *J. Mol. Biol.* *113*, 343-368.

Turner, D. C., Walliman, T., and Eppenberger, H. M. (1973). *Proc. Nat. Acad. Sci. U.S.A.* *70*, 702-705.

Motility in Cell Function
Proceedings of the First John M. Marshall Symposium in Cell Biology

THE MYOSIN FILAMENT: MOLECULAR STRUCTURE

Frank A. Pepe

Department of Anatomy
School of Medicine/G3
University of Pennsylvania
Philadelphia, Pennsylvania

The purpose in this presentation is to summarize briefly some information we have obtained recently, which is related to the backbone structure of the myosin filament. This consists of the following: a) Optical diffraction studies of the substructure spacing in the shaft of the myosin filament and the determination of differences in the structure of the backbone along the length of the filament (done with Peter Dowben and Francis Ashton). b) Determination of the myosin content of myosin filaments, the conditions for formation of synthetic filaments with length distributions comparable to those observed for natural filaments, and the effect of C-protein on synthetic filament formation (done with Barbara Drucker). c) Studies of the similarity of C-protein binding to LMM paracrystals and to myosin filaments (done with Barbara Drucker and Dan Safer).

A few years back, we observed substructure in relatively thin cross sections of the shaft of the myosin filament (Pepe and Drucker, 1972). These sections were less than 0.1 μm in thickness, and in some well preserved filaments, it was possible to observe 12 structural units hexagonally packed giving triangular cross sectional profiles with a structural unit missing at each apex of the triangle (Fig. 1a). This arrangement was particularly well visualized after Markham rotation printing, but it was also visible in the unrotated images. This arrangement in the shaft of the myosin filament is exactly that predicted by my model for the filament (Pepe, 1966, 1967). The center-to-center spacing of these structural units is about 3.7 nm which suggests that each structural unit may be made up of more than one myosin molecule, where the diameter of each myosin rod is about 2 nm. The model allows for this possibility.

ISBN 0-12-551750-5

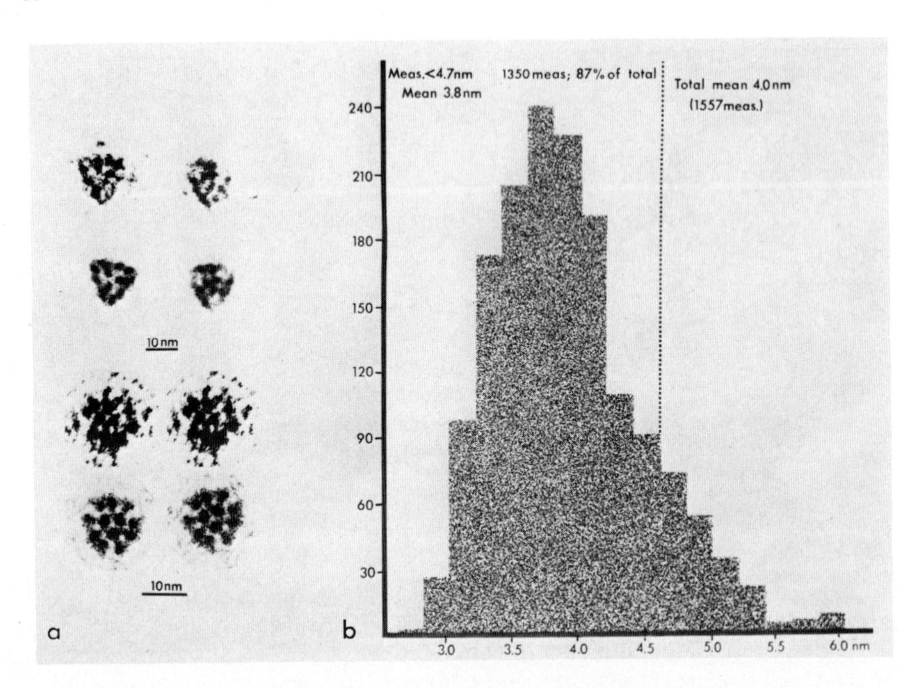

FIGURE 1. Substructure spacings in the backbone of individual myosin filaments. (a) Observations of substructure in thin (<0,1 μm thick) cross sections of individual myosin filaments in chicken pectoralis muscle. In the top half, the two upper images correspond to the two lower images. The lower images were formed by superimposing the top image, rotated in intervals of 120°. In the bottom half the same procedure was used. The center-to-center distance between structural units is approximately 3.7 nm and the packing is as predicted by the model (Pepe, 1966, 1967). (Modified from Pepe and Drucker, 1972). (b) Distribution of lattice plane spacings of the substructure in the backbone of fish muscle myosin filaments. Measurements were made on the optical diffraction patterns obtained from images of cross sections of individual myosin filaments.

As a result of these observations, we set to work trying to improve our measurements of the substructure spacing in cross sections of the filament. In the model, the structural units in the backbone of the filament are all parallel to the long axis of the filament. Therefore, we reasoned that by increasing the thickness of a cross section, we should improve the signal to noise ratio and obtain measurable substructure in more of the filaments. In doing this, we had to be sure that the filaments were well stained throughout the thickness of the section so we used alcoholic uranyl acetate followed by alcoholic lead citrate to facilitate penetration of the stain (Pepe and Dowben, 1977). In order to be able to observe these heavily stained thick sections, we used a 200 kV electron microscope beam to penetrate the sections. To make our measurements of substructure spacing less subjective and more rapid, we took optical diffraction patterns of the images of the cross sections of individual filaments. We found that, as expected for parallel structural units, with increasing section thickness the substructure spacing in cross sections of the myosin filament were more easily observed, the most clearly observed being obtained in sections 0.2 to 0.3 μm thick (Pepe and Dowben, 1977).

To improve the preservation of the substructure of the filament, we used a modification (unpublished) of the fixation technique described by Reedy (1976) in which the specimen is extensively cross linked prior to dehydration and embedding for electron microscopy. As a result of this we were able to observe substructure spacings clearly even in sections as thin as about 0.15 μm in thickness. The distribution of substructure spacings observed in this material is shown in Fig. 1b. This is a plot of 1557 measurements of the plane spacings calculated from the optical diffraction patterns and the peak corresponds to a plane spacing of about 3.8 nm or a center-to-center spacing of a little over 4 nm. This then confirms our earlier findings (Pepe and Drucker, 1972) suggesting that the center-to-center spacing between parallel structural units is large enough to include more than one myosin molecule per structural unit.

Recently, a general model for myosin filament structure has been proposed by Squire (1973). This general model is based on a packing scheme in which all molecules are equivalent; i.e., that each structural unit consists of a single myosin molecule, and requires that the molecules be tilted with respect to the long axis of the filament. Since the observations I have just described with sections 0.3 μm thick (Pepe and Dowben, 1977) clearly rule out a single myosin molecule per structural unit and a tilt of more than 1/3° assuming one myosin molecule per structural unit or more than 3/4° assuming a 4 nm spacing between structural units, I will not consider Squires model (1973) further.

The problem now becomes one of determining how many myosin molecules there are per structural unit. At 14 nm intervals along the filament model (Pepe, 1966, 1967), two of the structural units project from the surface of the filament as myosin cross bridges. Therefore, if there are two myosin molecules per structural unit there will be four myosin molecules per 14 nm interval, and if there are three molecules per structural unit there will be six molecules per 14 nm along the filament. There is some controversy in the literature about this figure.

In general, quantitative SDS gel electrophoresis has been used to obtain this figure by relating the weight of myosin to the weight of actin in the myofibril. In this way, Tregear and Squire (1973), and Potter (1974), obtained a myosin to actin weight ratio which corresponds to about three myosin molecules per 14 nm interval along the myosin filament and Morimoto and Harrington (1974) obtained a value of four molecules per 14 nm interval. Morimoto and Harrington (1974) also obtained the same value of four using a particle counting technique to determine the total molecular weight of myosin filaments isolated from muscle. In my laboratory, we have been able to obtain both of these values depending upon how extensively the fibrils were washed, the higher value of four molecules per 14 nm interval being obtained with more extensive washing of the fibrils in a large volume of buffer of buffered glycerol (*J. Mol. Biol.* in press). Since the higher value corresponds to a higher myosin to actin weight ratio, the required washing must be extracting either actin or some material with a chain weight similar to that of actin. Our studies were done using 6% polyacrylamide gels. More recently, we have repeated this work using 8% gels and we compared whole muscle to the extensively washed muscle. This is shown in Fig. 2. In whole muscle (Fig. 2b) there are three clearly resolved bands in the actin region, one with a slightly higher and the other with a slightly lower chain weight than actin. In 6% gels, these three bands are not resolved. In the extensively washed muscle (Fig. 2a) only the actin band remains. Regardless of whether whole muscle or extensively washed muscle was used, the myosin to actin weight ratio in these gels corresponds to a value of four myosin molecules per 14 nm interval along the myosin filament. This eliminates the possibility of loss of actin being responsible for the higher value obtained after extensive washing and, therefore, clearly indicates that the value of four myosin molecules per 14 nm interval along the myosin filament or a total of about 384 molecules per filament is correct.

In the work by Tregear and Squire (1973), gels consisting of 4 cm of 5% polyacrylamide layered over 5 cm of 10% polyacrylamide were used, and therefore, the migration in the 10% of the gel was not far enough to separate the actin from its neighboring bands. Although Potter (1974) used 12.5% polyacrylamide gels, he was primarily interested in the lower chain weight components, and therefore, again, the actin was not resolved from its closely neighboring bands. The material used by Morimoto and Harrington (1974) was comparable to our extensively washed muscle and they obtained the comparable value of four myosin molecules per 14 nm interval.

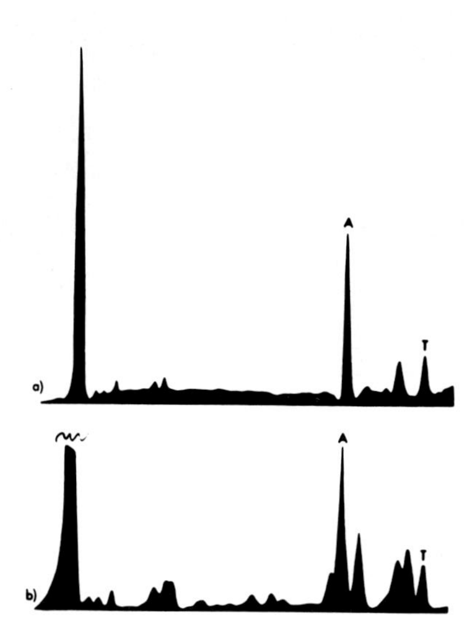

FIGURE 2. Determination of myosin content of rabbit psoas muscle myosin filaments. (a) Muscle extensively washed in 50% glycerine, 50 mM KCl, 0.5 mM $MgCl_2$ *5 mM phosphate buffer pH7.0. Scan is of an 8% polyacrylamide SDS gel run using a modification of the procedure described by Laemmli (1970) and stained with coomassie blue. (b) Whole muscle run under the same conditions as (a). A=band with chain weight of actin. T=band with chain weight of tropomyosin. The myosin to actin weight ratio obtained from both (a) and (b) is 2.6 corresponding to four myosin molecules per 14 nm interval along the filament.*

From the results I have discussed so far, we are now convinced that the structural units in the shaft of the myosin filament are arranged parallel to the long axis of the filament and that there are two myosin molecules per structural unit. If the two molecules are not tightly twisted in each structural unit then in cross section the two molecules side by side will give an elongated profile 4 nm long by 2 nm wide. In Fig. 3a these are placed in the model (Pepe, 1966, 1967) in two ways. In each case all the structural units are similarly oriented but each structural unit in one case is rotated by 30° on its long axis with respect to the structural units in the other case. The optical diffraction patterns from these two arrangements of the model (Fig. 3b) show the approximately 4 nm substructure spacings in three directions in one case and only in one direction in the other case. Next to these (Fig. 3c) are comparable optical diffraction patterns observed from cross sections of individual myosin filaments. We considered

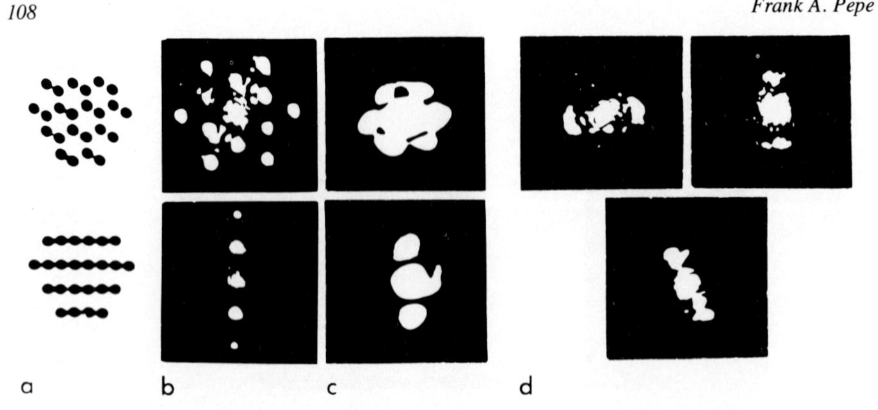

FIGURE 3. Optical diffraction patterns obtained from models and from images of cross sections of individual myosin filaments in fish muscle. (a) Two different arrangements in which each structural unit of the filament model (Pepe, 1966, 1967) consists of two myosin molecules. The center-to-center distance between structural units is the same in both cases. (b) The optical diffraction patterns observed from the corresponding models in (a). (c) Optical diffraction patterns observed from cross sections of individual filaments. (d) Optical diffraction patterns observed from three neighboring filaments in a cross section of the A-band. (Modified from Pepe and Dowben, 1977.)

the possibility that the optical diffraction patterns showing spacings in only one direction could result from tilting of the filament in the plane of the section. However, this is very unlikely because in neighboring filaments, which each show substructure spacing in one direction, the direction of the spacings is different (Fig. 3d). If tilt were responsible, neighboring filaments should all show the same direction. Similar differences in optical diffraction patterns (Fig. 3c) could be obtained with structural units having circular cross sectional profiles. However in that case the lattice arrangement of the 12 circular structural units would have to be different for the two optical diffraction patterns observed in Fig. 3c. The cross sectional profile of the structural units within a single filament has not been clearly observed.

The question now becomes whether these two different optical diffraction patterns (Fig. 3c) come from different parts of the same filament, or whether they represent filaments which have different internal structure. To determine if these different patterns come from different parts of the same filament we took serial cross sections along the length of the myosin filament and observed the changes in the optical diffraction pattern along the filaments. It was found that optical diffraction patterns with substructure spacing in three directions predominated toward each end of the filament and in the middle of the filament while those showing substructure spacing in only one direction predominated in the middle of each half of the myosin filament which is where C-protein binding to the filament occurs (Pepe and Drucker, 1975; Craig and Offer, 1976a). Therefore, the backbone structure which gives optical diffraction patterns showing substructure spacing in one direction can be related to the C-protein binding region of the filament (Fig. 7b,c,d).

C-protein localization has been shown to occur in seven stripes spaced 43 nm apart in the middle, one-third of each half of the myosin filament by antibody staining in electron microscopy (Pepe and Drucker, 1975; Craig and Offer, 1976a). Pepe and Drucker (1975) were further able to demonstrate that the C-protein is restricted to this portion of the filament by comparison of the antibody staining in fluorescence and in electron microscopy. In Fig. 4a, the fluorescent bands correspond to the stripes seen in the electron micrograph in Fig. 4b. This puzzling restricted binding of C-protein to only a portion of the A-band and therefore only along a portion of the myosin filament can now be related to a difference in the arrangement of the substructural units in the shaft of the myosin filament (Fig. 7b,c,d).

Another puzzling feature of C-protein binding to the myosin filaments is that it occurs at intervals of 43 nm whereas myosin cross bridges occur at intervals of 14 nm. In previous studies of C-protein binding to LMM paracrystals, Moos *et al.* (1975) were able to show that C-protein will bind to LMM paracrystals and enhance the 43 nm axial repeat of the paracrystal but they were not able to observe both the axial repeat of the paracrystal and the repeat of the bound C-protein in the same paracrystal leaving some doubt about whether the two were exactly the same. This doubt was erased by observations made by Pepe *et al.* (1975) and Chowrashi and Pepe (1977) where both the 43 nm paracrystal repeat and the added C-protein were observed simultaneously. In none of these studies was C-protein binding at 43 nm observed on LMM paracrystals having a 14 nm axial repeat; which is more analagous to the situation in the myosin filament where C-protein binding occurs at 43 nm intervals. We have now observed this in paracrystals, to which C-protein is bound, and which were positively stained with uranyl acetate, as can be seen in Fig. 4c. Therefore, the binding of C-protein to LMM paracrystals can now be directly related to the binding of C-protein to the myosin filament. The details of how the substructure arrangement of the myosin rods determines whether or not C-protein can bind and how binding at 43 nm intervals, instead of 14 nm intervals is achieved, remains to be determined.

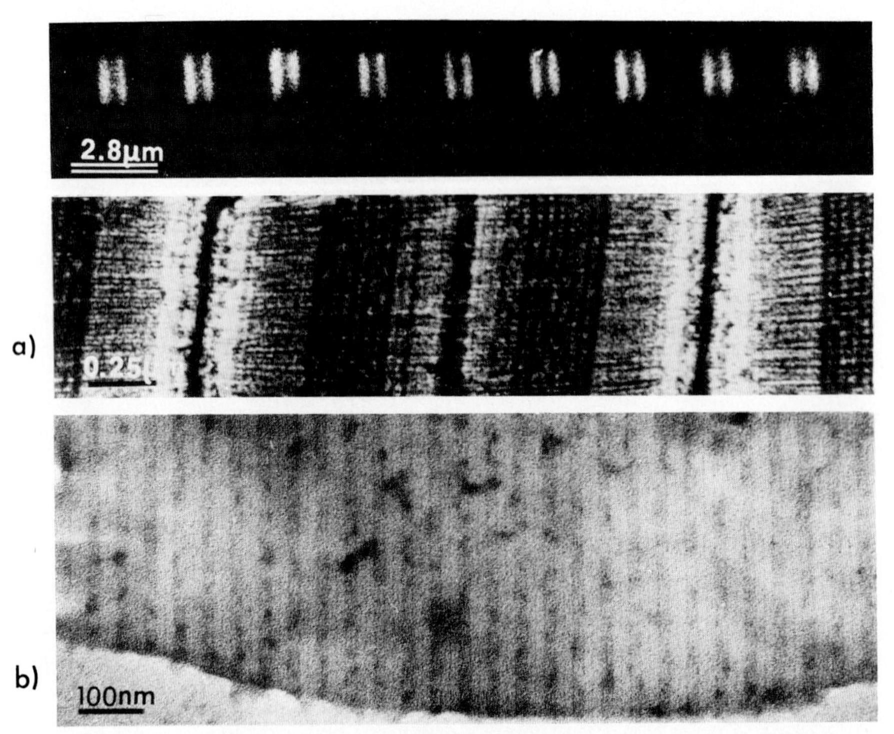

FIGURE 4. Restricted localization of C-protein along a portion of the length of the myosin filament and its relation to C-protein binding to LMM paracrystals. (a) Chicken pectoralis muscle myofibril stained with fluorescein labelled specific anti-C-protein (Pepe and Drucker, 1975). (b) Electron micrograph corresponding to (a). The seven dense stripes in each half of the A-band correspond to the bright fluorescent bands in (a) and represent specific staining of C-protein. The stripes occur at 43 nm intervals. (Pepe and Drucker, 1975.) (c) C-protein binding to an LMM paracrystal. The LMM paracrystal has a 14 nm axial repeat consisting of a narrow dark band every 14 nm. The C-protein binds at 43 nm intervals between two of the narrow dark bands of the LMM paracrystal. The presence of the C-protein increases the density of the space in which it is located. Positive stained with uranyl acetate. (Safer and Pepe, unpublished results.)

Another property related to the backbone structure of the myosin filaments is how the precise length of the filament is determined and whether all the information required for length determination is present in the myosin molecule or whether the presence of another protein is required. Questions such as these have been proposed because of the finding that synthetic myosin filaments grown from solutions of myosin by reducing the ionic strength have a structure which is similar to naturally occurring filaments but they vary greatly in length (Huxley, 1963). Recently, using highly purified myosin, we have been able to prepare synthetic myosin filaments comparable in length distribution to natural filaments. In Fig. 5 are some graphs of the length distributions obtained. The myosin was purified by column chromatography on DEAE Sephadex A-50 and the eluted myosin was collected in pools as it came off of the column. The first two graphs represent the length distri-

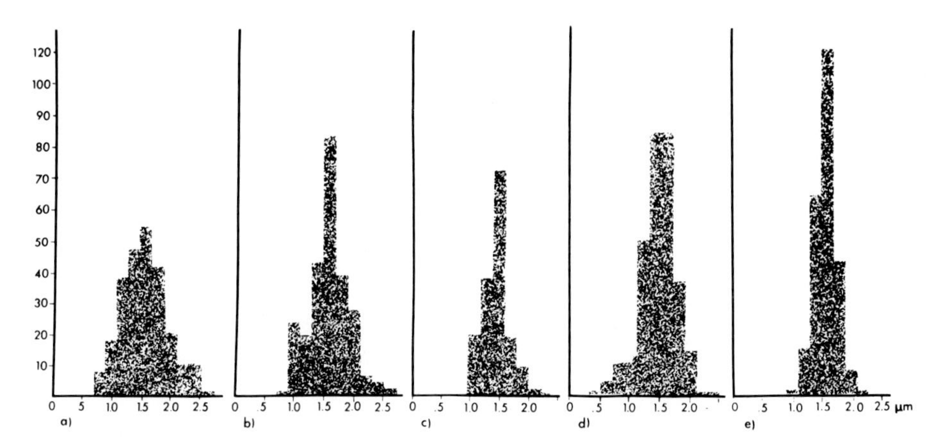

FIGURE 5. Length distribution of synthetic filaments grown from solutions of rabbit psoas muscle myosin. (a) and (b) Two pools of myosin coming from different portions of the elution curve during purification of myosin by DEAE Sephadex A-50 column chromatography. In (a) 51% of the filaments have lengths in the range 1.7 ± 0.2 μm and in (b) 63% have lengths in the same range. Filaments were formed with a starting protein concentration of 4 mg/ml by dilution from 0.6 M *KCl in 0.01* M *imidazole buffer pH 7.0. (c) Same pool as in (b) formed at pH 6.8 and with a starting protein concentration of 2 mg/ml. Now 77% of the filaments have lengths in the range 1.5 ± 0.2 μm. (d) Same as (c) except that C-protein is present in a molar ratio of 8:1 (myosin: C-protein). Now 65% of the filaments have lengths in the range 1.5 ± 0.2 μm. (e) The best length determination observed so far, with 84% of the filaments with lengths in the range 1.5 ± 0.2 μm, obtained with a starting concentration of 4 mg/ml and formed at 7.0.*

butions obtained with different myosin pools off the same column. The first two graphs represent the length distributions obtained with different myosin pools off the same column. In Fig. 5a, 51% of the filaments have lengths in the range 1.7 ± 0.2 μm, and in Fig. 5b, 63% of the filaments have lengths in the same range. The reason for these differences is not clear. These filaments were formed by dilution from 0.6 *M* KCl to 0.15 *M* in 0.01 *M* imidazole buffer at pH 7.0 and starting with a protein concentration of 4 mg/ml. Using the same myosin pool as in Fig. 5b and changing the pH to 6.8 and the starting protein concentration to 2 mg/ml, 77% of the filaments were in the range of 1.5 ± 0.2 μm (Fig. 5c). This is comparable to the lengths measured for isolated natural filaments by Morimoto and Harrington (1973) where 80% of the filaments were within ± 0.17 μm of the average length of 1.53 μm. The best length determination we have obtained was in a preparation where 84% of the filaments were in the range of 1.5 ± 0.2 μm (Fig. 5e).

Growing filaments in the presence of C-protein, we found little effect on the peak length but there was a broadening of the distribution and this increases with increasing amounts of C-protein present. At a molar ratio of myosin to C-protein of 8:1 the percent of filaments in the range 1.50 ± 0.2 μm decreased from 77% in the absence of C-protein (Fig. 5c) to 65% in the presence of C-protein (Fig. 5d). Therefore, not only is C-protein not required for length determination but it actually interferes with it.

Craig and Offer (1976b) have recently pointed out a significant aspect of the structure of the A-band which is related to myosin filament structure; and that is the presence of a low density gap along each edge of the A-band. Although they reported this observation only with muscle stained with anti-S1, the gap can also be seen in muscle without antibody staining (Fig. 6a) and after staining with anti-myosin (Fig. 6b). Therefore, the gap clearly represents a region in which myosin cross bridges are absent as evidenced both by antibody staining (Fig. 6b; Craig and Offer, 1976b) and by lack of protein density (Fig. 6a). In my model for the myosin filament (Pepe, 1966, 1977) this gap can be explained by the tapering of the filament model. The model, as shown in Fig. 7a, is made up of 12 structural units in the shaft of the filament. These are arranged in three equivalent groups of four structural units. Consider one group of four at the tapered end of the filament, i.e., *A* in Fig. 7a. This will provide myosin cross bridges at levels 6 and 5 separated by 14 nm. There will be no cross bridges at level 4, leaving a gap of 28 nm before the next cross bridges at level 3. Therefore, tapering of the filament as shown in the right hand portion of Fig. 7a will give a gap at each edge of the A-band as deduced by Craig and Offer (1976b) in myofibrils stained with anti-S1 and as is seen even in myofibrils which are not stained with antibody (Fig. 6a).

FIGURE 6. Incomplete axial cross bridge repeat at the edges of the A-band. (a) Normal fish muscle showing a clear less dense gap (see arrows) at each edge of the A-band. (b) Rabbit muscle stained with specific anti-myosin showing a clear less dense gap (see arrows) at each edge of the A-band.

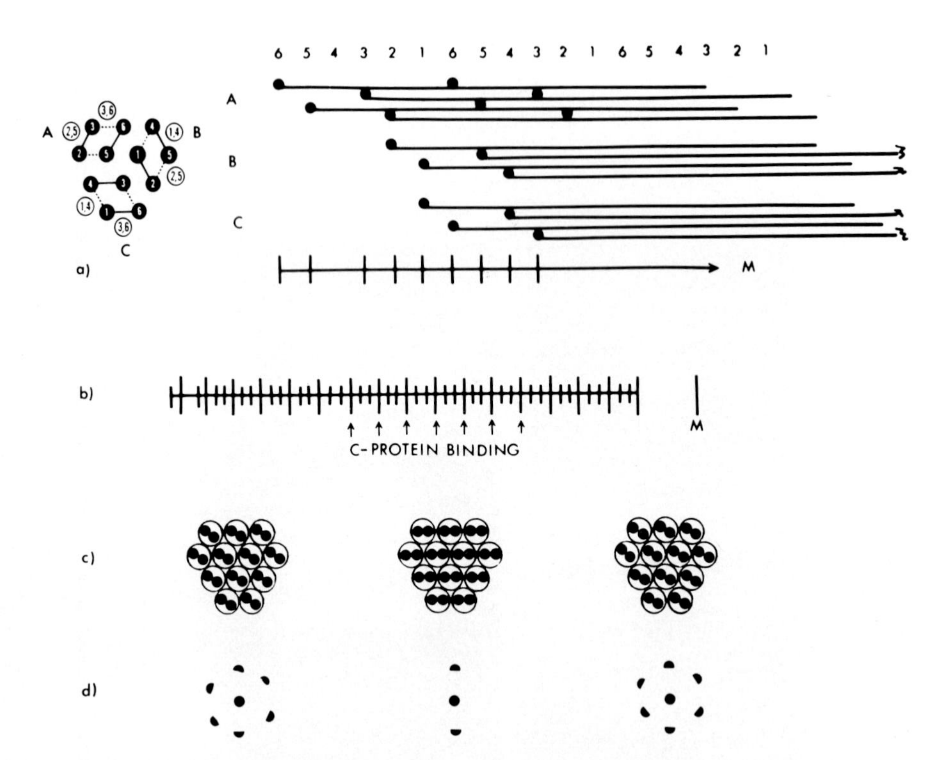

FIGURE 7. Summary of the relation between observations and the myosin filament model (Pepe, 1966, 1967). (a) A cross section of the filament model is on the left and a longitudinal representation of the tapered end of the model is at the right. Equivalent groups of four structural units are indicated by A, B, *and* C. *Below the longitudinal representation of the tapered end, the axial distribution of cross bridges is indicated and the source of the gap at each edge of the A-band (Fig. 6) is apparent. The numbers indicate intervals of 14 nm. (b) The positions of restricted location of C-protein along the length of the filament (Fig. 4) are indicated. (c) and (d) The possible structural differences, and the corresponding optical diffraction patterns observed (Fig. 3) are indicated in (c) and (d), respectively, relative to the portion of the filament model in (b) to which they correspond.*

The recent observations we have made, clearly are consistent with the model for the myosin filament (Pepe, 1966, 1967), and these are summarized as follows:

1) The observed increased visibility of substructure in cross sections of the shaft of the myosin filament with increasing section thickness (Pepe and Dowben, 1977) is expected by the parallel alignment of structural units along the long axis of the filament model.

2) The observation of twelve structural units close packed to give a triangular cross sectional profile with a structural unit missing at each apex of the triangular profile (Pepe and Drucker, 1972; Fig. 1a) is predicted by the model.

3) The observed center-to-center spacing of about 4 nm between structural units in cross sections of the myosin filament (Pepe and Drucker, 1972; Pepe and Dowben, 1977; Fig. 1) suggests that there is more than one myosin rod (with a diameter of 2 nm) in each structural unit. This is supported by X-ray diffraction studies presented by Wray (this volume) and Millman (this volume) in this symposium.

4) The unequivocal finding that there are four myosin molecules per 14 nm interval along the myosin filament (Fig. 2) is consistent with the presence of two myosin molecules per structural unit in the model.

5) The recent finding that the optical diffraction pattern produced by images of cross sections along the myosin filament is different in different portions of the filament (Pepe, Dowben and Ashton, unpublished results) indicates that the structure of the backbone changes along the length of the filament (Fig. 7c,d). This change can be related to the restricted binding of C-protein along only a portion of the myosin filament (Pepe and Drucker, 1975; Fig. 4a, Fig. 7b,c,d). The different optical diffraction patterns may be related either to the possibilities shown in Fig. 7c, where the structural units have elongated profiles, and the center-to-center distance between structural units does not change along the filament; or to the possibility that the structural units have circular profiles with the center-to-center distance between structural units changing along the length of the filament.

6) A fundamental characteristic of the model is that the information for precise length determination of the filaments is present on the myosin molecules and no other proteins are required. This is consistent with our finding that synthetic myosin filaments can be formed with precisely determined length

from highly purified myosin (Fig. 5).

7) The gap present along each edge of the A-band (Fig. 6) can be explained as resulting from the characteristics of the three equivalent groups of four structural units that make up the model (Fig. 7a).

ACKNOWLEDGMENT

This work was supported by USPHS Grant HL-15835 to the Pennsylvania Muscle Institute.

REFERENCES

Chowrashi, P. K., and Pepe, F. A. (1977). *J. Cell Biol. 74*, 136-152.

Craig, R., and Offer, G. (1976a). *Proc. R. Soc. Lond. B. 192*, 451-461.

Craig, R., and Offer, G. (1976b). *J. Mol. Biol. 102*, 325-332.

Huxley, H. E. (1963). *J. Mol. Biol. 7*, 281-308.

Laemmli, U. K. (1970). *Nature 227*, 680-685.

Moos, C., Offer, G., Star, R., and Bennett, P. (1975). *J. Mol. Biol. 97*, 1-9.

Morimoto, K., and Harrington, W. F. (1973). *J. Mol. Biol. 77*, 165-175.

Morimoto, K., and Harrington, W. F. (1974). *J. Mol. Biol. 83*, 83-97.

Pepe, F. A. (1966). *In* "Electron Microscopy" (R. Uyeda, ed.), Vol 2, pp. 53-54. Maruzen Co. Ltd., Tokyo.

Pepe, F. A. (1967). *J. Mol. Biol. 27*, 203-225.

Pepe, F. A., and Dowben, P. (1977). *J. Mol. Biol. 113*, 199-218.

Pepe, F. A., and Drucker, B. (1972). *J. Cell Biol. 52*, 255-260.

Pepe, F. A., and Drucker, B. (1975). *J. Mol. Biol. 99*, 609-617.

Pepe, F. A., Chowrashi, P. K., and Wachsberger, P. (1975). *In* "Comparative Physiology. Functional Aspects of Structural Materials" (L. Bolis, S. H. P. Maddrell, and K. Schmidt-Nielsen, eds.) pp. 105-120. Noord-Hollandsche Uitg. Mij., Amsterdam.

Potter, J. D. (1974). *Arch. Biochem. Biophys. 162*, 436-441.

Reedy, M. (1976). *Biophys. J. 16*, 126a.

Squire, J. (1973). *J. Mol. Biol. 77*, 291-323.

Tregear, R. T., and Squire, J. M. (1973). *J. Mol. Biol. 77*, 279-290.

Motility in Cell Function
Proceedings of the First John M. Marshall Symposium in Cell Biology

CYTOPLASMIC MYOSIN FILAMENTS

Thomas D. Pollard

Department of Anatomy
Harvard Medical School
Boston, Massachusetts

INTRODUCTION

The John M. Marshall Symposium in Cell Biology "Motility in Cell Functions" has given me this opportunity to review briefly the structure of the myosin filaments of nonmuscle cells. There are at least four distinct classes of myosin in various nonmuscle cells (Table 1) and each class forms different types of filaments. However, in spite of differences in size and shape, all of these myosin filaments seem to be built on common design principles.

VERTEBRATE MYOSINS

Virtually all vertebrate cells contain myosin molecules which resemble their muscle counterparts in most ways. Similar myosins have been found in other metazoan cells as well. These orthodox cytoplasmic myosins have a tail about 150 nm long with two heads at one end (Elliot *et al.*, 1976). Other physical properties, including the sizes of the constituent polypeptides (Pollard *et al.*, 1974), are also similar to muscle myosins. On the other hand, antibodies to muscle and nonmuscle myosins from the same species have only limited cross-reactivity (Pollard *et al.*, 1976), showing that the various myosins must have different amino acid sequences.

ISBN 0-12-551750-5

TABLE 1. *Myosins from Nonmuscle Cells*

Class	Mol. wt.	Composition
Orthodox myosin	*460,000*	*2 × 200,000*
Metazoa		*2 × 19,000*
		2 × 16,000
Macromyosin	*~460,000*	*2 × 230,000*
Carnivorous amoebae		*? light chains*
Slime moulds		
Midimyosin	*400,000*	*2 × 175,000*
Acanthamoeba		*1 × 17,500*
		2 × 16,500
Minimyosin	*~190,000*	*1 × 140,000*
Acanthamoeba		*1 × 16,000*
		1 × 14,000

A. *Cytoplasmic Myosin Filaments*

In light of the differences in primary structure, it was not surprising to find that nonmuscle myosins form filaments which differ from the thick filaments of striated muscles (Niederman and Pollard, 1975). Taking human platelet myosin as an example, it was found to form a homogeneous population of short, thin bipolar filaments under ionic conditions similar to those expected for the cytoplasm. The central bare zone is 150 nm long, just like the tail of the constituent molecules, while overall length of the filaments is twice the length of a myosin tail (Fig. 1). Thus the tails of the most distal molecules at each end of a filament would just meet in the middle of the bare zone. These platelet myosin filaments formed *in vitro* are very uniform in size, suggesting that the assembly process is regulated by some feature of the myosin molecule itself.

A rough two-dimensional model for the structure of the platelet myosin filament (Fig. 1) was determined from an analysis of electron micrographs. A longitudinal periodicity observed in aggregates of these platelet myosin filaments (Pollard, 1974), indicated that the heads of the myosin molecules project from the backbone of the filament in rows spaced 15 nm apart on each side of the bare zone. Given the dimensions of the bare zone and the size of the individual myosin

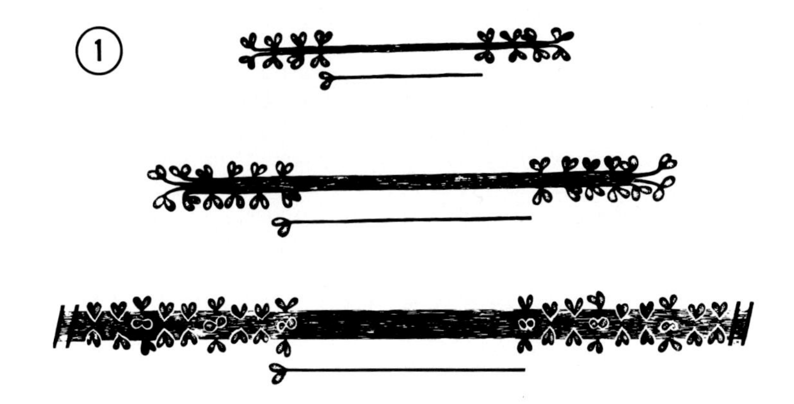

Figure 1. Line drawings of an artist's conception of myosin molecules and filaments. Top: an Acanthamoeba *myosin-II molecule and filament shown in two dimensions. Middle: a plately myosin molecule and filament shown in two dimensions. Bottom: a skeletal muscle myosin molecule and the central part of a thick filament having 6/4 symmetry, shown in three dimensions.*

molecules, it was calculated that there are two myosin molecules in each of the rows. Since the filaments are long enough for seven rows at each end, there must be (2 × 2 × 7) 28 myosin molecules in each filament (Niederman and Pollard, 1975).

B. *Muscle Myosin Filaments*

The bipolar thick filaments of striated muscle (Fig. 1) are about 18 nm wide, 1500 nm long, and contain 300 to 400 myosin molecules (see Pepe, this volume). The myosin heads project from the backbone at 15 nm intervals, but there is still controversy about whether there are 3 or 4 molecules in each row (see Pepe, this volume). The central bare zone is about 150 nm long.

When filaments are formed from purified skeletal muscle myosin, they resemble the natural thick filaments except that they vary considerably in length (Huxley, 1963).

The form of vertebrate smooth muscle myosin in the cell is also controversial (see Craig, this volume). These myosins form a number of different aggregates *in vitro*, including bipolar filaments (Kaminer *et al*., 1969) and side polar ribbons (Craig and Megerman, 1977).

C. *Hybrid Myosin Filaments*

In spite of the differences in the individual filaments that they form, the various types of vertebrate myosin seem to be capable of copolymerization (Pollard, 1975; Kaminer *et al.*, 1976). This has been shown indirectly by studying the size of filaments formed from mixtures of two different myosins. If, for example, platelet myosin and skeletal muscle myosin are mixed together at high ionic strength prior to filament formation at low ionic strength, the resulting filaments are all intermediate in size between the small platelet myosin filaments and the large skeletal muscle myosin filaments. The length of the hybrid filaments depends on the proportions of the two types of myosin. Even a small fraction of platelet myosin makes the copolymers much smaller than those of skeletal muscle myosin. It is possible that the strict filament length determining feature of the platelet myosin is responsible for restricting the size of the copolymers.

D. *Cytoplasmic Myosin* in vivo

The form of myosin within vertebrate nonmuscle cells remains a mystery. Providing that there are no inhibitors of filament formation in the cytoplasm, it is expected that the myosin is assembled into short, thin bipolar filaments like those formed from purified myosin *in vitro*. The apparent absence of these filaments from electron micrographs of vertebrate cells has been ascribed to their low concentration in these cells and the small probability of including all of one of these filaments in a single thin section (Niederman and Pollard, 1975).

In nonmuscle cells, the distribution of myosin changes with cell activity and myosin accumulates at sites of motile force generation such as the cleavage furrow of dividing cells (Fujiwara and Pollard, 1976). Since the cleavage furrow also contains a bundle of actin filaments called the contractile ring (Schroeder, 1973), it is thought that the interaction of the two contractile proteins produces the force which pinches the dividing cell in two.

ACANTHAMOEBA MYOSINS

Two different unique myosins have now been purified from the soil amoeba *Acanthamoeba castellanii* (Pollard and Korn, 1973a; Maruta and Korn, 1977b; Pollard *et al.*, 1977). The first myosin isolated, myosin-I, is a 190,000 molecular weight, more or less globular molecule. It is soluble at low ionic strength and does not form filaments. Myosin-II has a molecular weight of about 400,000 and is similar to other myosins in having two heads and a tail (Fig. 2). Its 90 nm tail is intermediate in length between the 150 nm tail of the orthodox myosins and the 50 nm tail of heavy meromyosin, a proteolytic fragment of myosin (Lowey *et al.*, 1969).

At low ionic strength myosin-II forms bipolar filaments that are remarkable for their small size. They are about 200 nm long and 6.6 nm wide (Fig. 3). Thus they are no wider than an actin filament and not much longer than the bare zone of other myosin filaments (Fig. 1). As in the case of platelet myosin filaments, the bare zone is the same length as the length of the myosin tail, and the overall length is twice as long. No more than 15 myosin tails could be packed into the 6.6 nm wide shaft of these minute filaments.

The occurence of two unusual myosins in *Acanthamoeba*, both of which are smaller than other types of myosin, raised the question: Are the two *Acanthamoeba* myosins related to each other, and it is possible that they are both products of the proteolytic cleavage of a larger, more conventional myosin? The physical properties and chemical compositions of the two myosins argue that they are unrelated to each other (Pollard *et al.*, 1978). Moreover, myosin-II is not broken down to fragments the size of myosin-I during cell homogenization, extraction and subsequent fractionation by gel permeation chromatog-

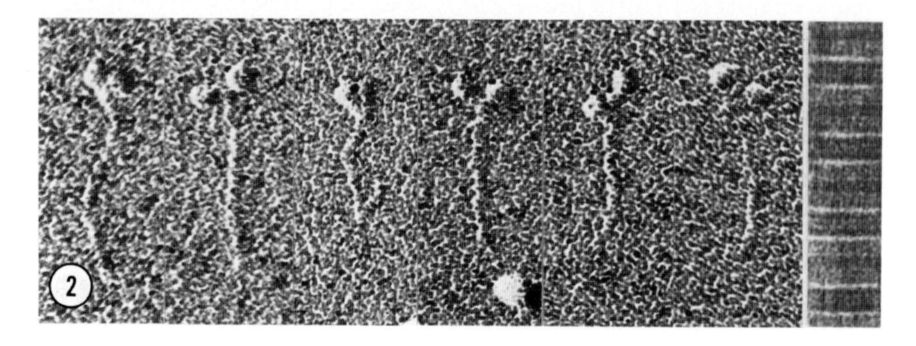

FIGURE 2. Electron micrographs of shadowed Acanthamoeba *myosin-II molecules. The right hand panel is a muscle tropomyosin paracrystal with a periodicity of 39.5 nm.*

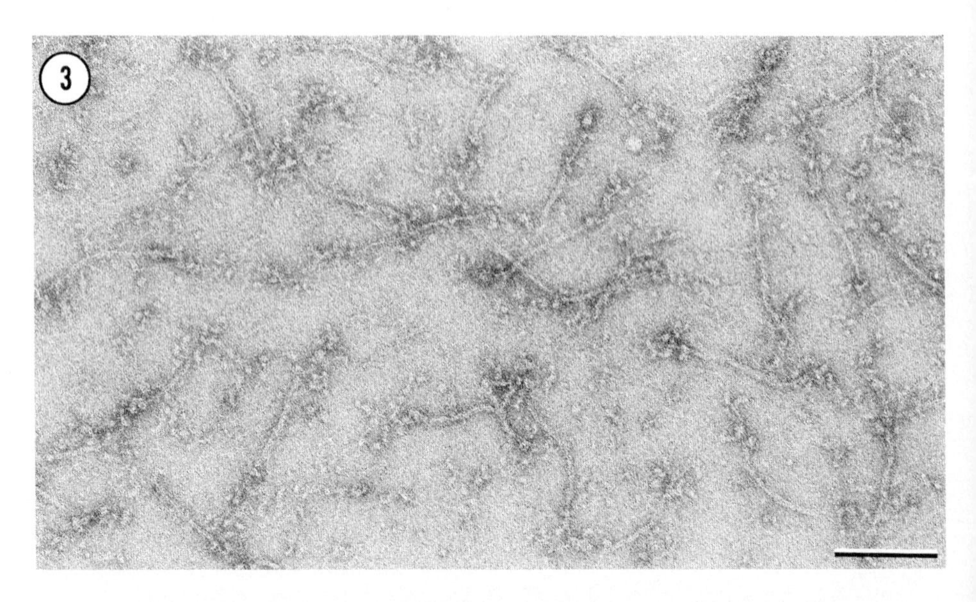

FIGURE 3. An electron micrograph of negatively stained Acanthamoeba *myosin-II filaments formed in 0.1* M *KCl, 10* mM *imidazole, pH 6.8. Magnification ×136,000, bar is 100 nm.*

raphy (Pollard *et al.*, 1978). The present evidence cannot rule out the possibility, however unlikely, that myosin-I is formed from myosin-II or a larger common precursor by a normal physiological process which does not occur in the homogenate.

It is not yet known why *Acanthamoeba* should have two types of myosin, but the differences between these myosins suggest that they must perform different functions in the cell. Myosin II, the more conventional myosin, probably has a contractile function: it forms polarized arrowhead-shaped complexes with actin (Pollard *et al.*, 1977) and experiments with both crude (Pollard, 1976) and partially purified (Maruta and Korn, 1977a) model contractile systems indicate that contraction results when it interacts with actin. Appreciable enzymatic interaction of myosin-II with actin has yet to be demonstrated. Myosin-I interacts strongly with actin in an enzyme assay (Pollard and Korn, 1973b) providing that a cofactor protein (Pollard and Korn, 1973b), recently shown to be a heavy chain kinase (Maruta and Korn, 1977c), is present. On the other hand, myosin-I does not form polarized complexes with actin (Pollard and Korn, 1973b), and it is less active than myosin-II in a model contractile system (Maruta and Korn, 1977a). The inability of myosin-I to form filaments suggests that it may have to be anchored to some cellular structure, conceivably microtubules or membranes, during force generation within the cell.

MACROMYOSINS

The myosins of slime moulds and carnivorous amoebae have two heads and a tail (Hatano and Takahashi, 1971), but their heavy chains are somewhat larger than those of other myosins (Table 1). These myosins form a variety of polymers *in vitro* including both bipolar filaments (Condeelis, 1977) and side polar filaments with rows of heads projecting at 15 nm intervals (D'Haese and Hinssen, 1974). Bipolar filaments are abundant in both fixed cells (Nachmias, 1968) and in motile extracts (Pollard and Ito, 1970), so that it is reasonably clear that these myosins are assembled into filaments in the living cells.

CONCLUDING REMARKS

Although myosin filaments come in a variety of sizes and shapes, most seem to be built on similar design principles. Most have bipolar symmetry which is necessary for a sliding filament mechanism of contraction. This bipolarity is established by a central bare zone containing an antiparallel array of myosin tails. Finding that the length of this bare zone is the same length as the myosin tail in filaments which differ as much as those in Fig. 1, suggests that the myosin tail length, rather than some other feature of the myosin tail, determines the bare zone length. A 15 nm longitudinal spacing of myosin molecules is a feature of filaments of muscle myosin, orthodox cytoplasmic myosins and macromyosins. It will be of great interest to determine whether the small *Acanthamoeba* myosin-II filaments also have this longitudinal spacing. If they do, it will suggest that this periodicity is independent of myosin tail length.

Under identical conditions, cytoplasmic myosins form much more homogeneous populations of filaments than muscle myosins. The length regulating mechanism must be intrinsic to the cytoplasmic myosin molecules. One possibility is that the cytoplasmic myosins cannot form purely parallel associations with other myosins within the backbone of the filament. Such associations with molecules of like polarity are necessary for a myosin filament to grow much longer than twice the length of the myosin tail. In short filaments extensive interactions with antipolar myosin molecules from the other side of the bare zone are possible for *all* of the myosin molecules.

The sizes of the various myosin filaments are probably related to their functions. Long, thick myosin filaments containing hundreds of myosin molecules are appropriate for striated muscle where large forces generating units are required and there is no need for the filaments to move about in the cytoplasm. Small myosin filaments are appropriate for nonmuscle cells which require little force for their movements and which continually redistribute the myosin within the cytoplasm. The small filaments must contain enough myosin to be an adequate force generating unit and certainly would diffuse through the viscous cytoplasm more rapidly than bulky thick filaments.

ACKNOWLEDGMENTS

This work was supported by N.I.H. Research Grant GM-19654.

REFERENCES

Condeelis, J. S. (1977). *J. Cell Sci. 25*, 387-402.
Craig, R., and Megerman, J. (1977). *J. Cell Biol. 75*, 990-996.
D'Haese, J., and Hinssen, H. (1974). *Cell Tiss. Res. 151*, 323-335.
Elliott, A., Offer, G., and Burridge, K. (1976). *Proc. Royal Soc. 193*, 45-53.
Fujiwara, K., and Pollard, T. D. (1976). *J. Cell Biol. 71*, 847-875.
Hatano, S., and Takahashi, K. (1971). *J. Mechanochem. Cell Motil. 1*, 7-14.
Huxley, H. E. (1963). *J. Mol. Biol. 7*, 281-308.
Kaminer, B. (1969). *J. Mol. Biol. 39*, 257-271.
Kaminer, B., Szonyi, E., and Belcher, C. D. (1976). *J. Mol. Biol. 100*, 379-386.
Lowey, S., Slayter, H. S., Weeds, A., and Baker, H. (1969). *J. Mol. Biol. 42*, 1-29.
Maruta, H., and Korn, E. D. (1977a). *J. Biol. Chem. 252*, 399-402.
Maruta, H., and Korn, E. D. (1977b). *J. Biol. Chem. 252*, 6501-6509.
Maruta, H., and Korn, E. D. (1977c). *J. Biol. Chem. 252*, 8329-8332.
Nachmias, V. T. (1968). *J. Cell Biol. 38*, 40-50.
Niederman, R., and Pollard, T. D. (1975). *J. Cell Biol. 67*, 72-92.
Pollard, T. D. (1974). *In* "Molecules and Cell Movement" (S. Inoue and R. E. Stephens, eds.), pp. 259-296. Raven Press, New York.

Pollard, T. D. (1975). *J. Cell Biol.* *67*, 93-104.
Pollard, T. D. (1976). *J. Cell Biol.* *68*, 579-601.
Pollard, T. D. and Ito, S. (1970). *J. Cell Biol.* *46*, 267-289.
Pollard, T. D., and Korn, E. D. (1973a). *J. Biol. Chem.* *248*, 4682-4690.
Pollard, T. D., and Korn, E. D. (1973b). *J. Biol. Chem.* *248*, 4691-4697.
Pollard, T. C., Thomas, S. M., and Niederman, R. (1974). *Anal. Biochem.* *60*, 258-266.
Pollard, T. D., Fujiwara, K., Niederman, R., and Maupin-Szamier, P. (1976). *In* "Cell Motility" (R. Goldman, T. Pollard, and J. Rosenbaum, eds.), pp. 689-724. Cold Spring Laboratory, New York.
Pollard, T. D., Porter, M. E., and Stafford, W. F. (1977). *J. Cell Biol.* *75*, 262a.
Pollard, T. D., Stafford, W. F., and Porter, M. E. (1978). *J. Biol. Chem.*, *253*, 4798.
Schroeder, T. E. (1973). *Proc. Nat. Acad. Sci. U.S.A.* *70*, 1688-1673.

REGULATION

Motility in Cell Function
Proceedings of the First John M. Marshall Symposium in Cell Biology

THE REGULATORY PROTEINS OF THE I FILAMENT AND THE CONTROL OF CONTRACTILE ACTIVITY IN DIFFERENT TYPES OF STRIATED MUSCLE

S. V. Perry, H. A. Cole, and G. K. Dhoot

Departments of Biochemistry and Immunology
University of Birmingham
Birmingham B15 2TT., United Kingdom

The contractile system of the muscle cell, the myofibril, is remarkably similar in structure in slow skeletal, fast skeletal, and cardiac muscle, yet each of these tissues responds to a stimulus in a manner that is characteristic for the muscle type. The differences in response of the various types of striated muscle reflect the nature of the regulatory systems present which can be regarded as consisting of three major components, namely: (1) the innervation; (2) the systems for controlling the Ca^{2+} concentration of the cytoplasm that are associated with the cell membrane and particularly with the sarcoplasmic reticulum; (3) the regulatory proteins of the myofibril that directly control the actomyosin ATPase.

This communication will be confined to discussing the ways in which the regulatory protein systems of the I filament consisting of the troponin complex and tropomyosin, vary in composition and function in different muscle types, and thus play an important role in determining the form of the contraction-relaxation cycle.

It is usually assumed that the only factor controlling the myofibrillar ATPase and hence contractile activity, is the Ca^{2+} concentration of the cytoplasm. Nevertheless this conclusion has been questioned, stimulated by the fact that *in vitro* it is difficult to reduce the ATPase activity of natural actomyosin or myofibrils to less than 5-10% of the fully activated level at Ca^{2+} concentrations similar to those in resting muscle, i.e., $<10^{-7}M$. Such levels of myofibrillar ATPase at these low Ca^{2+} concentrations are much higher than would be expected to occur in resting muscle. In view of this the question has been asked from time to time whether some other system, as yet unrecognized, is present in muscle that is able to inhibit the

ISBN 0-12-551750-5

Ca^{2+} insensitive component of the myofibrillar ATPase and thus reduce the enzyme activity in resting muscle to <1% than that of the fully activated muscle.

Recent reinvestigation of this aspect in our laboratory (Goodno *et al.*, 1978) has shown that a number of factors markedly affect the myofibrillar Mg^{2+}-stimulated ATPase at Ca^{2+} concentrations $<10^{-7}M$. In particular at these low Ca^{2+} concentrations, the ATPase activity is extremely low if the thiol groups of the myofibrillar proteins are kept fully reduced but increases on storage in the absence of thiol compounds. Also at ionic strength and ATP concentrations corresponding to those existing *in vivo* the ATPase activity at low Ca^{2+} concentrations is much suppressed. If all these factors are taken into consideration during *in vitro* enzymic studies the ATPase activity of myofibrils at Ca^{2+} concentrations $<10^{-7}M$ is usually about 1% of that when the system is fully activated at higher Ca^{2+} concentrations. Thus we may conclude that with myofibrils from rabbit fast muscle the change in ATPase activity obtained *in vitro* when the Ca^{2+} concentration increases from that of resting to the stimulated level approaches that presumed to apply *in vivo*.

It is therefore likely that the differences in the time course of the contractile responses of the various muscle types are due in part to the different effects that the changes in the intracellular Ca^{2+} concentration have on the protein system regulating the myofibrillar ATPase.

POLYMORPHISM OF THE REGULATORY PROTEINS IN STRIATED MUSCLE

The fact that each of the troponin components exist in several forms (see Perry, 1974; Cummins and Perry, 1978, for review) presumably reflects the way in which the regulatory protein system of the myofibril has evolved to accommodate the functional requirements of the different muscle types. The evidence available indicates that troponin I, troponin T and troponin C exist in three different forms, each of which is a different gene product in striated muscle. In view of the distribution of the different forms of troponin I in whole striated muscles, one might expect, as is the case with myosin (Gauthier and Lowey, 1977) that the different forms of the troponin components are associated with different cell types. For example in the heart the troponin complex is present as a single form different from the troponin complexes found in skeletal muscle. In the case of tropomyosin, there may not be a unique form of the protein in each of the three striated muscle cell types for as yet only two forms, the α and β tropomyosins, have been identified.

It is to be expected that the structural differences in the components of the regulatory protein system will be reflected in the response of the myofibrillar ATPase to changing Ca^{2+} concentrations in the sarcoplasm. The curve obtained by plotting ATPase activity of myofibrils isolated from fast skeletal muscle is very steep with the result that the ATPase rises rapidly with a relatively small change in Ca^{2+} concentration (Fig. 1). This enables a very rapid response to stimulation to take place and could be interpreted to reflect strong cooperativity between the Ca^{2+} binding sites of the troponin C. In marked contrast, the ATPase activity of myofibrils from slow muscles of the rabbit such as the soleus or crureus rises much more slowly with rising Ca^{2+} concentration and, therefore, a given increment of Ca^{2+} concentration (Fig. 1) produces a lower increase in ATPase activity. This feature is presumably of functional significance for slow skeletal muscles. The ATPase activity of cardiac myofibrils responds in a similar way to

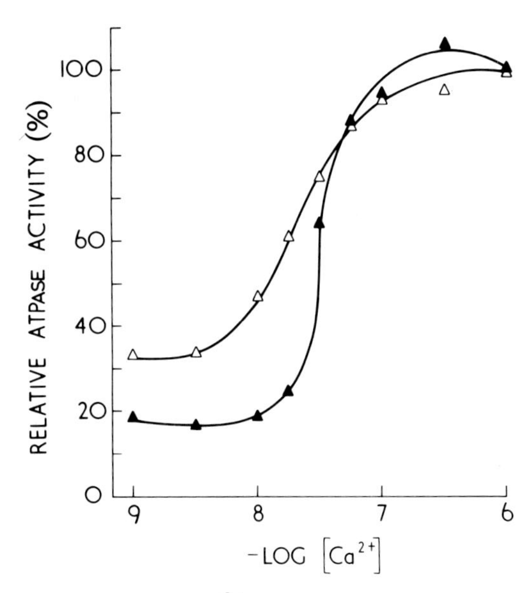

FIGURE 1. Effect of Ca^{2+} concentration on the Mg^{2+}-stimulated ATPase activity of myofibrils isolated from fast and slow muscles of the rabbit. Myofibrils from the combined soleus and crureus (0.9 mg/ml) or the L. dorsi muscles (0.22 mg/ml) incubated at 25°C for 5 min in 25 mM-Tris adjusted to pH 7.5 with 1 M-HCl/2.5 mM dithiothreitol/2.5 mM $MgCl_2$/2.5 mM-Tris-ATP/1 mM EGTA-Ca^{2+} buffer. ▲ , psoas myofibrils; △ , mixed soleus and crureus myofibrils.

increasing Ca^{2+} concentrations to that of myofibrils from slow skeletal muscle (Solaro and Briggs, 1974; Cole and Perry, 1978).

These properties no doubt reflect differences in the Ca^{2+} binding characteristics of the troponin C itself. Clear structural differences exist between the skeletal and cardiac muscle forms of this protein in a given species, based on the electrophoretic mobilities of the peptides obtained on cyanogen bromide digestion (Head *et al.*, 1977). They are implied by primary sequence studies on bovine cardiac troponin C (Van Eerd and Takahashi, 1975), and are reflected in the Ca^{2+} binding properties of the two forms (Potter and Gergely, 1975; Potter *et al.*, 1977). Interaction with troponin I also affects the Ca^{2+} binding properties of troponin C (Potter and Gergely, 1975) as probably does interaction with troponin T. Thus the precise Ca^{2+} binding properties of the myofibril will depend not only on the troponin C component, but also the nature of the troponin I and T with which it is associated. Although, as indicated above, the evidence would suggest that in each muscle cell type the components of the troponin complex are fixed, i.e., in fast skeletal muscle fibers the troponin complex consists of the fast troponin C, fast troponin I, and fast troponin T, i.e., the homocomplex, the possibility exists that, for example, in some fast skeletal muscle cells the slow form of one of the components might be replaced by the fast form. In this way a series of heterocomplexes might be formed with their Ca^{2+} binding properties different from the homocomplex characteristic of the fiber type. If this were the case the maximum number of different troponin complexes that could exist in skeletal muscle is eight, each probably with slightly different Ca^{2+} binding characteristics and thereby regulating the ATPase of the myofibrillar in different ways in response to the changing Ca^{2+} concentration.

To investigate whether in fact homo or heterocomplexes of troponin do exist in striated muscle cells we have undertaken a study of the distribution of the proteins of the regulatory system in mammalian striated muscle using the immunoperoxidase technique. In the first place the distribution of the polymorphic forms of troponin I has been studied (Dhoot *et al.*, 1978) as the fast and slow skeletel and cardiac muscle forms of the protein have been well characterized and can be readily isolated by affinity chromatography (Syska *et al.*, 1974). In general the antibodies raised against the polymorphic forms of troponin I are specific for the polymorphic form of troponin I, but not for the species from which it is derived. By staining of cross sections of adult skeletal muscle using the immunoperoxidase technique, it has been shown that the fast skeletal form of troponin I is localized in cells which stain strongly for myosin ATPase at pH 9.4, i.e., type II fibers (Fig. 2).

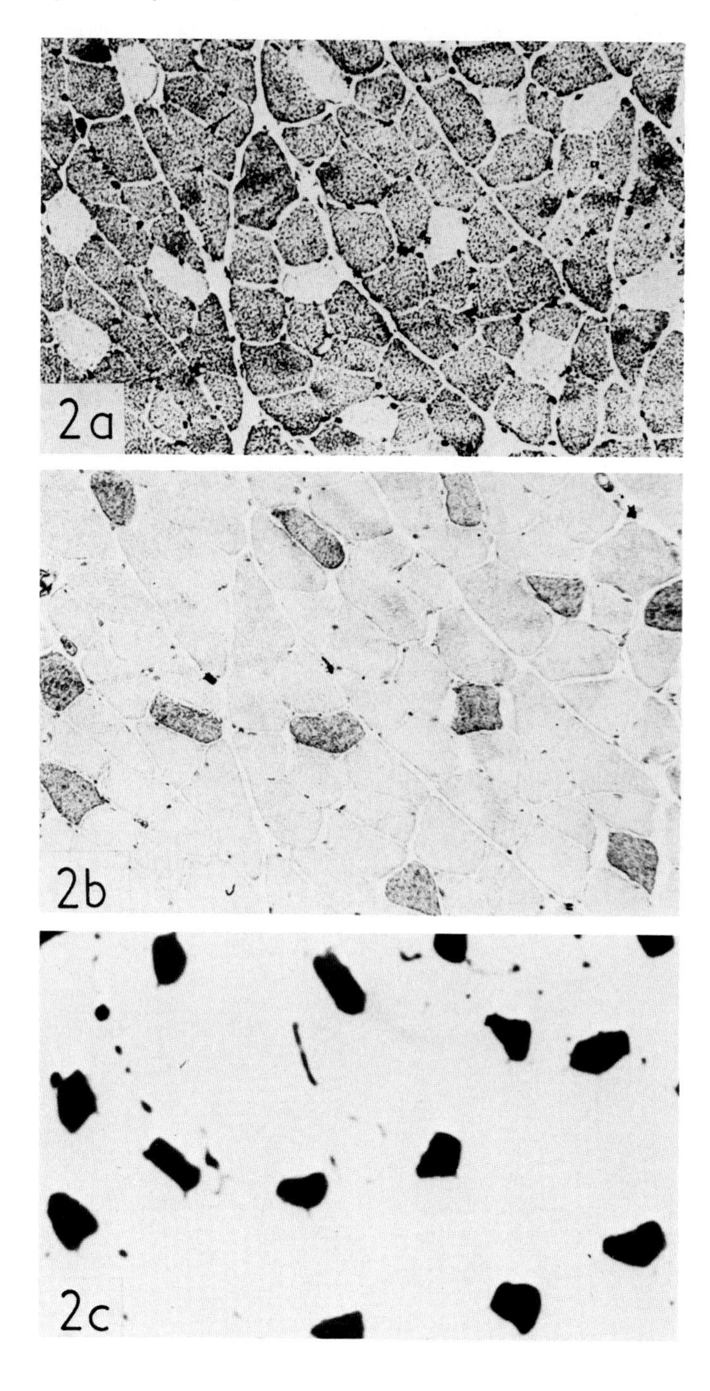
2a
2b
2c

Likewise the slow form of skeletal troponin I is located in the cells staining for myosin ATPase measured after incubation at pH 4.2, type I fibers. This distribution was observed in all normal adult muscles studied from man, rat, rabbit, hamster, guinea pig, baboon, and rhesus monkey. Thus it appears that fast troponin I is absent or present in very low amounts which are undetectable by the immunoperoxidase staining procedure in type I cells. Similarly, slow troponin I is absent from type II cells. In the adult fiber, therefore, only one of the genes controlling troponin I synthesis is expressed. Intermediate type cells were not normally present and when they were observed usually there was evidence for a change from the normal environment of the muscle. Staining for both polymorphs of troponin I in a given cell was observed in adult muscle that had undergone pathological change, after alterations in innervation or during development.

Studies with antibodies to the fast and slow skeletal muscle forms of troponin T and the α form of tropomyosin have not yet been so extensive, but support the general premise that only one polymorphic form of a component of the troponin complex is expressed at one time in the normal adult muscle and that the homocomplex is normally present in a given normal cell type (Fig. 3).

The results obtained to date clearly indicate that fast troponin I is associated with fast troponin T in type II fibers and complementarily slow troponin I is associated with slow troponin T in type I fibers. The α form of tropomyosin appears to be restricted to type II cells.

These studies indicate that a predetermined association of the proteins of the regulatory system occurs in the adult muscle cell. For example, in type II fibers containing the fast isoenzymic forms of myosin the I filament regulatory proteins are fast I and fast T and probably fast C existing as a homocomplex and α-tropomyosin. These observations suggest that some type of cooperative control exists for the expression of the genes controlling the synthesis of most, if not all, of the proteins of the myofibril.

FIGURE 2. Localization of fast and slow forms of troponin I in different skeletal muscle cell types identified by myosin ATPase staining. Serial sections of rat gastrocnemius muscle prepared and stained as described by Dhoot et al., *(1978). Antibodies raised in the guinea pig. (a) Stained with antibody to rabbit fast skeletal muscle troponin I by immunoperoxidase procedure. (b) Stained with antibody to rabbit slow troponin I by immunoperoxidase procedure. (c) Stained for myosin ATPase after incubation at pH 4.2, i.e., type I cells.*

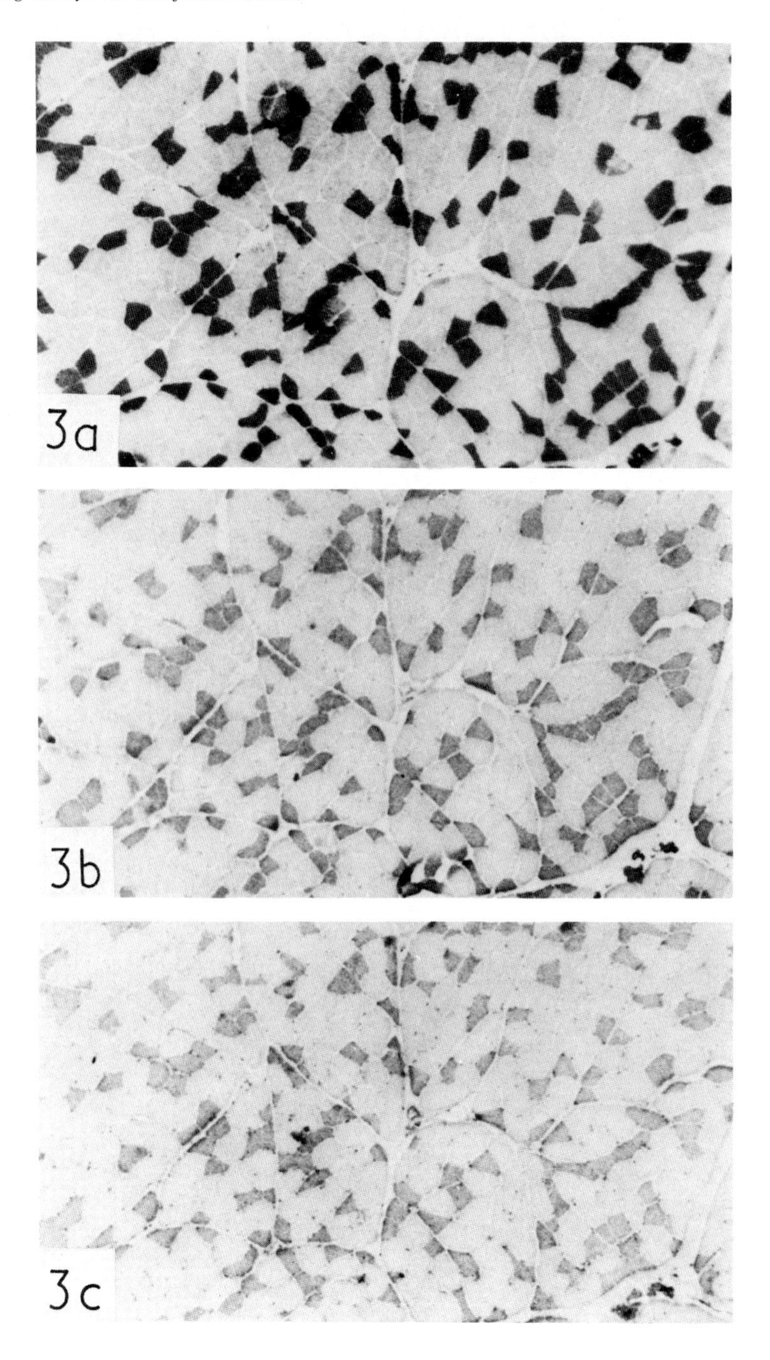
3a
3b
3c

It appears also that in a given fiber of adult striated muscle the pattern of response of the myofibrillar ATPase to normal muscle only the genes responsible for the synthesis of the components of the troponin homocomplex characteristic for the fiber type are expressed. Only in extreme conditions, such as changes in innervation, or as a consequence of disease or regeneration of the tissue, is this genetic control disturbed and do the genes responsible for the expression of the other polymorphic forms of the regulatory proteins become active. It is not yet clear whether this leads to the existence of a series of heterocomplexes, but the fact that fast and slow forms of all three components are produced under these conditions suggests that fast and slow troponin homocomplexes are present.

ROLE OF PHOSPHORYLATION OF REGULATORY PROTEINS IN CHANGING THE RESPONSE OF THE CONTRACTILE SYSTEM TO Ca^{2+}

It is now established that phosphorylation of three of the four proteins of the regulatory protein system in the I filament, namely troponin I (Bailey and Villar-Palasi, 1971; Stull *et al.*, 1972; Perry and Cole, 1974), troponin T (Perry and Cole, 1973; Cole and Perry, 1975; Moir *et al.*, 1977) and tropomyosin (Ribolow and Barany, 1977) can occur. Of these systems only in the case of troponin I is there evidence of a physiological role for this process.

The phosphorylation sites of troponin I from fast skeletal muscle (Moir *et al.*, 1974; Huang *et al.*, 1974) and cardiac muscle (Moir and Perry, 1977) of the rabbit have been determined and presumptions about those of troponin I from slow muscle can be made by comparison of its amino acid sequence with that of the other two polymorphic forms (Wilkinson and Grand, 1978) (Fig. 4).

FIGURE 3. Localization of fast troponin I, fast troponin T and α-tropomyosin in human semispinalis capitis muscle. Immunoperoxidase staining procedure on serial sections using antibodies to rabbit proteins raised in the guinea pig. (a) Antibody to fast skeletal muscle troponin I. (b) Antibody to fast skeletal muscle troponin T. (c) Antibody to α-tropomyosin.

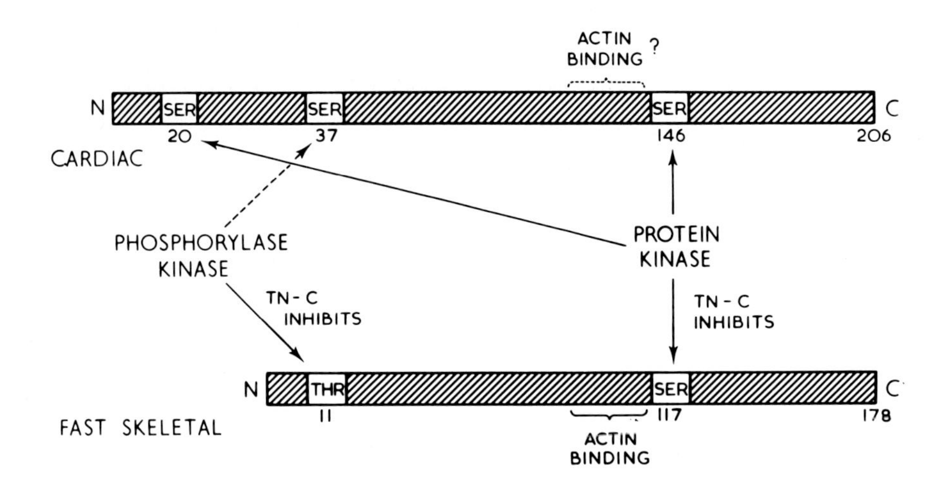

FIGURE 4. Scheme illustrating the phosphorylation sites and primary sequences of troponin I from cardiac and fast and slow skeletal muscles of the rabbit (Moir et al., *1974; Moir and Perry, 1977; Wilkinson and Grand, 1978).*

Fast skeletal troponin I has two major sites of phosphorylation, threonine 11, the principal site phosphorylated by phosphorylase kinase, and serine 117, the principal site for 3':5'-cyclic AMP-dependent protein kinase. Although the majority of the amino acid sequence of rabbit cardiac troponin I is homologous with the fast skeletal form of the protein, it has an additional 26 residue peptide at the N terminus (Grand *et al.*, 1976). This peptide contains a phosphorylation site at serine 20 which is, therefore, unique for the cardiac form. Although this residue is the main site for phosphorylation catalyzed by 3':5'-cyclic AMP-dependent protein kinase, phosphorylation also occurs at ser 146, a site analagous to serine 117 of fast skeletal troponin I. The phosphorylase kinase site would, by analogy with fast skeletal troponin I, be expected to be ser 37 but this has not yet been demonstrated directly (Fig. 4).

Direct comparison of the rates of phosphorylation of fast skeletal and cardiac troponin I and the effects of troponin C on them reveals some differences that are of significance for the role of this process in the regulation of contraction in the two types of muscle.

1) If isolated directly by homogenization of fresh muscle in strong urea solution followed by isolation by affinity chromatography on a column of Sepharose-troponin C (Syska *et al.*,1974) both polymorphic forms of troponin I contain covalently bound phosphate. Whereas the fast skeletal form usually has 0.5 moles P/mole, the cardiac form contains 1.2-2.0 moles P/mole, the precise value being variable and depending on the conditions of the muscle immediately before the troponin I was isolated (Cole and Perry, 1975).

2) The phosphorylation of isolated rabbit fast skeletal troponin I catalyzed by phosphorylase kinase and by 3':5'-cyclic AMP-dependent protein kinase is completely inhibited by equimolar amounts of troponin C with cardiac troponin I as substrate, whereas only the phosphorylation by phosphorylase kinase is inhibited under similar conditions (Cole and Perry, 1975).

3) Isolated cardiac troponin I is phosphorylated 20-30 times faster than fast skeletal troponin I by 3':5'-cyclic AMP-dependent protein kinase (Cole and Perry, 1975).

Although these results are obtained from *in vitro* studies, they suggest that the phosphorylation of troponin I catalyzed by 3':5'-cyclic AMP-dependent protein kinase has special significance in cardiac muscle. These conclusions are further supported by investigation of the changes in phosphorylation of troponin I occurring *in situ* in the perfused heart. During perfusion by the Langendorff procedure the bound phosphate content of rabbit cardiac troponin I remains constant at about 1.2 moles P/mole troponin I (Jones *et al.*, 1978). On treatment with adrenaline the phosphate content in the perfused rat heart (England, 1975) and rabbit heart (Solaro *et al.*, 1976; Jones *et al.*, 1978) rises, in the latter case from 1.2 moles P/mole to about 1.8 moles P/mole troponin I. The increase in phosphate does not correlate precisely with the increase in force, as the rise in phosphate content of the troponin I occurs some seconds after the increase in force. It also falls at a slower rate than the force when the adrenaline is removed from the perfusate. Thus in the perfused heart phosphorylation of troponin I occurs in response to the effect of adrenaline on the myocardium, but does not correlate exactly with the force changes.

By isolation of the N terminal peptide consisting of residues 1 to 48 obtained on digestion of troponin I with cyanogen bromide, the increased phosphorylation of troponin I that occurs during adrenaline treatment can be shown to be almost entirely associated with serine 20 (Solaro *et al.*, 1976; Moir *et al.*, 1978).

When myofibrils or natural actomyosin prepared from cardiac muscle are incubated with purified preparations of 3':5'-cyclic AMP-dependent protein kinase, phosphorylation occurs mainly on the troponin I component. This protein can be isolated from the actomyosin preparations by affinity chromatography and shown to be phosphorylated by the incorporation of ^{32}P when $[\gamma-^{32}P]ATP$ is used as substrate for the protein kinase. Phosphorylation of the troponin I in this way, which is mainly confined to serine 20 (Moir and Perry, 1977), leads to a small, but significant decrease in sensitivity of the ATPase activity of actomyosin or myofibrils to Ca^{2+}, i.e., slightly higher concentrations of Ca^{2+} are required to produce 50% of maximal activation of the myofibrillar ATPase (Fig. 5 and Table 1).

In the experiments illustrated in Table 1 the Ca^{2+} concentration required to give 50% of maximal activation of actomyosin ATPase is increased by a factor of about two as a result of phosphorylation of the troponin I which only represents 1-2% of the total protein present in the system. Ray and England (1977) have also obtained similar results and shown that after treatment of cardiac myofibrils with phosphatase the Ca^{2+} concentration required for 50% activation falls. Although it was originally claimed that phosphorylation of cardiac troponin I made the actomyosin system more sensitive to Ca^{2+} (Rubio *et al.*, 1975), the balance of experimental evidence from a number of other laboratories suggests that the converse is the case (Reddy and Wyborny, 1976; Bailin, personal communication).

Similar experiments in which myofibrils or natural actomyosin from fast skeletal muscle are incubated with 3':5'-cyclic AMP-dependent protein kinase do not lead to much additional phosphorylation of the troponin I or any significant change in the effect of Ca^{2+} on the ATPase activity. If cardiac troponin is used to sensitize skeletal actomyosin, phosphorylation of the troponin I by 3':5'-cyclic AMP-dependent protein kinase leads to a decrease in sensitivity as is the case with natural actomyosin from cardiac muscle. The ability to change the sensitivity of the ATPase system to Ca^{2+} is therefore a property of the troponin rather than the actomyosin.

TABLE 1. Effect of Phosphorylation by Protein Kinase on Calcium Sensitivity of the Mg^{2+}-Stimulated ATPase of Bovine Cardiac Natural Actomyosin.

P content of actomyosin (mole P/100,000g)			Δ pCa^{2+} for 50% activation after phosphorylation
Control	Exp.	Δ P	
0.04	0.20	0.16	0.20
0.04	0.20	0.16	0.32
--	--	--	0.40
0.02	0.09	0.07	0.30
0.06	0.30	0.24	0.25
--	--	--	0.25

Incubations as described in legend to Fig. 1. Electrophoresis of extracts of actomyosin in sodium dodecyl sulphate indicated that most of the radioactivity when the actomyosin was incubated with [γ-^{32}P]ATPase and 3':5'-cyclic AMP-dependent protein kinase was located in the troponin I band.

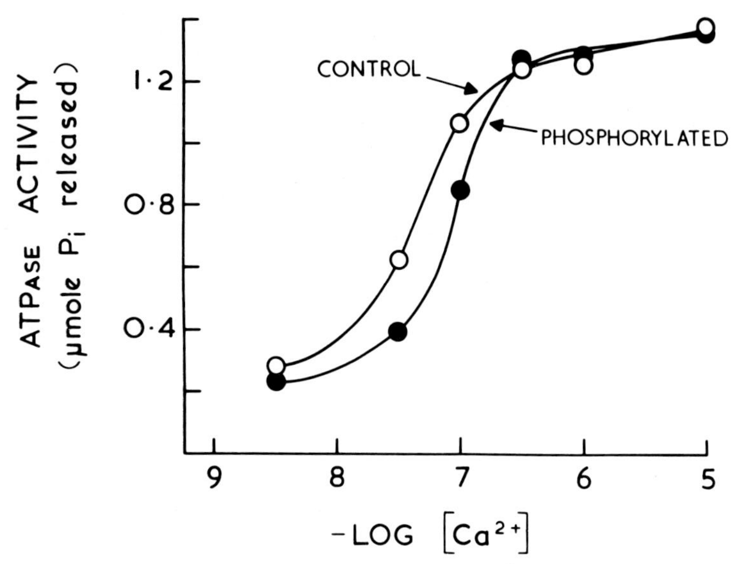

FIGURE 5. Effect of phosphorylation of troponin I of cardiac natural actomyosin on the Mg^{2+}-stimulated ATPase activity at different Ca^{2+} concentrations. Bovine cardiac actomyosin (3.8 mg/ml) was preincubated with 3':5'-cyclic AMP-dependent protein kinase in the presence of 3':5'-cyclic AMP. ATPase activity of control and phosphorylated actomyosin determined as described in legend to Fig. 1. O, control; ●, phosphorylated.

Thus it appears that due to the special properties of cardiac troponin the myofibrillar ATPase of cardiac muscle can change its response to the Ca^{2+} concentration inside the cell. This property is regulated by adrenaline and depends directly on the unique phosphorylation site of cardiac troponin I, serine 20.

In the scheme illustrated in Fig. 6 for the mode of action of adrenaline on the contractile response of cardiac muscle, the phosphorylation of troponin I forms a negative feedback loop regulating the response of the myofibrillar ATPase to Ca^{2+}. The phosphorylation of troponin I and consequent reduction in Ca^{2+} sensitivity represents a means of smoothing out the response to the increased Ca^{2+} concentration that occurs in cardiac muscle after stimulation of the β-receptors by adrenaline. The immediate consequent of the rise in Ca^{2+} concentration is the rapid increase in contractile force that is characteristic of the inotropic response. Activation of the 3':5'-cyclic AMP-dependent kinase which also occurs leads to increased phosphorylation at serine 20, usually 20-30% phosphorylated in

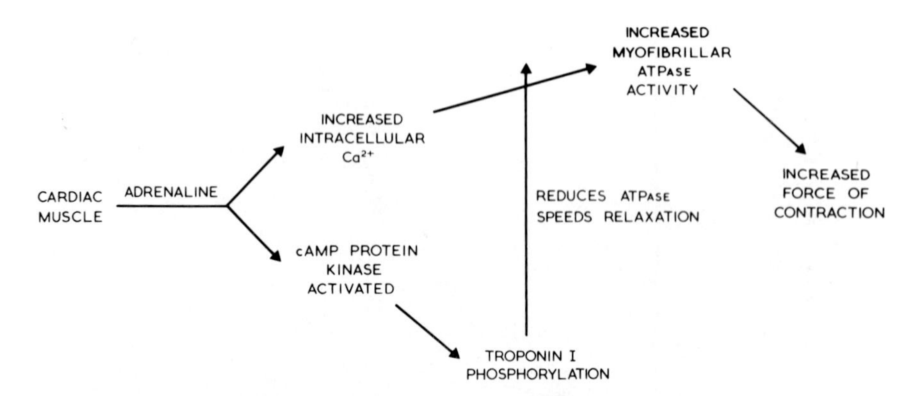

FIGURE 6. Scheme outlining the changes in myofibrillar response to Ca^{2+} *concentration that occur in cardiac muscle in the presence of adrenaline.*

the normal beating perfused heart before intervention with adrenaline (Moir *et al.*, 1978). As a consequence changes occur in the binding characteristics of the cardiac troponin C which lead to a higher Ca^{2+} concentration being required to produce the same level of activation of the myofibrillar ATPase. This fact explains the increased rate of relaxation associated with the inotropic effect of adrenaline, for as the Ca^{2+} concentration decreases in diastole the fall in ATPase activity and hence tension will be more rapid when troponin I is phosphorylated above the normal level.

Thus we can distinguish between the two main mechanisms that are effective at the myofibrillar level in striated muscle for regulating the characteristic contractile response to the changes in Ca^{2+} concentration on stimulation.

These are illustrated diagrammatically in Fig. 7. The first mechanism is under genetic control and in the normal adult muscle cell is responsible for the expression of one of the types of the troponin homocomplex. This form of control is relatively invariable and characteristic of the muscle cell type and is changed only by extreme conditions which lead to changes at the genetic level. Whereas each skeletal muscle cell is under an all or none nervous control and increase in force developed by the muscle is obtained by increased recruitment of cells, the cardiac cell is able to change the force developed per cell. This is achieved by developing a more flexible system of regulation in cardiac muscle, a process that involves changing the rate of Ca^{2+} flux into the cytoplasm and which can be controlled by external factors such as β-adrenergic agents. The development of this flexibility in regulating the Ca^{2+} concentration in the cardiac cytoplasm requires further

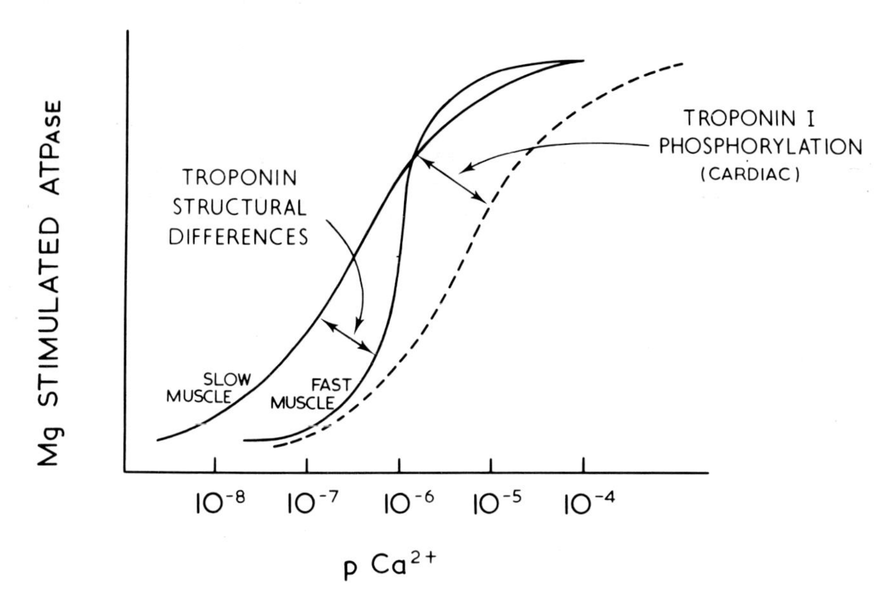

FIGURE 7. Scheme to represent mechanisms operating at the I filament level for changing the response of the myofibrillar ATPase to Ca^{2+} concentration. The thick lines represent the way in which the ATPase activity changes in response to changing Ca^{2+} in fast and slow skeletal muscle. The dotted line represents the change in response to Ca^{2+} produced by phosphorylation of troponin I of cardiac muscle.

adaptation of the control at the myofibrillar level to smooth out the contractile response. To this end the troponin I phosphorylation system has been evolved to change the sensitivity of the myofibrillar ATPase to Ca^{2+} and also enabling relaxation to be speeded up.

ACKNOWLEDGMENTS

This work was supported from grants by the Medical Research Council and the Muscular Dystrophy Group of Great Britain.

REFERENCES

Bailey, C., and Villar-Palasi, C. (1971). *Fed. Proc. 30,* 1147.

Cole, H. A., and Perry, S. V. (1975). *Biochem. J. 149,* 525-533.

Cole, H. A., and Perry, S. V. (1978). to be published.

Cummins, P., and Perry, S. V. (1978). *Biochem. J. 171,* 251-259.

Dhoot, G. K., Gell, P. G. H., and Perry, S. V. (1978). *Experimental Cell Research, 117,* 357-370.

England, P. J. (1975). *FEBS Letts. 50,* 57-60.

Gauthier, G. E., and Lowey, S. (1977). *J. Cell. Biol. 74,* 760-779.

Goodno, C. C., Wall, C. M., and Perry, S. V. (1978). *Biochem. J.,* in press.

Grand, R. J. A., Wilkinson, J. M., and Mole, L. E. (1976). *Biochem. J. 159,* 633-641.

Head, J. F., Weeks, R. A., and Perry, S. V. (1977). *Biochem. J. 161,* 371-382.

Huang, T. S., Bylund, D. B., Stull, J. T., and Krebs, E. G. (1974). *FEBS Letts. 42,* 249-252.

Jones, C. M., Solaro, R. J., and Perry, S. V. (1978). to be published.

Moir, A. J. G., and Perry, S. V. (1977). *Biochem. J. 167,* 333-343.

Moir, A. J. G., Wilkinson, J. M., and Perry, S. V. (1974). *FEBS Letts. 42,* 253-256.

Moir, A. J. G., Cole, H. A., and Perry, S. V. (1977). *Biochem. J. 161,* 371-382.

Moir, A. J. G., Solaro, R. J., and Perry, S. V. (1978). To be published.

Perry, S. V. (1974). *In* "Exploratory Concepts in Muscular Dystrophy, II" (A. T. Milhorat, ed.), pp. 319-328. Excerpta Medica, Amsterdam.

Perry, S. V., and Cole, H. A. (1973). *Biochem. J. 131,* 425-428.

Perry, S. V., and Cole, H. A. (1974). *Biochem. J. 141,* 733-743.

Potter, J. D., and Gergely, J. (1975). *J. Biol. Chem. 250,* 4628-4633.

Potter, J. D., *et al.* (1977). *In* "Calcium Binding Proteins and Calcium Function" (R. H. Wasserman *et al.*, eds.) pp. 239-250. North-Holland, New York.

Ray, K. P., and England, P. J. (1976). *FEBS Letts. 70,*11-16.

Reddy, Y. S., and Wyborny, E. (1976). *Biochem. Biophys. Res. Comm. 73,* 703-709.

Ribelow, H., and Barany, M. (1977). *Arch. Biochem. Biophys. 179,* 718-720.

Rubio, R., Bailey, C., and Villar-Palasi, C. (1975). *J. Cyclic Nucleotide Res. 1,* 143-150.

Solaro, R. J., and Briggs, F. N. (1974). *In* "Calcium Binding Proteins" (W. Drabikowski, H. Strzelecka-Golaszewska, and E. Carafoli, eds.) pp. 587-607. Elsevier, Amsterdam.

Solaro, R. J., Moir, A. J. G., and Perry, S. V. (1976). *Nature 262,* 615-617.

Stull, J. T., Brostrom. C. O., and Krebs, E. G. (1972). *J. Biol. Chem. 247,* 5272-5274.

Syska, H., Perry, S. V., and Trayer, I. P. (1974). *FEBS Letts. 40,* 253-257.

Van-Eerd, J. P., and Takahashi, K. (1975). *Biochem. Biophys. Res. Com. 64,* 122-127.

Wilkinson, J. M., and Grand, R. J. A. (1978). *Nature 271,* 31-35.

Motility in Cell Function
Proceedings of the First John M. Marshall Symposium in Cell Biology

PHOSPHORYLATION OF MYOSIN: A POSSIBLE REGULATORY MECHANISM IN SMOOTH MUSCLE

R. Dabrowska,[1] *J. M. F. Sherry,*[2]
and
D. J. Hartshorne[2]

[1]Department of Biochemistry of Nervous System and Muscle
Nencki Institute of Experimental Biology
Warsaw, Poland

[2]Departments of Biological Sciences and Chemistry
Carnegie-Mellon University
Pittsburgh, Pennsylvania

INTRODUCTION

It is the general concensus of opinion that contraction in smooth muscle is due to the cyclic interactions of the myosin cross-bridges with actin, in fundamentally the same process as in skeletal muscle. It is also accepted that these interactions are subject to control by the intracellular Ca^{2+} concentration; at relatively high levels of ionized Ca^{2+} (in the order of $5 \times 10^{-6}M$) contraction occurs and at lower concentrations of Ca^{2+} relaxation occurs. There is, however, no universal agreement on the regulatory mechanism that recognizes these Ca^{2+} transients and then transmits the relevant signal to the contractile apparatus.

One of the tenets that is established is that regulation must be achieved via a Ca^{2+}-dependent *activation* of actomyosin ATPase activity. The Mg^{2+}-ATPase activity of pure smooth muscle myosin is not activated by actin, unlike the situation with skeletal muscle myosin. This point was established experimentally several years ago (Bárány *et al.*, 1966; Yamaguchi *et al.*, 1970) and its importance to the regulatory mechanism of smooth muscle has been stressed repeatedly by Ebashi and his colleagues (Ebashi *et al.*, 1975a,b; Mikawa *et al.*, 1977b). In its simplest concept, the regulatory proteins effect an activation of actomyosin ATPase

ISBN 0-12-551750-5

activity (and hence tension development, *in situ*) but only in the presence of Ca^{2+}; in the absence of Ca^{2+} the myosin is not activated and the hydrolysis of ATP is prevented.

In the search for candidates to fulfill the regulatory function one is therefore limited to activators of smooth muscle myosin, and this is a point that is sometimes overlooked. The results that have been accumulated over the last decade may be considered in three categories. These have postulated that the regulatory mechanism in smooth muscle is due to a) a troponin-like system, b) phosphorylation of the myosin light chains, and c) an enzymic process that is not yet identified, but excludes phosphorylation. Each of these will be considered in more detail.

A. *Troponinlike mechanisms*

Historically the biochemistry of smooth muscle has depended to a large extent for its development on the better understood skeletal muscle system. With respect to the regulatory process this is no exception. Following the discovery of troponin by Ebashi and his co-workers in skeletal muscle (see Ebashi and Endo, 1968) a similar system was claimed in smooth muscle (Ebashi *et al.*, 1966; Carsten, 1971). More recently Head *et al.* (1977) have isolated proteins from various smooth muscles which are similar to the skeletal troponin subunits, namely troponin I (TpI) and troponin C (TpC). (A subunit similar to troponin T has not been found). In a wide variety of cells a protein similar to TpC, but not identical to it, is present (the modulator protein, to be discussed in more detail below) and there is the possibility that the TpC isolated by Head *et al.* (1977) is in fact the TpC-like modulator protein. This contention is supported by the recent findings of Drabikowski *et al.* (1977, 1978) who could find only the modulator protein in chicken gizzards and rabbit uterus and no evidence for the existence of TpC. Earlier experiments had cast doubt on the involvement of troponin, in its conventional sense, in smooth muscle regulation as troponinlike subunits were not present in Ca^{2+}-sensitive actomyosin (Driska and Hartshorne, 1975; Sobieszek and Bremel, 1975), thin filaments (Driska and Hartshorne, 1975; Sobieszek and Small, 1976) and partially purified actin (Hartshorne *et al.*, 1977). Thus, if the troponinlike subunits are indeed functional in the regulatory mechanism it would probably be the TpI-modulator complex that is involved (the substitution of TpC for modulator is a significant alteration, as the TpI-modulator complex does not require TpT for its Ca^{2+}-dependent response (Amphlett *et al.*, 1976)). In our opinion even this system is not reasonable, since the primary requirement is for an activation process and the established role of troponin is to inhibit an active state in the absence of Ca^{2+} and not to activate a dormant state in the presence of Ca^{2+}. It is possible that the troponin-like system might operate in tandem with another regulatory mechanism and this remains to be established.

V. *Phosphorylation of the Myosin Light Chains*

This mechanism is the most popular of the three alternatives, and is accepted by many investigators in the field (Sobieszek, 1977; Aksoy *et al.*, 1976; Gorecka *et al.*, 1976; Chacko *et al.*, 1977). The system requires the concerted action of two enzymes, a myosin light chain kinase (phosphotransferase) and a phosphatase. The critical points are as follows: the 20,000 dalton light chains of myosin, in the presence of Ca^{2+}, are phosphorylated by the kinase, this allows the subsequent activation by actin of the Mg^{2+}-ATPase activity. As long as Ca^{2+} is present hydrolysis of ATP will proceed (leading to contraction or tension development in whole muscle). When Ca^{2+} is removed the kinase is not active and the phosphatase removes the phosphate groups from the myosin light chains; this prevents the actin-activation of ATPase activity and the actomyosin system is turned off (i.e., the muscle relaxes). Some aspects of this cycle are not established, for example the role of Ca^{2+} binding to the myosin light chains, but the general outline is accepted. It should also be pointed out that this system fulfills the requirement for a Ca^{2+}-dependent activation of the Mg^{2+}-ATPase activity of actomyosin.

C. *80,000 Dalton Activator*

Subsequent to the early observations of a native tropomyosinlike protein in smooth muscle (Ebashi *et al.*, 1966) Ebashi and his co-workers have isolated a protein factor which they consider to be the essential regulatory component in smooth muscle (Mikawa *et al.*, 1977b; Hirata *et al.*, 1977). This has a molecular weight of 80,000 daltons, and requires tropomyosin for its action. In this respect it is similar to troponin, but it is present in significantly lower amounts than a conventional troponinlike protein would ($<1:1$ *M* ratio with tropomyosin; Mikawa *et al.*, 1977b). For this reason Ebashi and colleagues do not consider it to be analagous to skeletal muscle troponin. Its mechanism of action, however, is not known, although it is reported not to be a myosin light chain kinase (Mikawa *et al.*, 1977a; Mikawa *et al.*, 1977b). It is worth noting in the context of this discussion that a purified protein would need to have a dual function in order to cope with both the activation of myosin leading to contraction, and the deactivation of myosin leading to relaxation. Further it is assumed that the reversibility of this mechanism (presumably enzymic) is regulated by Ca^{2+}. The identification of the mechanism and its reversibility remain to be established in Ebashi's scheme.

D. Experimental Summary and Rationale

Since there is obviously a wide range of divergent opinion on the regulatory mechanism in smooth muscle, our goal was to purify the myosin light chain kinase and to examine its activity at different levels of purity. We reasoned that any non-kinase, but functional components, would be lost during the isolation procedures. Our results indicated that the myosin light chain kinase was responsible for the Ca^{2+}-dependent activation of ATPase activity, and this confirmed our earlier hypothesis. The kinase was composed of two components, one of molecular weight 105,000 daltons (referred to as 105K) and one of 17,000 daltons. The smaller protein was remarkably similar to TpC and was identified as the modulator protein (also called the Ca^{2+}-dependent regulatory protein). Besides being involved in the regulation of contractile activity in smooth muscle the latter has several functions in the cell, including the activation of cAMP phosphodiesterase activity (Cheung, 1970; Kakiuchi *et al.*, 1970), the activation of adenylate cyclase activity (Cheung *et al.*, 1975; Brostrom *et al.*, 1975), the regulation of Ca^{2+}-dependent Mg^{2+}-ATPase activity of erythrocyte membranes (Gopinath and Vincenzi, 1977; Jarrett and Penniston, 1977), and the stimulation of phosphorylation of brain membrane proteins (Schulman and Greengard, 1978). It is significant also that the modulator protein is found in a wide variety of cell types (see review, Wang, 1977), and this prompted the idea that its partner in the kinase function, the 105K, might also be of widespread occurrence. We have therefore begun a search for the larger of the two kinase subunits in different snimal cell types, and in this presentation we report on our findings with human blood platelets.

MATERIALS AND METHODS

The following procedures were used: chicken gizzard myosin (Hartshorne *et al.*, 1977), rabbit skeletal muscle actin (Driska and Hartshorne, 1975), rabbit skeletal muscle tropomyosin (Hartshorne and Mueller, 1969), Mg^{2+}-activated ATPase activity and ^{32}P incorporation (Aksoy *et al.*, 1976), amino acid analyses (Dabrowska *et al.*, 1978). The isolation of 105K and the modulator protein from the native tropomyosin fraction (Ebashi *et al.*, 1966) was as described by Dabroska *et al.* (1977).

The crude platelet kinase was prepared as follows: Outdated platelets were obtained from the blood bank, no less than 12 units were used per preparation. The platelets were washed three times to remove serum proteins with 0.15 *M* KCl, 15 m*M* tris-HCl (pH 7.5), 1 m*M* ethyleneglycol bis-(β-aminoethyl ether) -N,N'-tetraacetic acid (EGTA), 0.2 m*M* phenylmethylsulfonyl fluoride. The platelet pellet after the final wash was suspended in 5 volumes of buffer A (consisting of: 0.8 *M* KCl, 1 m*M* EGTA, 1 m*M* $MgCl_2$, 10 m*M* tris-HCl (pH 7.5), 0.2 m*M* dithiothreitol, 0.2 m*M* phenylmethylsulfonyl fluoride) sonicated for 15 sec at frequency 5 (using a Sonifier, Bronson Ultrasonics Corp., Model W185) and centrifuged at 100,000 × g for 30 min. The supernatant was subject to ammonium sulfate fractionation and the precipitate obtained between 35 and 75% saturation, was collected and dialyzed versus buffer A. This solution (> 10 mls) was applied to a Sepharose 4B column (100 × 2.5cm), equilibrated with buffer A, and fractions collected at a flow rate of approximately 20 ml per hour. The leading edge of the retarded peak was found to contain the kinase activity.

RESULTS AND DISCUSSION

A. *Biochemical Characterization of the Kinase Components*

It was discovered that the myosin light chain kinase from chicken gizzard was composed of two distinct proteins, the 105K component and the modulator protein (Dabrowska *et al.*, 1977). Neither protein alone had any effect on either the Mg^{2+}-ATPase activity of gizzard myosin plus skeletal actin or, the incorporation of phosphate from γ-labelled [^{32}P] ATP into gizzard myosin. However, when the two components were mixed both the ATPase activity and the extent of phosphorylation was increased. This is shown in Fig. 1. For the ATPase assays gizzard myosin, skeletal actin and tropomyosin, and modulator were held constant and the amount of 105K was varied: for the phosphorylation assays, gizzard myosin and modulator were held constant, and 105K was varied. Each curve was composed from four separate preparations of 105K. The important points to be derived from Fig 1 are that the two kinase components together are essential for activity, that the activation of ATPase activity follows approximately the extent of myosin phosphorylation, and that in the absence of Ca^{2+}, whatever components are present, there is no effect.

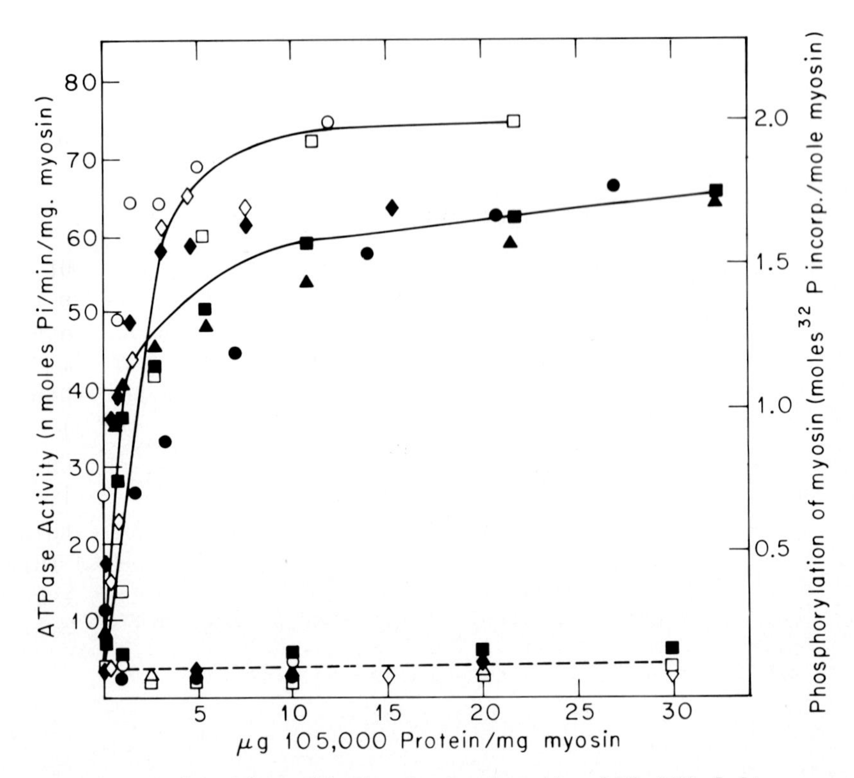

FIGURE 1. The effect of varying the 105,000 dalton component at constant modulator concentration on the Mg^{2+}-ATPase activity of gizzard myosin plus skeletal muscle actin, and the phosphorylation of gizzard myosin. ATPase assay conditions (closed symbols): 10 mM $MgCl_2$, 50 mM KCl, 2.5 mM Tris-HCl (pH 7.6), 25°C, gizzard myosin approximately 0.5 mg/ml, skeletal muscle actin, 0.25 mg/ml, tropomyosin 0.075 mg/ml. Modulator (17,000 dalton component) constant at 2.4 μg/ml. Phosphorylation assay conditions (open symbols): Solvent, myosin, and modulator concentrations as above. (Actin and tropomyosin were not added.) Dashed line indicates both ATPase and phosphorylation assays done in the presence of 1 mM EGTA, other conditions as above. Different symbols indicate different preparations of the 105,000 component. (Figure adapted from Dabrowska et al. *(1977), reprinted courtesy of Academic Press, Inc.)*

The modulator from a variety of sources appeared to function identically. This is shown in Fig. 2, where the effect of a varying amount of modulator from adrenal medulla, brain, and chicken gizzard are compared. Each was equally effective in activating the Mg^{2+}-ATPase activity and the extent of myosin

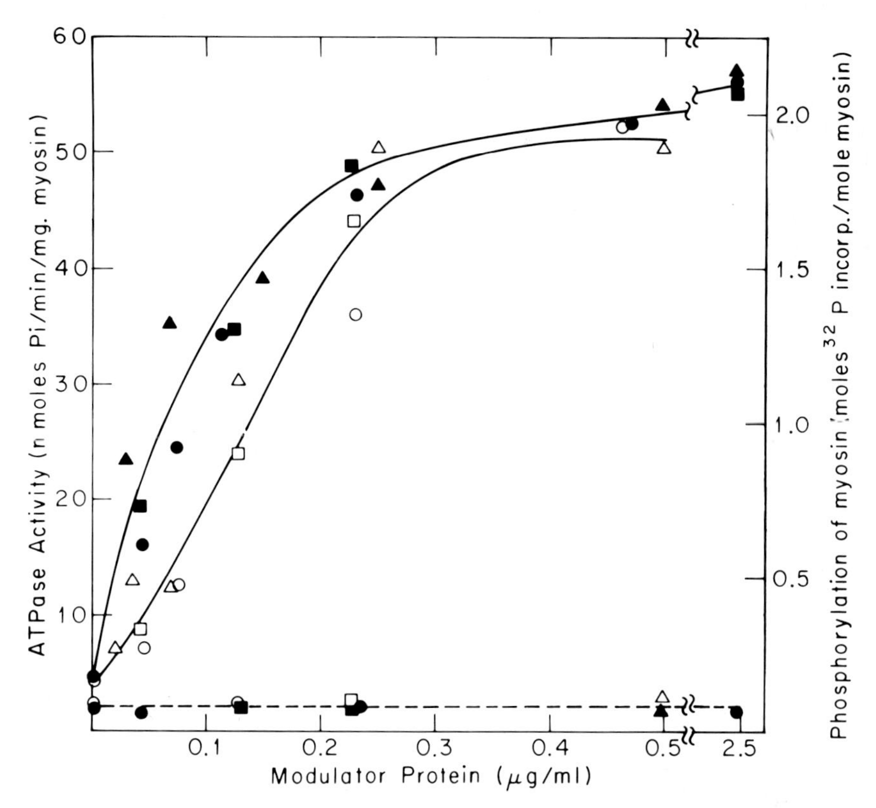

FIGURE 2. The effect of the 17,000 dalton component and other modulator proteins at a constant concentration of the 105,000 dalton component on the Mg^{2+}-ATPase activity of gizzard myosin plus skeletal muscle actin, and the phosphorylation of gizzard myosin. ATPase assay conditions (solid symbols) as in Fig. 1, except that the concentration of the modulator was varied, and that of the 105,000 component was constant at 6.4 μg/ml. Phosphorylation assay conditions (open symbols) as in Fig. 1, at varied modulator and constant 105,000, as above. Dashed line indicates both ATPase and phosphorylation assays done in the presence of 1 mM *EGTA. The modulators from chicken gizzard, (O, O), adrenal medulla (Δ Δ), and brain ( ) were used. (Figure adapted from Dabrowska* et al. *(1978), reprinted courtesy of the American Chemical Society.)*

phosphorylation. Once again in the absence of Ca^{2+} no activation was detected. When the modulator was substituted by TpC, up to 30 μg/ml, no activation of ATPase activity occurred. In an assay of phosphodiesterase activity (results not shown) Dabrowska *et al.* (1978) showed that the modulator from adrenal medulla and gizzard were also about equally effective.

Several points remain to be established. For example, the stoichiometry of the two proteins in the active complex has not been established, and the nature and site of the kinase interaction with myosin is unknown. It is also not meaningful at this stage to compare kinetic parameters with those obtained for other protein kinases, and in particular myosin light chain kinases (Daniel and Adelstein, 1976; Pires and Perry, 1977). We have not determined the K_m for the kinase, with respect to the light chain, nor do we have sufficient data to calculate a rate for the phosphorylation reaction. Until these are established comparisons are pointless.

B. *Physical Properties of the Two Components*

The amino acid analyses of the modulator and 105K are given in Table 1. The analysis of the smooth muscle modulator is very similar to those reported for adrenal medulla (Kuo and Coffee, 1976), bovine brain (Watterson *et al*., 1976), bovine heart (Stevens *et al*., 1976), and rat testis (Dedman *et al*., 1977). The presence of 1 mole of trimethyllysine per mole protein is an unusual feature and serves as a useful index of identification. Tryptophan was not present in the modulator from smooth muscle as indicated by the lack of tryptophan emission fluorescence and the uv absorption spectrum. Scanning for tryptophan fluorescence served as a useful criteria for estimating the purity of column fractions. A protein with a similar mobility to the 105K on SDS-electrophoresis is α-actinin (subunit molecular weight about 100,000 daltons [Suzuki *et al*., 1976]). The amino acid composition of α-actinin, from either skeletal muscle (Robson and Zeece, 1973) or chicken gizzard (R. M. Robson, personal communication) is, however, quite distinct from the 105K.

In many of its physical properties the modulator protein resembles TpC. The two proteins, however, are different. This has been illustrated above in terms of biological activity as it was found that TpC did not activate the 105K in ATPase and phosphorylation assays. In physical properties a difference may be detected on the urea-polyacrylamide gels as the modulator proteins had a slightly slower mobility than TpC (Amphlett *et al*., 1976; Dabrowksa *et al*., 1978).

TABLE 1. *Amino Acid Composition of the Two Components of the Chicken Gizzard Myosin Light Chain Kinase*

Amino acid	*17,000 component (modulator)*[a]		*105,000 component*[b]	
	moles/mole	*moles/10^5g protein*	*moles/mole*	*moles/10^5g protein*
Lysine	7	41	102	97
Histidine	1	6	16	15
Trimethyllysine	1	6	0	0
Arginine	6	35	38	37
Aspartic acid	23	135	92	87
Threonine[c]	12	71	61	58
Serine[d]	5	29	86	82
Glutamic acid	28	165	139	132
Proline	2	12	47	45
Glycine	12	71	71	68
Alanine	11	65	73	70
Valine	7	41	52	49
Methionine	9	53	20	19
Isoleucine	8	47	44	42
Leucine	9	53	59	56
Tyrosine	2	12	23	22
Phenylalanine	8	47	27	26

[a]*Taken from Dabrowska* et al. *(1978).*
[b]*Taken from Dabrowska* et al. *(1977).*
[c]*Corrected assuming a 6% loss during 22 hr hydrolysis.*
[d]*Corrected assuming a 10% loss during 22 hr hydrolysis.*

C. Comparison with Other Myosin Light Chain Kinases

There are only a few reports in which a reasonable homogeneity of a light chain kinase has been established. In two of these instances a kinase of about 80,000 daltons has been isolated, one from blood platelets (Daniel and Adelstein, 1976) and one from skeletal muscle (Pires and Perry, 1977). The two differ in that a Ca^{2+}-dependent phosphorylation of myosin was found only for the skeletal muscle kinase. There is no indication that either kinase required additional protein components for activity. Recently Yazawa and Yagi (1977) partially purified a skeletal muscle light chain kinase and found it to consist of two components, of approximate molecular weights 100,000 and 200,000, the latter being the Ca^{2+}-binding component. These results are similar to ours (reported above). Thus in the case of skeletal muscle one is faced with the possibility that two distinct myosin light kinases are present, and this suggestion may be extended to include other tissues. The validity of this generalization, however, can be accepted only when the two types of kinases are isolated from several sources. An alternative explanation is that the smaller kinase is derived from the larger subunit (for example, following proteolysis). Although this is largely speculative our preference is for the second possibility. We have shown that following proteolysis of the crude kinase Ca^{2+} regulation is lost (Hartshorne *et al.*, 1977). More recently Dabrowska and Hartshorne (in preparation) found that proteolysis of the 105K resulted in the appearance of smaller proteins, including a component about 80,000 daltons, and that this was correlated with the loss of Ca^{2+}-sensitivity. Thus in the case of the platelet kinase our suggestion is that it is derived from a larger and Ca^{2+}-dependent kinase, presumably as a result of proteolysis. (It should also be emphasized that when single column fractions are assayed for kinase activity only the modulator independent kinase will be detected.) In the case of skeletal muscle we have no adequate explanation, since the kinase of Pires and Perry (1977), although of molecular weight about 80,000, retained Ca^{2+}-sensitivity. It is possible that certain proteolytic processes could retain Ca^{2+}-sensitivity in the hydrolytic fragments, although this has not been demonstrated. The requirement for modulator protein in our system and not in that of Pires and Perry (1977) is also a feature that cannot be resolved at this time.

D. 105K and Modulator Protein in Other Cell Types.

Myosin and actin are present in a wide variety of eucaryotic cells, and since in most nonmuscle cells the regulatory mechanism(s) is not established it was decided to test the myosin light chain kinase (i.e., 105K plus modulator protein) as a candidate. Our initial experiments were done with blood platelets, and the rationale for this choice and for initiating the search in general, was based on the following facts: 1) It is known that the modulator protein has a widespread distribution (see Wang, 1977; and Dabrowska *et al.*, 1978, for references) and since it has been shown to be involved in one process regulating actin-myosin interactions (in smooth muscle) it is not unreasonable to look for this mechanism in other cell types. It is of specific interest that the modulator protein has been shown to be present in platelets (Muszbek *et al.*, 1977). 2) A myosin light chain kinase was found in blood platelets which had the characteristics of a "degraded" kinase, and thus this system was appropriate for us to test our proteolysis theory.

E. Preliminary Results with the Platelet Kinase

So far we have not purified the 105K component from blood platelets and the results that are presented here are done with a partially purified kinase (see Materials and Methods). The kinase preparation that was used, however, did contain a component comigrating with 105K on SDS-electrophoresis.

The active kinase component was eluted from Sepharose 4B in the leading edge of the retarded peak (see Materials and Methods). ATPase and phosphorylation assays done with this fraction are shown in Fig. 3. In each case gizzard myosin was used. The important features of these results are 1) a kinase component which requires modulator protein can be obtained from blood platelets; 2) this protein in the presence of Ca^{2+} and modulator activates the actin-moderated ATPase activity of gizzard myosin and catalyzes the incorporation of phosphate into the myosin molecule; 3) there appears to be a correlation between the activation of ATPase activity and the incorporation of phosphate; 4) in the absence of Ca^{2+} the ATPase activity was not activated and the incorporation of phosphate was drastically reduced.

These results do not eliminate the possibility that a Ca^{2+}-independent kinase is also present in blood platelets, but they do demonstrate that a Ca^{2+}-dependent myosin light chain kinase is a potential candidate for the regulation of actin-myosin mediated processes in blood platelets. In the future we need to confirm these results and to extend our search to include other cell types.

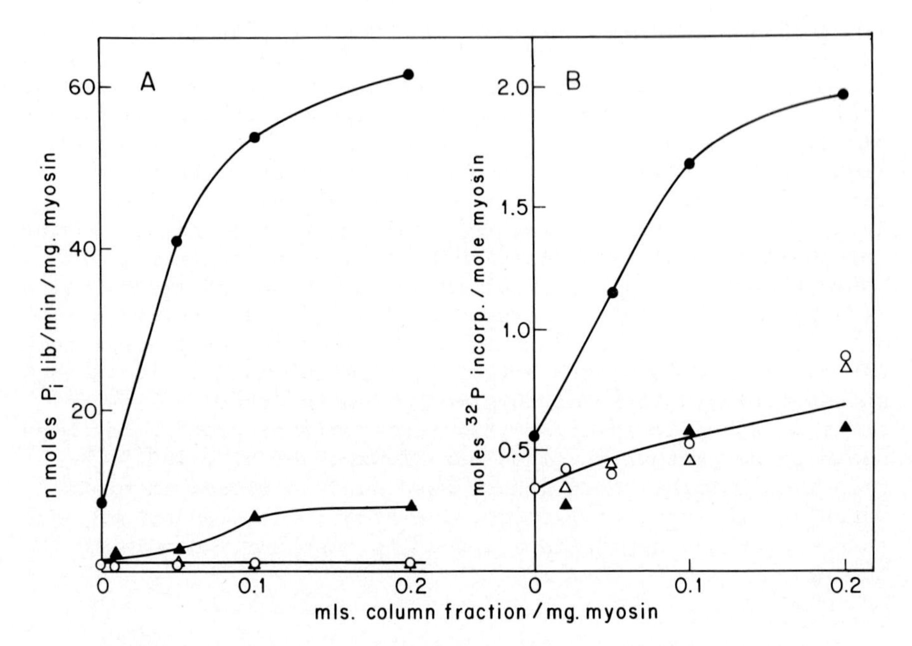

FIGURE 3. The effect of a crude platelet kinase on the Mg^{2+}-ATPase activity of gizzard myosin plus skeletal actin, and the phosphorylation of gizzard myosin. ATPase assays shown in A; solvent conditions, myosin, actin and tropomyosin concentrations as in Fig. 1. Assays done in presence of gizzard modulator, 2.4 μg/ml (0 , 0) in the absence of modulator (Δ , Δ), and in the presence of 1 mM ETGA (0 , Δ). Phosphorylation assays shown in B; solvent conditions and myosin concentration as in Fig. 1. Assays done in the presence and absense of modulator (as above) and in the presence and absence of EGTA (as above). The absorbance, at 278 nm, of the column fraction was 0.7; however, 105K was only a minor component of this fraction.

ACKNOWLEDGMENTS

The authors wish to thank S. Abmayr for valuable assistance. This work was supported by Grant HL-09544 from the National Institutes of Health.

REFERENCES

Aksoy, M. O., Williams, D., Sharkey, E. M., and Hartshorne, D.J. (1976). *Biochem. Biophys. Res. Commun. 69,* 35.

Amphlett, G. W., Vanaman, T. C., and Perry, S. V. (1976). *FEBS Lett. 72,* 163.

Bárány, M., Bárány, K., Gaetjens, E., and Bailin, G. (1966). *Arch. Biochem. Biophys. 113,* 205.

Brostrom, C. O., Huang, Y. C., Breckenridge, B. McL., and Wolff, D. J. (1975). *Proc. Natl. Acad. Sci. U.S.A. 72,* 64.

Carsten, M. (1971). *Arch. Biochem. Biophys. 147,* 353.

Chacko, S., Conti, M. A., and Adelstein, R. S. (1977). *Proc. Natl. Acad. Sci. U.S.A. 74,* 129.

Cheung, W. Y. (1970). *Biochem. Biophys. Res. Commun. 38,* 533.

Dabrowska, R., Aromatorio, D., Sherry, J. M. F., and Hartshorne, D. J. (1977). *Biochem. Biophys. Res. Commun. 78,* 1263.

Dabrowska, R., Sherry, J. M. F., Aromatorio, D. K., and Hartshorne, D. J. (1978). *Biochemistry 17,* 253.

Daniel, J. L., and Adelstein, R. S. (1976). *Biochemistry 15,* 2370.

Dedman, J. R., Potter, J. D., Jackson, R. L., Johnson, J. D., and Means, A. R. (1977). *J. Biol. Chem. 252,* 8415.

Drabikowski, W., Kuźnicki, J., and Grabarek, Z. (1977). *In* "Calcium Binding Proteins and Calcium Function" (R. H. Wasserman *et al.*, eds.), pp. 270-272. Elsevier North-Holland, Amsterdam, New York, Oxford.

Drabikowski, W., Kuźnicki, J., and Grabarek, Z. (1978). *Comp. Biochem. Physiol.*, in press.

Driska, S., and Hartshorne, D. J. (1975). *Arch. Biochem. Biophys. 167,* 203.

Ebashi, S., and Endo, M. (1968). *Progr. Biophys. Mol. Biol. 18,* 123.

Ebashi, S., Iwakura, H., Nakajima, H., Nakamura, R., and Ooi, Y. (1966). *Biochem. Z. 345,* 201.

Ebashi, S., Nonomura, Y., Kitazawa, T., and Toyo-oka, T. (1975a). *In* "Calcium Transport in Contraction and Secretion" (E. Carofoli *et al.*, eds.) pp. 405-414. North Holland, Amsterdam.

Ebashi, S., Toyo-oka, T., and Nonomura, Y. (1975b). *J. Biochem. 78,* 859.

Gopinath, R. M., and Vincenzi, F. F. (1977). *Biochem. Biophys. Res. Commun. 77,* 1203.

Gorecka, A., Aksoy, M. O., and Hartshorne, D. J. (1976). *Biochem. Biophys. Res. Commun. 71,* 325.

Hartshorne, D. J., *et al.* (1977). *In* "The Biochemistry of Smooth Muscle" (N. L. Stephens, ed.) pp 513-532. University Park Press, Baltimore.

Hartshorne, D. J., Gorecka, A., and Aksoy, M. O. (1977). *In* "Exitation-Contraction Coupling in Smooth Muscle" (R. Casteels, T. Godfraind, and J. C. Rüegg, eds.) pp. 377-384. Elsevier North-Holland, Amsterdam, New York, Oxford.

Hartshorne, D. J., and Mueller, H. (1969). *Biochim. Biophys. Acta 175,* 301.

Head, J. F., Weeks, R. A., and Perry, S. V. (1977). *Biochem. J. 161,* 465.

Hirata, M., Mikawa, T., Nonomura, Y., and Ebashi, S. (1977). *J. Biochem. 82,* 1793.

Jarrett, H. W., and Penniston, J. T. (1977). *Biochem. Biophys. Res. Commun. 77,* 1210.

Kakiuchi, S., Yamazaki, R., and Nakajima, H. (1970). *Proc. Japan Acad. 46,* 587.

Kuo, I. C. Y., and Coffee, C. J. (1976). *J. Biol. Chem. 251,* 1603.

Mikawa, T., Nonomura, Y., and Ebashi, S. (1977a). *J. Biochem. 82,* 1789.

Mikawa, T., Toyo-oka, T., Nonomura, Y., and Ebashi, S. (1977b). *J. Biochem. 81,* 273.

Muszbek, L., Kuźnicki, J., Szabó, T., and Drabikowski, W. (1977). *FEBBS Lett. 80,* 308.

Pires, E. M. V., and Perry, S. V. (1977). *Biochem. J. 167,* 137.

Robson, R. M., and Zeece, M. G. (1973). *Biochim. Biophys. Acta 295,* 208.

Schulman, H., and Greengard, P. (1978). *Nature 271,* 478.

Sobieszek, A. (1977). *In* "The Biochemistry of Smooth Muscle" (N. L. Stephens, ed.), pp. 413. University Park Press, Baltimore.

Sobieszek, A., and Bremel, R. D. (1975). *Eur. J. Biochem. 55,* 49.

Sobieszek, A., and Small, J. V. (1976). *J. Mol. Biol. 102,* 75.

Stevens, F. C., Walsh, M., Ho, H. C., Teo, T. S., and Wang, J. H. (1976). *J. Biol. Chem. 251,* 4495.

Suzuki, A., *et al.* (1976). *J. Biol. Chem. 251,* 6860.

Wang, J. H. (1977). *In* "Cyclic Nucleotides: Mechanism of Action" (H. Cramer and J. Schultz, eds.) pp. 37-56. Wiley, New York.

Watterson, D. M., Harrelson, Jr., W. G., Keller, P. M., Sharief, F., and Vanvman, T. C. (1976). *J. Biol. Chem. 251,* 4501.

Yamaguchi, N., Miyazawa, Y., and Sekine, T. (1970). *Biochim. Biophys. Acta 216,* 411.

Yazawa, M., and Yagi, K. (1977). *J. Biochem. 82,* 287.

Motility in Cell Function
Proceedings of the First John M. Marshall Symposium in Cell Biology

STUDIES ON MYOSIN LIGHT CHAIN KINASE IN MACROPHAGES, SCALLOPS, AND PLATELETS

Robert S. Adelstein, John A. Trotter,
David R. Hathaway, Robert Heinen,
and
Mary Anne Conti

Section on Molecular Cardiology
National Heart, Lung, and Blood Institute,
National Institutes of Health
Bethesda, Maryland

INTRODUCTION

Phosphorylation of myosin regulates the contractile proteins isolated from platelets, proliferative myoblasts, *Acanthamoeba* and a number of different smooth muscles (Adelstein and Conti, 1975; Scordilis and Adelstein, 1977; Maruta and Korn, 1977; Gorecka *et al.*, 1976; Chacko *et al.*, 1977a; Sobieszek and Small, 1977). Phosphorylated myosin undergoes actin-activation of its Mg^{2+}-ATPase activity but nonphosphorylated myosin is not activated by actin. Myosin phosphorylation, which in most cases occurs on the 20,000 dalton light chain of the myosin molecule, is catalyzed by a specific kinase (Daniel and Adelstein, 1976; Pires and Perry, 1977). Dephosphorylation of myosin is mediated by a phosphatase, which has been purified from skeletal muscle and partially purified from platelets (Morgan *et al.*, 1976; Barylko *et al.*, 1977).

In this paper we report on three new aspects of myosin phosphorylation: 1) Rabbit alveolar macrophages contain a myosin phosphorylating-dephosphorylating system. 2) Preliminary data indicate the presence of a myosin light chain kinase in scallops. 3) The enzyme platelet myosin light chain kinase, which previously was isolated in a Ca^{2+}-independent form has now also been isolated in a Ca^{2+}-dependent form.

ISBN 0-12-551750-5

MACROPHAGE MYOSIN PHOSPHORYLATION

Macrophages exhibit a number of functions such as motility, secretion, and phagocytosis, that may be controlled by the contractile proteins actin and myosin. Since previous work from this laboratory had indicated that phosphorylation of human platelet myosin (Adelstein and Conti, 1975) and rat proliferative myoblast myosin (Scordilis and Adelstein, 1977) was necessary for actin-myosin interaction, it was important to determine whether a similar type of regulatory mechanism existed in macrophages. Regulation of macrophage contractile proteins was of additional interest because of the finding by Hartwig and Stossel (1975) that the presence of a third protein, called "cofactor" was necessary for actin-activation of myosin ATPase activity. The studies reported here (JAT and RSA) establish the presence of a myosin phosphorylating-dephosphorylating system in macrophages.

Methods

Rabbits were stimulated by a single intravenous injection of complete Freund's adjuvant. Following sacrifice 2-4 weeks later, alveolar macrophages were harvested by intratracheal lavage with normal saline and collected by centrifugation.

The preparation of a fraction containing actomyosin and myosin light chain kinase was similar to that previously outlined for platelets (Adelstein and Conti, 1976) with the exception that the cells were broken in 0.34 *M* sucrose-10 pyrophosphate m*M* 15 m*M* Tris·HCl (pH 7.5) - 10 m*M* dithiothreitol (DTT) - 1 m*M* EDTA 1 m*M* PMSF (phenylmethylsulfonylfluoride) in a Dounce homogenizer. Following extraction for 45 min the suspension was sedimented at 45,000 × g × 1 hr. The supernatant was made 10 m*M* in $MgCl_2$ and ATP and fractionated by addition of saturated ammonium sulfate-10 m*M* EDTA. The 35-55% saturated ammonium sulfate fraction was dissolved in 0.6 *M* KCl - 15 m*M* Tris·HCl (pH 7.4) - 1 m*M* EDTA -5 m*M* DTT and dialyzed overnight in the same buffer with 0.06 *M* KCl to precipitate actomyosin. The pH of the solution was lowered to 6.3 following dialysis and the precipitate sedimented and redissolved in the high ionic strength buffer at pH 7.4.

Phosphorylation was carried out using the above solution following adjustment of the KCl concentration by dialysis to 0.2 *M*. The conditions for phosphorylation were as described (Adelstein and Conti, 1976) with the exception that 5 m*M* ATP was used in the incubation mixture. Phosphorylation was terminated by chilling the sample on ice, raising the KCl concentration to 0.6 *M* and applying the sample to a 1.5 × 90 cm column of Sepharose 4B equilibrated with 0.5 *M* KCl - 15 m*M* Tris·HCl (pH 7.4) - 1 m*M* EDTA - 2.5 m*M* DTT. Nonphosphorylated controls were treated in a similar manner but pyrophosphate was used in place of ATP. An isolated fraction of myosin light chains was prepared from skeletal and smooth muscle myosin by the method of Perrie and Perry (1970).

Results

Evidence for Macrophage Myosin Kinase: Fig. 1 (top) shows a Sepharose 4B gel filtration profile following chromatography of the incubation mixture used for phosphorylation of macrophage myosin. Three peaks of activity labeled AM, M, and K are depicted. The first two peaks (AM and M) indicate the K^+-EDTA-activated ATPase activity and represent actomyosin (AM) and myosin (M), respectively. Figure 2 is a photograph of 1% SDS-7 1/2% polyacrylamide gels of a pooled fraction of the actomyosin (left) and myosin (right), eluted from the Sepharose 4B gel filtration column.

When γ-labeled $AT^{32}P$ was included in the phosphorylating mixture both actomyosin and myosin coeluted with peaks of ^{32}P-radioactivity. In order to identify the radioactive proteins each peak (actomyosin and myosin) was pooled and subjected to SDS-polyacrylamide gel electrophoresis. After staining with Coomassie brilliant blue (see Fig. 2) the gels were sliced into 2 mm segments and ^{32}P-radioactivity determined. The bottom left graph in Fig. 1 is the radioactivity profile of a gel of macrophage myosin. The major peak of ^{32}P-radioactivity comigrated with the 20,000 dalton light chain of macrophage myosin. A similar pattern of radioactivity was found for the actomyosin peak (not shown). These experiments indicate that the 20,000 dalton light chain of macrophage myosin is phosphorylated. Quantitation of the extent of phosphorylation based on ^{32}P-incorporation showed that approximately 1 mole of phosphate could be incorporated per mole of myosin light chain.

Further evidence for the presence of macrophage myosin kinase in the incubation mixture used for phosphorylation is indicated by the presence of peak K in the upper part of Fig. 1. This peak was obtained by assaying the fractions eluted from Sepharose 4B for their ability to transfer ^{32}P from γ-labeled $AT^{32}P$ to the isolated 20,000 dalton light chain of turkey gizzard smooth muscle myosin. The panel on the bottom-right of Fig. 1 is the radioactivity profile of an SDS-polyacrylamide gel showing that the only protein phosphorylated during this assay was the 20,000 dalton light chain of smooth muscle myosin. If the smooth muscle myosin light chain fraction was omitted from the assay no peak of radioactivity was found on the SDS-polyacrylamide gel.

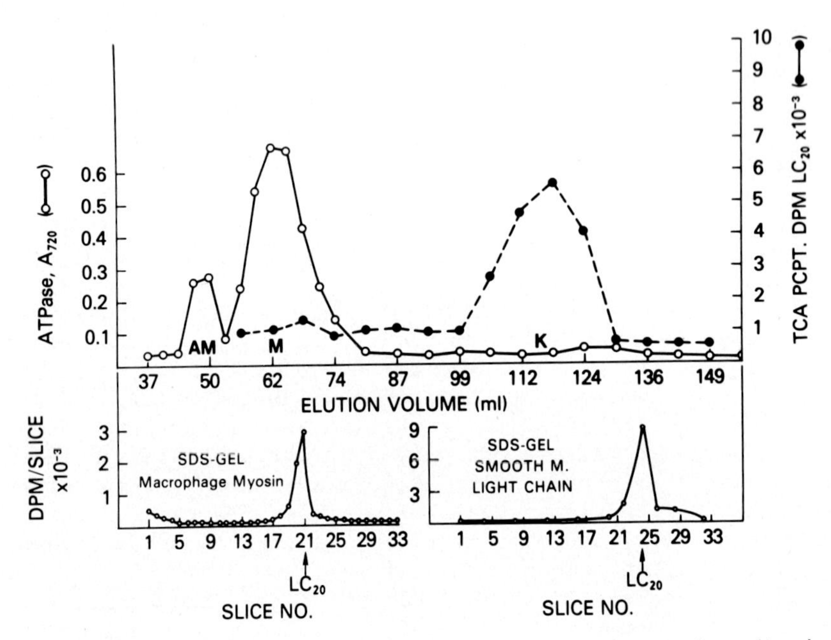

FIGURE 1. Profile of Sepharose 4B gel filtration (top) and distribution of ^{32}P *after SDS-polyacrylamide gel electrophoresis (bottom).* Sepharose 4B gel filtration: *A 2-ml sample of crude macrophage actomyosin prepared as outlined in Methods, was applied to a 1.5 × 90 cm column equilibrated and eluted at 21 ml/hr with 0.5* M *KCl-15* mM *Tris·HCl (pH 7.5)-1* mM *EDTA - 2.5* mM *DTT. Samples of 2.4 ml were collected, of which 0.1 ml was used for ATPase assay (solid line) and 0.1 ml for myosin light chain kinase assay (dashed line). The three peaks are actomyosin (AM), myosin (M), and Kinase (K). SDS-polyacrylamide gel electrophoresis: The elution diagram on the bottom left shows the distribution of* ^{32}P *following gel electrophoresis of peak M, which had been labeled with γ-labeled* $AT^{32}P$*. The elution diagram bottom right shows the distribution of* ^{32}P *following gel electrophoresis of an incubation mixture containing peak K, smooth muscle light chain and γ-labeled* $AT^{32}P$*.* LC_{20} *= 20,000 dalton light chain.*

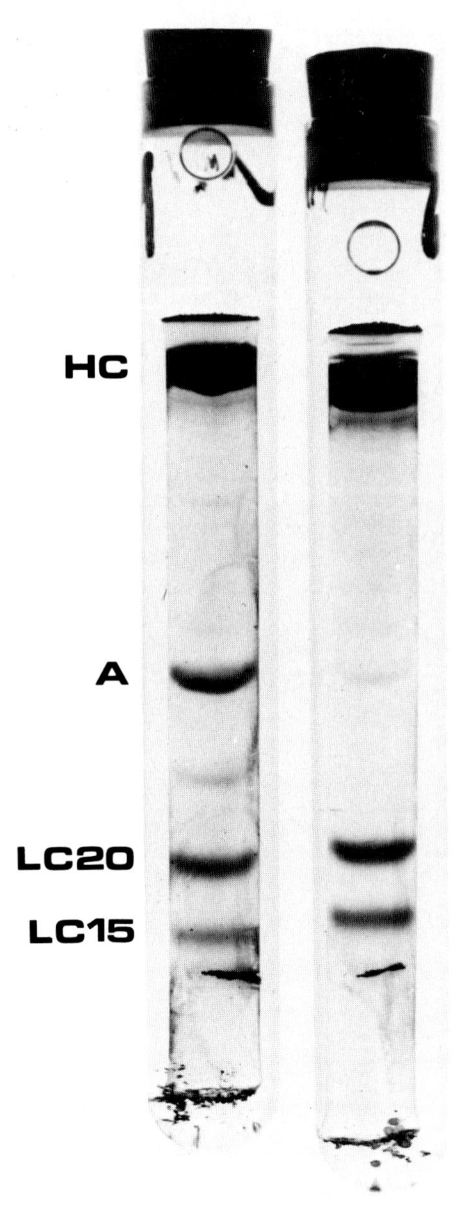

FIGURE 2. Photograph of SDS-polyacrylamide gels following electrophoresis of macrophage actomyosin (left) and myosin (right). Pooled peaks AM and M (see Fig. 1) were applied to 1% SDS-7 1/2% polyacrylamide gels. Electrophoresis was from top to bottom. The stained bands are myosin heavy chain (HC), actin (A), the 20,000 dalton light chain of myosin (LC20), and the 15,000 dalton light chain of myosin (LC15). The dye marker is seen just below LC15.

Phosphatase Activity: Evidence for macrophage myosin phosphatase activity was obtained utilizing the pooled peak of kinase activity (peak K, Fig. 1) following Sepharose 4B gel filtration. Previous experiments had shown that both skeletal muscle myosin phosphatase (Morgan *et al.*, 1976) and platelet myosin phosphatase (Adelstein *et al.*, 1977) were eluted together with their respective kinases from Sepharose 4B.

The 18,500 dalton light chain of rabbit skeletal muscle myosin was used as substrate in these experiments. Following incubation with γ-labeled $AT^{32}P$ and the pooled kinase fraction from Sepharose 4B, phosphorylation of the light chain was allowed to proceed for 1 hr (see Fig. 3). At this time an aliquot was removed from the incubation mixture, passed through a Sephadex G-25 column to remove ATP and monitored for ^{32}P-incorporation at 90, 120, and 150 min. As can be seen from Fig. 3, there is a gradual but significant dephosphorylation of the light chains following the removal of ATP indicating the presence of a phosphatase. Readdition of ATP to an aliquot at 120 min resulted in rephosphorylation of the myosin light chain, indicating that the myosin kinase was still active.

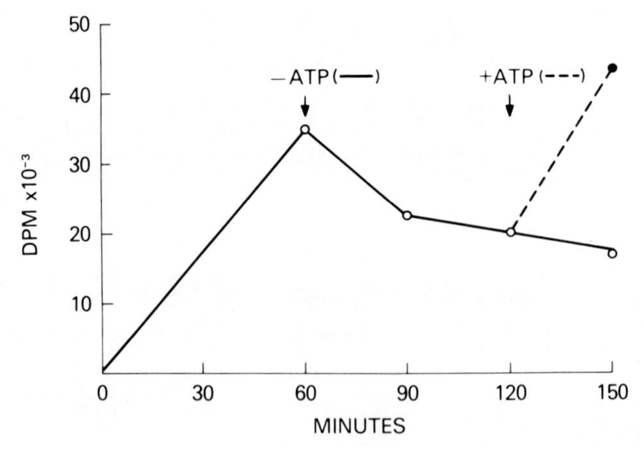

FIGURE 3. Time course of phosphorylation and dephosphorylation of skeletal muscle myosin light chains catalyzed by macrophage myosin kinase and phosphatase. Skeletal muscle myosin light chains were incubated with pooled peak K from Fig. 1 and γ-labeled $AT^{32}P$. At 60 min an aliquot of the incubation mixture was applied to Sephadex G-25 to remove ATP. At 120 min γ-labeled $AT^{32}P$ was added back to an aliquot of the incubation mixture from which $AT^{32}P$ had been removed. The amount of incorporation into skeletal muscle myosin light chains was determined at the times indicated.

Discussion

Although it is possible that the cofactor protein described by Hartwig and Stossel (1975) might be the myosin light chain kinase, further work is required to establish if this is the case. It is interesting that the cofactor found by Pollard and Korn (1973) to be necessary for actin-activation of *Acanthamoeba* myosin I has recently been reported (Maruta and Korn, 1977) to be a myosin kinase which phosphorylates the heavy chain of *Acanthamoeba* myosin I.

The presence of a myosin kinase and phosphatase in macrophages suggests that myosin phosphorylation-dephosphorylation may regulate actin-myosin interaction in these cells. Recent experiments (Trotter and Adelstein, manuscript in preparation) indicate that phosphorylated macrophage myosin has a much higher actin-activated ATPase activity than nonphosphorylated myosin. While this would indicate that phosphorylation of macrophage myosin plays a major role in regulating macrophage contractile proteins, the relevance of these biochemical findings to macrophage physiology is yet to be determined.

SCALLOP MYOSIN KINASE

Unlike skeletal and cardiac muscle myosin, scallop myosin does not require a separate system of regulatory proteins (i.e., troponin-tropomyosin) to mediate the effect of Ca^{2+} on actin-activation. Ca^{2+}-regulation of scallop myosin ATPase activity is mediated by a light chain of myosin which can be removed with EDTA (Szent-Györgyi *et al.*, 1973; and Szent-Györgyi, see this volume). When the regulatory light chain is bound to scallop myosin actin-activation can only occur in the presence of Ca^{2+}. Removal of 0.5 moles of this light chain/mole of myosin results in a form of myosin that can be activated by actin in the presence and absence of Ca^{2+} (Kendrick-Jones *et al.*, 1976).

In collaboration with Andrew Szent-Györgyi, we (RH, MAC, RSA) have addressed two questions involving scallop myosin. The first is can the regulatory light chains of scallop myosin be phosphorylated? The second is does scallop muscle possess the enzyme myosin light chain kinase?

Methods

Scallop, as well as other isolated mollusc light chains and scallop myofibrils were a gift of Dr. Andrew Szent-Györgyi. Smooth muscle myosin light chains were prepared by the method of Perrie and Perry (1970).

Preparation of the kinase was as follows. Washed myofibrils were extracted in 0.5 *M* NaCl - 10 m*M* pyrophosphate - 4 m*M* Imidazole·HCl (pH 7.4) - 5 m*M* DTT following brief homogenization. The extract was sedimented at 45,000 × g for 10 min and the supernatant fractionated with solid ammonium sulfate following addition of ATP to 5 m*M* and $MgCl_2$ to 15 m*M*. The 50-70% ammonium sulfate fraction was solubilized in 0.5 *M* NaCl-10 m*M* Imidazole·HCl (pH 7.4) - 5 m*M* DTT and an aliquot applied to a Sepharose 4B column equilibrated with the same buffer.

Results

Jakes *et al.* (1976) showed that the regulatory light chain of scallop myosin contains a serine residue that is homologous in sequence to the phosphorylated serine in the skeletal and smooth muscle myosin light chain. Previous efforts to phosphorylate this light chain with skeletal muscle myosin kinase were unsuccessful. We, therefore, chose platelet myosin kinase in our attempt to catalyze phosphorylation of the regulatory light chain.

Table 1 summarizes the results of incubating platelet myosin kinase with scallop regulatory light chains. Approximately 0.18 mole of phosphate is incorporated per mole of myosin. Table 1 also shows that under similar conditions only insignificant amounts of phosphate are incorporated into the nonregulatory light chains prepared by treatment with guanidine hydrochloride. The 1% incorporation found may reflect a small amount of contamination of this fraction with regulatory light chain.

Partial phosphorylation of the scallop myosin regulatory light chain suggested that scallops might contain a myosin light chain kinase. Therefore an ammonium sulfate fraction was prepared as outlined in the Methods section and applied to a Sepharose 4B column. Figure 4 is the elution profile of this column showing a peak of myosin ATPase activity (at 64 ml) due to the presence of scallop myosin and a peak (at 100ml) of myosin kinase activity.

TABLE 1. Mollusc Light Chains as Substrates for Platelet Myosin Kinase

Scallop	*moles ^{32}P/mole light chain*
Regulatory light chain	*0.18*
Guanidine - HCl light chain	*0.01*

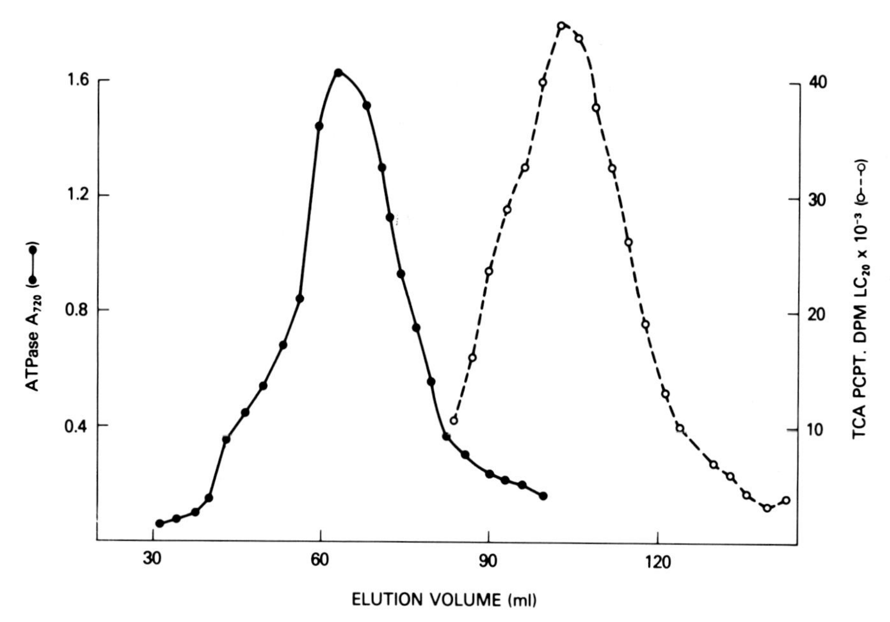

FIGURE 4. Profile of Sepharose 4B gel filtration of scallop myosin and myosin light chain kinase. A 2-ml sample of the 50-70% ammonium sulfate fraction of scallop myosin was applied to a 1.5 × 88 cm column equilibrated and eluted with 0.5 M NaCl- 10 mM Imidazole-HCl (pH 7.4) - 5 mM DTT. Samples of 3 ml were collected of which 0.05 ml was used for the myosin ATPase assay (solid line) and 0.1 ml was used for scallop myosin kinase assay (dashed line).

The kinase activity was located by assaying the column fractions with the 20,000 dalton light chain of smooth muscle myosin and γ-labeled $AT^{32}P$. Figure 5 is a scan of an SDS-polyacrylamide gel showing the relative migration of the various proteins and the location of ^{32}P-radioactivity following incubation of scallop kinase and the smooth muscle myosin light chain fraction. The only peak of radioactivity eluted from the gel is seen to comigrate with the 20,000 dalton light chain of myosin. Using scallop kinase and smooth muscle myosin light chains we have been able to incorporate up to 0.3 moles of phosphate per mole of light chain.

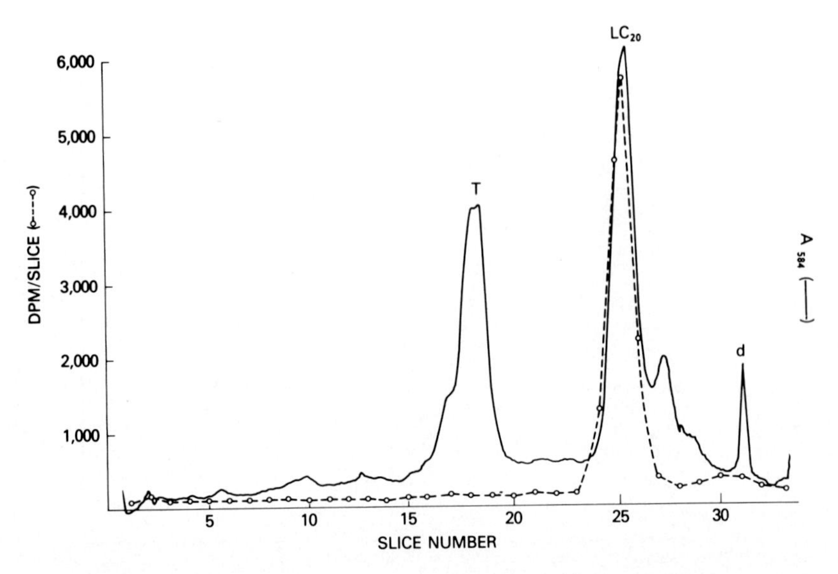

FIGURE 5. 1% SDS - 7 1/2% polyacrylamide gel scan and ^{32}P-elution profile of a smooth muscle myosin light chain fraction phosphorylated by scallop myosin kinase. The solid line is a scan at 584 nm of the Coomassie brilliant blue stained gel showing a peak at smooth muscle tropomyosin (T), the 20,000 dalton light chain (LC20), and the dye marker (d). The dashed line indicates the ^{32}P eluted from each 2 mm slice. Direction of electrophoresis was from left to right.

Discussion

Having found evidence for the existence of a myosin light chain kinase in the scallop, the next substrate to assay was scallop regulatory light chains. To date experiments similar to those shown in Fig. 5 have resulted in only low levels of incorporation into these light chains. The possibility exists that these light chains have been isolated in a phosphorylated form, which would preclude their phosphorylation until they are first dephosphorylated.

The major questions still to be answered are whether the kinase present in scallops catalyzes phosphorylation of scallop myosin, and if it does, whether phosphorylation alters any property (e.g., the actin-activated ATPase activity, Ca^{2+}-sensitivity) of the scallop myosin.

PLATELET MYOSIN KINASE

The kinase that catalyzes phosphorylation of platelet myosin has previously been purified and found to exist in a Ca^{2+}-independent form (Daniel and Adelstein, 1976). We (DRH and RSA) now report that platelet myosin light chain kinase can be isolated in a Ca^{2+}-dependent form by altering the method used for enzyme isolation.

Methods

Platelet myosin light chain kinase was purified from fresh platelet concentrates which were washed as outlined previously (Adelstein *et al.*, 1971). Major alterations in the procedure for preparing the kinase included: a) use of fresh platelets only; b) lysis of platelets by a N_2-decompression bomb in place of freeze-thawing and butanol; c) using 5 m*M* EDTA in the extracting solution in place of 1 m*M* EDTA. Avoiding freeze-thawing and increasing the EDTA concentration to 5 m*M* appears to be essential for isolation of a Ca^{2+}-dependent kinase.

Extraction of the lysed platelets in a high ionic strength buffer, precipitation of actomyosin and kinase by lowering the KCl concentration to 0.1 *M* and the pH to 6.3, and ammonium sulfate fractionation were carried out as previously outlined (Adelstein and Conti, 1976).

Results

The 35-50% ammonium sulfate fraction showed Ca^{2+}-dependent incorporation of ^{32}P from γ-labeled $AT^{32}P$ into the 20,000 dalton light chain of platelet myosin present in the same fraction. Partial purification of the Ca^{2+}-dependent kinase was carried out by gel filtration on Sepharose 4B. The kinase eluted from this column was found to be more active in the presence of Ca^{2+} (Table 2). The substrate for the isolated kinase was the 20,000 dalton light chain of smooth muscle myosin.

The pooled fraction from Sepharose 4B was rechromatographed on Sephadex G-100 in the presence of 5 m*M* EGTA and 1 *M* KCl. Figure 6 is an elution profile of kinase activity from such a column. The enzyme became more Ca^{2+}-dependent following elution from Sephadex G-100 (See Table 2). For comparison, Table 2 also shows a preparation of kinase that was isolated in the presence of 1 m*M* EDTA, which yielded a Ca^{2+}-independent kinase.

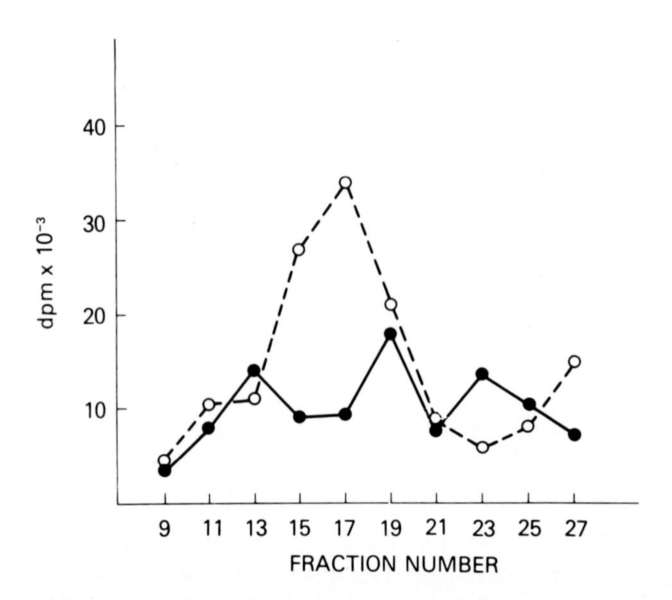

FIGURE 6. Profile of Sephadex G-100 gel filtration of platelet myosin kinase. A 0.5 ml sample of the kinase peak eluted from Sepharose 4B was applied to a 0.9 × 60 cm column equilibrated and eluted with 1 M KCl-5 mM EGTA-5 mM EDTA-20 mM Imidazole·HCl (pH 7.4). 0.8 ml samples were collected of which 0.1 ml was used to assay for platelet myosin kinase in the presence of 0.1 mM Ca^{2+} (dashed line) or in the presence of 1 mM EGTA (solid line). Smooth muscle myosin light chains were used as substrate.

Figure 7 shows the ^{32}P-elution pattern from an SDS-polyacrylamide gel following electrophoresis of an incubation mixture containing the Ca^{2+}-dependent, Sephadex G-100 purified platelet kinase and smooth muscle myosin light chain. This figure not only illustrates the difference between including 0.1 m*M* Ca^{2+} and 1 m*M* EGTA in the incubation mixture, but it substantiates that the 20,000 dalton myosin light chain was the substrate for the kinase.

TABLE 2. Ca^{2+} Requirement of Platelet Myosin Kinases[a]

Kinase	moles ^{32}P/mole light chain $+Ca^{2+}$	$+EGTA$	Ratio $Ca^{2+}/EGTA$
Ca^{2+}-independent			
Sepharose 4B	0.48	0.64	0.75
Sephadex G-100	0.20	0.29	0.68
Ca^{2+}-dependent			
Sepharose 4B	0.50[b]	0.44[b]	1.14
Sephadex G-100	0.19	0.006	31.7
Ca^{2+}-dependent			
Sepharose 4B	0.60	0.26	2.31
Sephadex G-100	0.21	0.02	10.5

[a]Data is from three different preparations.
[b]Data from S. P. Scordilis, this laboratory.

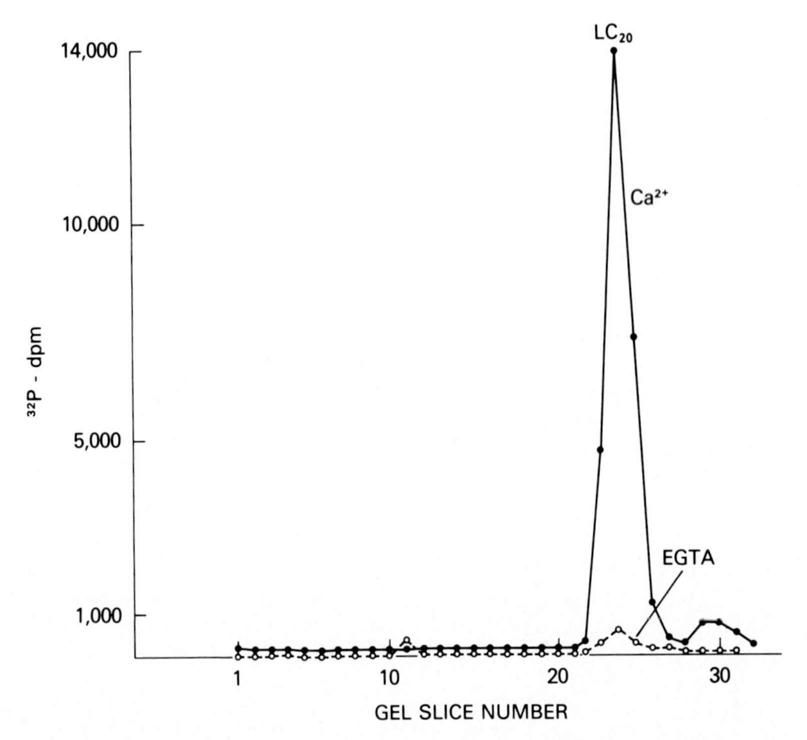

FIGURE 7. ^{32}P-elution profile from 1% SDS-7 1/2% polyacrylamide gels of smooth muscle myosin light chains phosphorylated by Ca^{2+}-dependent platelet myosin kinase. The elution profile from two equally loaded gels are superimposed, showing the amount of ^{32}P eluted following electrophoresis of a smooth muscle myosin light chain fraction, platelet kinase and γ-AT^{32}P incubated in the presence of 0.1 mM $CaCl_2$ (solid line) and 1 mM EGTA (dashed line). LC20 indicates the location of the 20,000 dalton smooth muscle myosin light chain. Migration was from left to right.

Discussion

A major question to be resolved is the difference between the Ca^{2+}-dependent and Ca^{2+}-independent kinases isolated from platelets. Among the possibilities are a) the Ca^{2+}-independent kinase is produced by proteolysis of the Ca^{2+}-dependent kinase, b) Ca^{2+}-sensitivity is not an intrinsic property of the kinase but is conferred by another set of proteins, which can alter the Ca^{2+} requirements of the enzyme, and c) more than one form of kinase can coexist in the platelet.

The possibility that the Ca^{2+}-independent kinase is a proteolytic product of the Ca^{2+}-dependent enzyme is difficult to rule out. The platelet contains multiple proteolytic enzymes, and a number of proteins, including myosin, have been shown to undergo digestion in this type of preparation (Adelstein *et al.*, 1971). On the other hand Ca^{2+}-independent kinases have also been isolated from a number of other cells, including macrophages (Adelstein *et al.*, 1977) proliferative myoblasts (Scordilis and Adelstein, 1977) and smooth muscle cells grown in culture (Chacko *et al.*, 1977b). In the case of smooth muscle cells, using identical conditions for preparation of the myosin kinase, a Ca^{2+}-dependent enzyme was prepared from intact aorta cells while a Ca^{2+}-independent enzyme was prepared from aortic cells grown in culture (Chacko *et al.*, this volume).

Is a separate protein such as the Ca^{2+}-dependent regulatory protein (CDR) responsible for the Ca^{2+}-dependence of the platelet kinase? Recent experiments (Hathaway and Adelstein, 1979) have shown that the Ca^{2+}dependence of the platelet myosin light chain kinase is mediated by the Ca^{2+}-binding protein CDR. Thus this enzyme is similar to 3':5'-cyclic nucleotide phosphodiesterase (Cheung 1970; Kakiuchi *et al.*, 1970), the Ca^{2+}-Mg^{2+} ATPase of erythrocytes (Gopinath and Vincenzi, 1977; Jarrett and Penniston, 1977) and smooth muscle myosin light chain kinase (Dabrowska *et al.*, 1977) in the manner that Ca^{2+} regulates its activity. Similar to the smooth muscle kinase (Dabrowska *et al.*, 1977), the platelet kinase is inactive in the absence of CDR.

Current experiments are being directed toward converting Ca^{2+}-dependent kinases, such as those found in platelets, into Ca^{2+}-independent kinases. The possibility exists that similar to the troponin-tropomyosin regulatory system in muscle more than one protein may be required to confer Ca^{2+}-dependence on a Ca^{2+}-independent kinase. The ultimate goal of this research is to understand the manner in which Ca^{2+} regulates nonmuscle and smooth muscle contractile processes.

ACKNOWLEDGMENTS

We wish to acknowledge the valuable editorial assistance of Mrs. Exa Murray in the preparation of this manuscript.

JAT is a postdoctoral fellow of the Muscular Dystrophy Association of America, Inc.

REFERENCES

Adelstein, R. S., and Conti, M. A. (1975). *Nature 256*, 597-598.

Adelstein, R. S., and Conti, M. A. (1976). *In* "Cell Motility, Cold Spring Harbor Conferences on Cell Proliferation," Vol. 3, pp. 725-738, Cold Spring Harbor Laboratory, Cold Spring Harbor, New York.

Adelstein, R. S., Pollard, T. D., and Kuihl, W. M. (1971). *Proc. Nat. Acad. Sci. U.S.A. 68*, 2703-2707.

Adelstein, R. S., *et al.* (1977). *In* "Exitation-Contraction Coupling in Smooth Muscle" (R. Casteels, T. Godfraind, and J. C. Ruegg, eds.), pp. 359-366. Elsevier/North Holland Biomedical Press, Amsterdam.

Barylko, B., Conti, M. A., and Adelstein, R. S. (1977). *Biophys. J. 17*, 270a.

Chacko, S., Conti, M. A., and Adelstein, R. S. (1977a). *Proc. Nat. Acad. Sci. U.S.A. 74*, 129-133.

Chacko, S., Blose, S. H., and Adelstein, R. S. (1977b). *In* "Exitation-Contraction Coupling in Smooth Muscle" (R. Casteels, T. Godfraind, and J. C. Ruegg, eds.), pp. 367-375. Elsevier/North Holland Biomedical Press, Amsterdam.

Cheung, W. Y. (1970). *Biochem. Biophys. Res. Commun. 38*, 533-538.

Dabrowska, R., Aromatorio, D., Sherry, J. M. F., and Hartshorne, D. J. (1977). *Biochem. Biophys. Res. Comm. 78*, 1263-1272.

Daniel, J. L., and Adelstein, R. S. (1976). *Biochemistry 15*, 2370-2377.

Gopinath, R. M. and Vincenzi, F. F. (1977). *Biochem. Biophys. Res. Commun. 77*, 1203-1209.

Gorecka, A., Aksoy, M. O., and Hartshorne, D. J. (1976). *Biochem. Biophys. Res. Commun. 71*, 325-331.

Hartwig, J. H., and Stossel, T. (1975). *J. Biol. Chem. 250*, 5696-5705.

Hathaway, D. R., and Adelstein, R. S. (1979). *Proc. Nat. Acad. Sci. U.S.A. (in press)*.

Jakes, R., Northrop, F., and Kendrick-Jones, J. (1976). *FEBS. Lett. 70*, 229-234.

Jarrett, H. W., and Penniston, J. T. (1977). *Biochem. Biophys. Res. Commun. 77*, 1210-1216.

Kakiuchi, S., Yamazaki, R., and Nakajima, H. (1970). *Proc. Japan Acad. 46*, 587-592.

Kendrick-Jones, J., Szentkiralyi, E. M., and Szent-Györgyi, A. G. (1976). *J. Molec. Biol. 104*, 747-775.

Maruta, H., and Korn, E. D. (1977). *J. Biol. Chem. 252*, 8329-8332.

Morgan, M., Perry, S. V., and Ottaway, J. (1976). *Biochem. J. 157*, 687-697.

Perrie, W. T., and Perry, S. V. (1970). *Biochemical J. 119,* 31-38.

Pires, E. M. V., and Perry, S. V. (1977). *Biochem. J. 167,* 137-146.

Pollard, T. D., and Korn, E. D. (1973). *J. Biol. Chem. 248,* 4691-4697.

Scordilis, S. P., and Adelstein, R. S. (1977). *Nature 268,* 558-560.

Sobieszek, A., and Small, J. V. (1977). *J. Molec. Biol. 112,* 559-576.

Szent-Györgyi, A. G., Szentkiralyi, E. M., and Kendrick-Jones, J. (1973). *J. Molec. Biol. 74,* 179-203.

Motility in Cell Function
Proceedings of the First John M. Marshall Symposium in Cell Biology

ASPECTS OF MYOSIN LINKED REGULATION OF MUSCLE CONTRACTION

Andrew G. Szent-Györgyi

Department of Biology
Brandeis University
Waltham, Massachusetts

The criteria for demonstrating thick filament control in a particular muscle are simple. To obtain good evidence one must isolate a species of myosin that depends on calcium for its actin activated Mg-ATPase activity in the presence of pure actin. One also has to establish the presence of specific high affinity calcium binding sites on these regulated myosins that will allow the calcium switch to operate at about a micromolar calcium concentration range in the presence of millimolar amounts of magnesuum.

We have proposed the existence of thick filament control in molluscan muscles by isolating such regulated myosins and, in addition, finding that the thin filaments of these muscles were not regulatory when tested with pure rabbit myosin (Kendrick-Jones *et al.*, 1970). Specific calcium binding by myosin and the calcium dependence of the ATPase in the presence of pure actin are important criteria that should be experimentally demonstrated for proposing myosin linked regulation in muscles. These considerations appear to be of particular relevance in vertebrate striated muscles where thick filament regulation has been suggested on the basis of indirect evidence. One notes, however, that the regulatory switch in muscles controls the interaction of the myosin intermediate (M**ADP,Pi) with actin. More indirect calcium effects may play a physiologically important role in modulating this interaction possibly at a longer time scale.

Myosin linked regulation is a common phenomenon. Regulated myosins have been prepared from over 15 different muscles and such control mechanism has been deduced by the mere indirect competitive actin activation assay in many other invertebrate muscles (Lehman and Szent-Györgyi, 1975). Myosin control ap-

ISBN 0-12-551750-5

pears to be lacking only in vertebrate striated muscles and in the fast muscles of some decapods.

The task to demonstrate that thick filament regulation is a subunit regulation of the myosin molecule is also a straightforward one. It requires the reversible dissociation of the light chain from myosin with the concurrent loss of calcium binding and calcium requirement of the actin activated ATPase activity. Moreover, upon recombination of the light chains with myosin these calcium dependent functions ought to be restored. Although myosin control is widespread, the removal and readdition of the regulatory light chain has been achieved only with scallop myosin (Szent-Györgyi *et al.*, 1973; Kendrick-Jones *et al.*, 1976). The dissociation of the regulatory light chains from most myosins requires procedures that irreversibly abolish the ATPase activity of myosin.

The *in vitro* evidence for control is based on the measurement of the calcium requirement of the actin activated ATPase activity. Since this ATPase activity is a measure of cross-bridge cycling it is the best available *in vitro* measure of contractile activity using isolated proteins. The role of the control proteins is to prevent cross link formation in the absence of calcium. The ATPase activity, however, is a measure of all the steps involved in the cycle and although it depends on actomyosin formation it can be influenced by additional parameters. It is therefore useful to examine the role of myosin control in tension generation.

In *Placopecten magellanicus*, Dr. R. M. Simmons of the University College, London, and I have obtained rather direct evidence that calcium regulation of tension is mediated by regulatory light chains. (Simmons and Szent-Györgyi, 1978). Scallop muscle is an obvious choice for a study of calcium regulation of tension for several reasons. Biochemical evidence indicates that scallop muscles lack actin control and that regulatory light chains can be reversibly removed. The muscle is a striated muscle and is sufficiently ordered to make it suitable for X-ray diffraction and electron microscopic studies (Wray *et al.*, 1975; Bennett and Millman, 1976).

Fiber bundles of Placopecten can be chemically skinned with a relaxing solution (50 m*M* KCl, 5m*M* $MgCl_2$, 5 m*M* ATP, 5 m*M* EGTA, 20 m*M* imidazole-HCl, pH 7.0 that contains 1/2% Brij 35. The light chain content of fiber bundles of the size used for tension studies (50-200 µm wide, 50-500 µm thick, and 5mm long) can be readily determined by urea acrylamide gel electrophoresis (Szent-Györgyi *et al.*, 1973; Kendrick-Jones *et al.*, 1976). Once regulatory light chains are removed the muscle develops full tension in relaxing solutions in the absence of calcium although the tension development is slower. Transferring the fibers into activating solutions does not produce larger tensions (Fig. 1). Removal of one or more of the

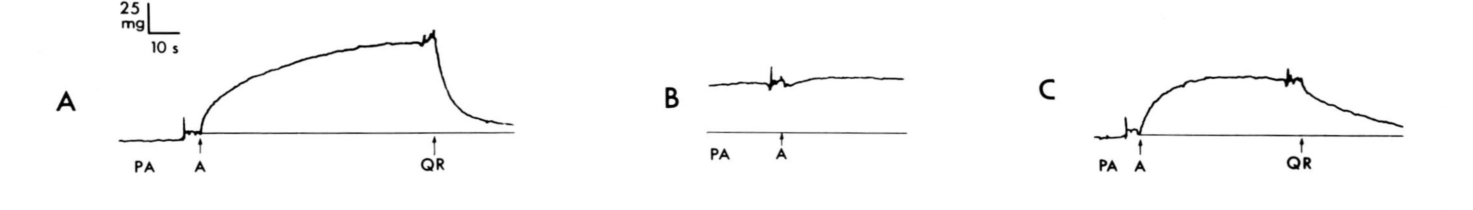

FIGURE 1. Tension records from chemically skinned muscle fiber bundles from the striated adductor of Placopecten magellanicus. *Lower lines are tension baselines (because of drift in the RCA 5734 transducer used, these are in some cases approximate). A simple rapid activation technique was used in which a fiber bundle was first placed in a relaxing solution while a low EGTA concentration ("preactivating" solution) to allow a large buffered* Ca^{2+} *gradient when the bundle was transferred to an activating solution. Rapid relaxation was achieved using a high EGTA concentration ("quick relaxing" solution). Symbols for solutions: PA, preactivating; QR, quick relaxing; R, relaxing; Rig., rigor. The tension and time scales in* a *apply to all the records. Temperature 20-25°C. First fiber bundle,* a-c; a, *activation-relaxation procedure before desensitisation;* b, *after desensitisation, the tension developed in relaxing solution is not much enhanced in activating solution;* c, *after resensitisation, the activation-relaxation response is similar to that of* a. *This fiber bundle was not analyzed for light chain content. Second fiber bundle,* d *relaxation of tension when freshly skinned fiber bundle in rigor is placed in relaxing solution;* e, *development of tension when desensitised bundle in rigor is placed in relaxing solution. (From Simmons and Szent-Györgyi, 1978.)*

regulatory light chains per myosin leads to a complete loss of calcium sensitivity of tension generation. Fibers recombine with regulatory light chains fully in about 20-30 min in rigor solution as determined by urea acrylamide gel electrophoresis. Such regenerated fiber bundles now relax partially or fully in relaxing solution, indicating a regain of calcium requirement for tension generation (Fig. 1). The results indicate thus that light chains regulate the calcium sensitivity of tension generation in a similar fashion to ATPase activity. The result further shows that in Placopecten striated adductor muscle regulation is myosin linked since EDTA treatment does not affect frog muscles that are regulated by troponin. Thus tension measurements support well the interpretation based on ATPase activity studies (Simmons and Szent-Györgyi, 1978). Removal of light chains from these muscles lead to well defined changes in the X-ray diffraction pattern in the rigor state. The change in the 385 Å reflection indicates that the way myosin and actin interact can be modulated by the regulatory light chains (Vibert *et al.*, 1978).

The removal of light chains affect specifically the actin activated ATPase of scallop myosin. The ATPase activity no longer requires calcium and is maximum in EGTA at free calcium concentrations of about 10^{-8} *M* or less. (Szent-Györgyi *et al.*, 1973). Such myosin preparations lost calcium control and are desensitized. Losses of light chains lead to proportional losses in the specific calcium binding. Recombination of the regulatory light chains with the desensitized scallop myosin restores fully calcium dependent functions, the calcium dependence of the ATPase, and calcium binding of the recombined and untreated preparations are similar (Szent-Györgyi *et al.*, 1973).

One out of the two regulatory light chains of scallop myosin are removed by cold EDTA treatment. Such treatments leave the "SH-" or "essential" light chains attached to myosin. Although the myosin preparations retain one of the two specific calcium binding sites, the actin activated ATPase activity is not calcium sensitive. It appears, therefore, that calcium sensitivity requires some type of interaction between the two halves of myosin. (Kendrick-Jones *et al.*, 1976).

The regulatory light chains modify only the actin activated ATPase activity, while the K-EDTA and the high calcium (10m*M* Ca^{2+}) ATPase are unaffected by the regulatory light chains. It appears that the regulatory light chains do not interact directly with the ATPase site but modify ATPase activity indirectly by interfering with actin combination in the absence of calcium (Lehman *et al.*, 1972; Szent-Györgyi *et al.*, 1973). In that sense thick filament regulation is analogous to thin filament regulation. The regulating components in both systems inhibit actomyosin formation by preventing the myosin product to combine with actin in the absence of calcium (Koretz *et al.*,

1972; Parker *et al.*, 1970; Eisenberg and Kielley, 1970).

Regulatory light chains have been found in all myosins so far tested. These regulatory light chains can be isolated in pure form, although the procedures needed to detach the light chains inactivate most myosins irreversibly. The light chains of the different myosins hybridize with desensitized scallop myofibrils and restore calcium sensitivity. (Kendrick-Jones, 1974; Kendrick-Jones *et al.*, 1976). However, not all regulatory light chains contribute to calcium binding. Molluscan regulatory light chains behave like scallop regulatory light chains and restore both calcium binding and calcium sensitivity. The light chains of myosins from rabbit or frog skeletal myosin, bovine cardiac myosin, and fast lobster myosin do not restore the second specific calcium binding site lost during desensitization as do the molluscan light chains, nevertheless the hybrid regains calcium sensitivity. We have interpreted this apparent paradox by assuming that in these hybrids the added light chains resensitized scallop myofibrils indirectly by cooperation with the remaining scallop regulatory light chain during the "off" state (Kendrick-Jones *et al.*, 1976). Accordingly, triggering by calcium in the hybrids would always occur by a reaction mediated by the scallop light chain. The lack of effect on calcium binding by the regulatory light chains of vertebrate skeletal, cardiac, and lobster myosins reflect the inability of these myosins to bind calcium specifically. Interestingly gizzard myosin that appears to be regulatory (Bremel, 1974; Sobieszek and Small, 1976) has light chains that function like molluscan regulatory light chains (Kendrick-Jones *et al.*, 1976).

The finding that magnesium and manganese ions bind competitively to the divalent cation sites of the rabbit regulatory light chain but do not compete for the high affinity calcium sites of molluscan myosin (Bagshaw, 1977) is in line with these observations. The regulatory light chains of the myosins from rabbit skeletal, beef cardiac, chicken gizzard, and lobster tail muscles can be phosphorylated and the activity of some of these myosins depend on the phosphorylated state (Pires *et al.*, 1974; Adelstein *et al.*, 1976; Chaco *et al.*, 1976; Sobieszek, 1976; Gorecka *et al.*, 1976) and may also have a long term modulating influence. In contrast the scallop regulatory light chain is not phosphorylated (Jakes *et al.*, 1976) and there is no evidence as yet for requirement of phosphorylation for the regulation and for the activity of molluscan myosins.

The hybridization experiments indicate that regulatory light chains in general retained a common attachment site. This may be identical or near to the general divalent cation binding site. In addition there appear to be a specific calcium binding site that is required for calcium regulation. Little is known about the position of this site, and why the specific calcium binding requires the association of the regulatory light chain with the heavy chain of myosin. It may also be significant that regulatory light chains are elongated structures exceeding in length 100 Å (Stafford and Szent-Györgyi, 1978). It is not clear why regulation requires the presence of both regulatory light chains in the intact myosin. Study of myosin from which both regulatory light chains have been removed may give further information of the mechanism of regulation. Recently we have found that at elevated temperatures (30°-35°C), EDTA dissociates reversibly both regulatory light chains, (Chantler and Szent-Györgyi, 1978) and it is possible now to check some of the interpretations on the mechanism of the subunit regulation of myosin more directly.

One also notes that there is no certainty that all the components of myosin controls have been identified. Little is known of the function of the other class of light chains. There is evidence that in rabbit myosin the "alkali" light chains modify the actin activated ATPase activity but do not directly act on the enzymatic center (Wagner and Weeds, 1977). It is clear that for a direct information about their function one has to find conditions to reversibly dissociate these light chains from scallop myosin.

ACKNOWLEDGMENTS

This research was supported PHS Grant Am 15963 and by a grant from the Muscular Dystrophy Association.

REFERENCES

Adelstein, R. S., Chaco, S., Barylko, B. and Scordilis, S. P. (1976). *In* "Contractile Systems in Non-Muscle Tissues" (S. V. Perry *et al.*, eds.), p. 153. Elsevier North Holland Biomedical Press.

Bagshaw, C. R. (1977). *Biochemistry 16,* 59.

Bremel, R. D. (1974). *Nature 252,* 405.

Chaco, S., Conti, M. A. and Adelstein, R. S. (1977). *Proc. Natl. Acad. Sci. U.S.A. 74,* 129.

Chantler, P. D. and Szent-Györgyi, A. G. (1978). *Biophys. Soc. Abstract. 45a.*

Eisenberg, E. J., and Kielley, W. W. (1970). *Biochem. Biophys. Res. Commun. 40,* 50.

Gorecka, A., Aksoy, M. D., and Hartshorne, D. J. (1976). *Biochem. Biophys. Res. Commun. 71,* 325.

Jakes, R., Northrop, F., and Kendrick-Jones, J. (1976). *FEBS. Lett. 70,* 229-234.

Kendrick-Jones, J. (1974). *Nature 249,* 631.

Kendrick-Jones, J., Lehman, W., and Szent-Györgyi, A. G. (1970). *J. Mol. Biol. 54,* 313.

Kendrick-Jones, J., Szentkiralyi, E. M., and Szent-Györgyi, A. G. (1976). *J. Mol. Biol. 104,* 747.

Koretz, J. C., Hunt, T., and Taylor, E. W. (1972). *Cold Spring Harbor Symp. Quant. Biol. 37,* 179.

Lehman, W., and Szent-Györgyi, A. G. (1975). *J. Gen. Physiol. 66,* 1.

Millman, B. M., and Bennett, P. M. (1976). *J. Mol. Biol. 103,* 439.

Parker, L., Pyun, H. Y., and Hartshorne, D. J. (1970). *Biochem. Biophys. Acta 223,* 453.

Pires, E., Perry, S. V., and Thomas, M. A. (1974). *FEBS. Lett. 41,* 292.

Simmons, R. M., and Szent-Györgyi, A. G. (1978). *Nature 273,* 62.

Sobieszek, A. (1977). *Eur. J. Biochem. 73,* 477.

Sobieszek, A., and Small, J. V. (1976). *J. Mol. Biol. 102,* 75.

Stafford, W. F., III., and Szent-Györgyi, A. G. (1978). *Biochemistry 17,* 607.

Szent-Györgyi, A. G., Szentkiralyi, E. M., and Kendrick-Jones, J. (1973). *J. Mol. Biol. 74,* 179.

Vibert, P., Szent-Györgyi, A. G., Craig, R., Wray, J., and Cohen, C. (1978). *Nature 273,* 64.

Wagner, P. D., and Weeds, A. G. (1977). *J. Mol. Biol. 109,* 455.

Wray, J. S., Vibert, D. J., and Cohen, C. (1975). *Nature* 561.

CELLULAR CONTRACTILITY

Motility in Cell Function
Proceedings of the First John M. Marshall Symposium in Cell Biology

PHARMACOLOGY: A RESOURCE TO DETERMINE REGULATION OF EXCITATION-CONTRACTION COUPLING IN SKELETAL MUSCLE

C. Paul Bianchi

Department of Pharmacology
Thomas Jefferson University

Cell pharmacology is the study of drugs and their biological action on cellular function. It provides a resource to determine the nature of the cellular control processes.

Our knowledge of excitation-contraction coupling in skeletal muscle fibers has been based largely on the use of drugs that modify or alter the regulation of muscle contraction. The following drugs are examples of those whose biological action reveals important information concerning the regulation of calcium release from the sarcoplasmic reticulum of skeletal muscle fibers. Caffeine, procaine, lidocaine, benzocaine, and Dantrolene are drugs that modify the regulation of muscle contraction as a result of their interaction with cellular receptive sites.

Caffeine is classified as a central nervous system stimulant. It also exerts effects on smooth muscle, cardiac muscle, and in general on most cells either by acting as a phosphodiesterase inhibitor or by increasing calcium influx or internal release of calcium from cellular stores. Procaine, tetracaine, benzocaine, and lidocaine are classified as local anesthetics (Fig. 1), but each alters the release of calcium from the sarcoplasmic reticulum, while lidocaine enhances its release.

Caffeine is a derivative of xanthine (2,6 dioxypurine) in which the 1,3,7 positions are methylated (Fig. 1). Other methylxanthines are theophylline (1,3 dimethylxanthine) and theobromine (3,7 dimethylxanthine). Theophylline is the most potent of the three methylxanthines as a competitive inhibitor of phosphodiesterase, while caffeine is the most potent in causing contractures of skeletal muscle. Smooth muscle is relaxed by methylxanthines with theophylline the most active and caffeine the least. Two of the most powerful regulators of

ISBN 0-12-551750-5

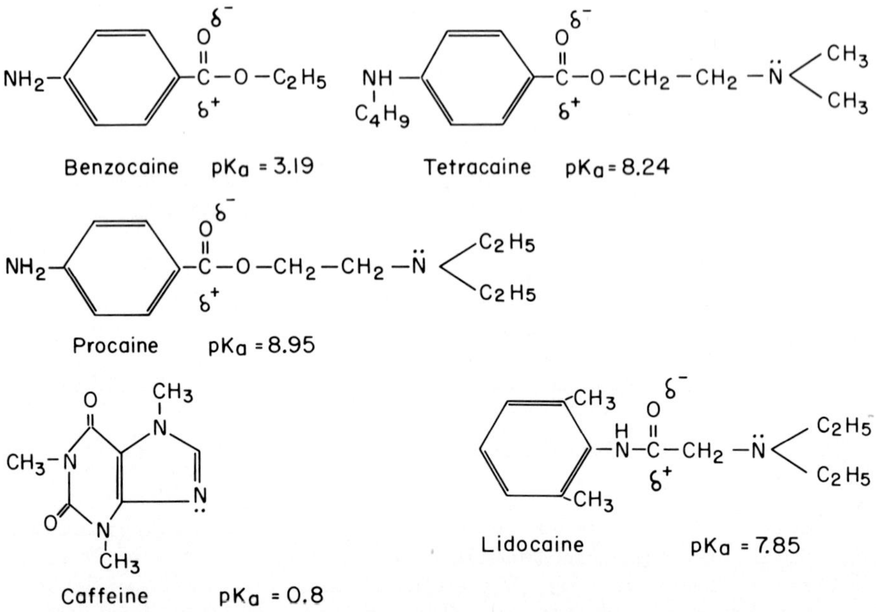

Benzocaine $pK_a = 3.19$ Tetracaine $pK_a = 8.24$

Procaine $pK_a = 8.95$

Caffeine $pK_a = 0.8$ Lidocaine $pK_a = 7.85$

FIGURE 1. Structural formalae of drugs that alter excitation-contraction of skeletal muscle. The PKa values are given and the dipolar nature of the carbonyl oxygen is shown.

cellular processes (calcium and cyclic AMP) are modified by caffeine. Caffeine crosses the cell membrane more rapidly than theophylline or theobromine and quickly gains access to intracellular sites (Bianchi, 1962). In skeletal muscle caffeine increased both influx and efflux of calcium. Efflux increases due to a release of calcium from the sarcoplasmic reticulum. Release of calcium is due to enhancement of a calcium-induced release of calcium (Endo, 1977). Procaine and tetracaine, both tertiary amine local anesthetics, block the release of calcium from the sarcoplasmic reticulum. They also block a caffeine-induced contracture due to a competitive antagonism of caffeine on muscle (Feinstein, 1963).

Procaine exists in two molecular forms at physiological pH of 7.2. The predominant form is the charged form (RH^+); the uncharged form is approximately 1% of the total. The charged form of procaine acts as a competitive antagonist of caffeine. The presence of the proton on the tertiary amine group masks the basicity of the tertiary amine group and allows the carbonyl group to compete with caffeine for receptor sites in the sarcoplasmic reticulum. Benzocaine (Fig. 1) exists as an uncharged molecule at pH 7.2 and lacks the tertiary amine group; benzocaine does block the caffeine-induced contracture indicating that it is the carbonyl group that is necessary to block the action of caffeine. Lidocaine, which contains a sterically hindered carbonyl group, does not block a caffeine-induced contracture. It can cause contracture of skeletal muscle and release of calcium from internal stores. The PKa of lidocaine is less than the PKa of either tetracaine or procaine and therefore exists to a greater extent as the free base at pH 7.2. It is the free base form of lidocaine that is responsible for causing contracture of muscle and release of calcium from the sarcoplasmic reticulum (Bianchi, 1975).

Depolarization of frog sartorius muscle fibers between -70 and -50 mV can cause calcium release from the sarcoplasmic reticulum without causing contracture. The free myoplasmic Ca^{2+} concentration rises to 10^{-8}M (Taylor and Godt, 1976) and is associated with a marked increase in oxygen uptake and resistance to stretch (Bianchi, *et al.*, 1975). The enhanced oxygen uptake can be blocked by procaine, tetracaine, and benzocaine, but it is stimulated by lidocaine.

Glycerol "shock" treatment of frog sartorius muscle uncouples potassium depolarization from increased oxygen uptake and contraction. Such treatment does not prevent stimulation of calcium influx or efflux by potassium depolarization. The biochemical lesion produced by glycerol "shock" treatment must be related to the step between calcium release from the sarcoplasmic reticulum and stimulation of the energy sink responsible for ATP utilization and enhanced oxygen uptake. Vos and Frank (1972) demonstrated that depolarization of sartorius muscle to levels below mechanical threshold (- 60 mV) increased the resistance of the muscle to stretch and potentiated subsequent potassium contractures. Their results suggest that activation of the actomyosin ATP'ase and cross bridge formation may be responsible for ATP utilization. Glycerol "shocked" muscle primed with 0.5 m*M* caffeine, 10^{-6} *M* dibutyryl c-AMP or 10^{-10}*M* cyclic GMP restores the stimulation of oxygen uptake by potassium depolarization (Bianchi *et al.*, 1975).

The cyclic AMP level of frog sartorius muscle (N=6, 107±7 mols/g wet weight) shows no significant decrease in the glycerol "shocked" muscle (N=6, 90±4 p mols/g wet weight); however cyclic GMP decreases from 15.0±0.6 to 11.0±0.4 pmols/g wet weight (N=6). Cyclic GMP may be required to increase the sensitivity of the troponin C to myoplasmic-free calcium, thus allowing low levels of Ca^{2+} (10^{-7}M) to permit cross bridge formation and ATP utilization.

Dantrolene is a relatively new drug, a muscle relaxant with a long action. It depresses the coupling of the muscle action potential to contraction in skeletal muscle fibers. Dantrolene does not alter the action potential, but the threshold for coupling of the action potential to contraction shifts from +18 mV to +40 mV, reducing the effectiveness of the muscle action potential. Dantrolene is a highly lipid soluble drug that forms a sodium salt in 0.1N NaOH; in neutral solutions it is predominantly unionized. The depression of the muscle twitch occurs in two phases. In the initial phase the time constant of depression is 4 min; in the second phase the depression develops at a slower rate (time constant of 25 min). The depression of the muscle twitch by Dantrolene can be overcome by drugs that lower the electrical threshold for excitation-contraction coupling. Dantrolene does not block caffeine-induced contracture (Putney and Bianchi, 1974).

Desmedt and Hainaut (1977) demonstrated that Dantrolene inhibited the intracellular release of calcium in the barnacle giant muscle fiber. It did not significantly affect the sequestration of myoplasmic calcium.

These pharmacological data suggest that there are two types of calcium channels or carriers in the sarcoplasmic reticulum. Depolarization activates one type which is inhibited by Dantrolene but not by 10 m*M* procaine. Increased calcium levels in the myoplasm activate a second type. Caffeine (2 m*M*) increases the sensitivity of this type to myoplasmic-free calcium, 10^{-8} *M* Ca^{2+} being sufficient to cause the release of calcium from the sarcoplasmic reticulum in sufficient quantity to cause contraction. The calcium-induced release of calcium is inhibited by procaine but not by Dantrolene.

The two sites for calcium release may be regulated under physiological conditions by the muscle action potential. Depolarization of the transverse tubular membrane may cause calcium release from the terminal cisternae by activation of carriers or channels of the first type. With the increase of calcium in the myoplasm the sites for calcium-induced release of calcium are activated and calcium release from the sarcoplasmic reticulum is amplified. Stephenson (1978) has demonstrated, in elegant experiments on skinned muscle fibers, that chloride-induced release of calcium from the sarcoplasmic reticulum can be blocked by the presence of 5 m*M* EGTA in the myofibrillar-free space. She considers that there is a calcium dependence for chloride stimulation of calcium release.

Additional knowledge of the molecular interaction of Dantrolene, caffeine, and procaine with receptive sites in the sarcoplasmic reticulum should lead to a better understanding of the processes which regulate excitation-contraction coupling in skeletal muscle.

REFERENCES

Bianchi, C. P. (1962). *J. Pharm. Exp. Therap. 138*, 41-47.

Bianchi, C. P. (1975). *In* "Cellular Pharmacology of Excitable Tissues" (T. Narahashi, ed.), pp 485-519. Charles C. Thomas, Springfield, Illinois.

Bianchi, C. P., Narayan, S., and Lakshminarayanaiah, N. (1975). *In* "Calcium Transport in Contraction and Secretion" (E. Caracoli *et al.*, eds.), pp 503-515. North-Holland/American Elsevier, New York.

Desmedt, J. E., and Hainaut, K. (1977). *J. Physiol. 265*, 565-585.

Endo, M. (1977). *Physiological Reviews 57*, 71-108.

Feinstein, M. B. (1963). *J. Gen. Physiol. 47*, 151-172.

Putney, J., and Bianchi, C. P. (1974). *J. Pharm. Exp. Therap. 189*, 202-212.

Stephenson, E. W. (1978). *J. Gen. Physiol. 71*, 411-430.

Taylor, S., and Godt, R. E. (1976). *In* "Symposia of the Soc. for Exp. Biol. Calcium in Biological Systems," 361-380.

Motility in Cell Function
Proceedings of the First John M. Marshall Symposium in Cell Biology

THE COMPOSITION OF THE SARCOPLASMIC RETICULUM OF STRIATED MUSCLE: ELECTRON PROBE STUDIES

Avril V. Somlyo, Henry Shuman
and
Andrew P. Somlyo

Pennsylvania Muscle Institute,
Presbyterian-University of Pennsylvania Medical Center
Philadelphia, Pennsylvania

INTRODUCTION

The activation of contraction in fast striated muscles involves the translocation of Ca from the sarcoplasmic reticulum (SR) to the myofilaments by, some as yet undefined, signal traversing the triadic junction between the T-tubule and the terminal cisternae (TC) (Ebashi, 1976; Endo, 1977; Fuchs, 1974). The purpose of our studies was to obtain further information about the composition and ultrastructure of the triadic junction and its components. Specifically, the combination of cryoultramicrotomy and electron probe analysis enabled us to test the hypothesis, based on flux data, that the SR is a compartment in ionic communication with the extracellular space (Conway, 1957; Harris, 1963; Keynes and Steinhardt, 1968; Makino and Page, 1975; Rogus and Zierler, 1973). Our findings show this *not* to be the case, and suggest that electromechanical coupling must involve a triggering mechanism at the triad that does not involve ionic communication between the extracellular and SR compartments. In view of the reported swelling of the SR in muscles incubated in hypertonic solutions and examined after chemical fixation (Birks and Davey, 1969), we also examined the ultrastructure and, with electron probe analysis, the composition of cryo sections of hypertonically treated frog skeletal muscle. We shall also review here the alterations in T-tubule elements in fatigued muscle (Gonzales-Serratos *et al.*, 1978). Detailed accounts of these studies have been published (Somlyo *et al.*, 1977a,b; Gonzales-Serratos *et al.*, 1978; Franzini-Armstrong *et al.*, 1978). We

ISBN 0-12-551750-5

shall also present some preliminary observations made with the aid of tannic acid during fixation used to obtain better preservation of the triadic structures.

METHODS AND MATERIALS

The extensor digitorum longus IV, the musculi lumbricales of the toe, or small bundles of the semitendinosus from *Rana pipiens* or bundles of fibers of the swimbladder from the toadfish *Opsanus tau* were used.

For cryoultramicrotomy and subsequent electron probe analysis muscles mounted on low mass stainless steel holders were frozen by shooting them into supercooled (-164 ± 2°C) Freon 22. Sections 900 Å to 2000 Å were cut at -130°C in a modified LKB cryokit. The sections were picked up from the dry knife onto Cu grids with C foils, dried at $<10^{-6}$ Torr overnight, carbon-coated, and stored dessicated. A detailed description of the freezing and cryoultramicrotomy is given elsewhere (Somlyo *et al.*, 1977b).

Electron probe analysis was done on a Philips EM 301 or EM 400 transmission electron microscope having goniometer stages modified to accept a 30 mm^2 Kevex Si (Li) X-ray detector interfaced with a Kevex 5100 (Burlingame, California) multichannel analyzer and a Tracor Northern (Middleton, Wisconsin) NS 880 computer and TN 1,000 magnetic tape recorder. Modified Philips liquid nitrogen cooled holders, operated at -100° to -110°C in the EM 301 and set at approximately -160°C in the EM 400 were used to minimize contamination and mass loss. The detailed characteristics of the system have been described previously (Shuman *et al.*, 1976). The SR was probed in regions adjacent to the T-tubules, where these could be identified and for paired comparison the regions of the cytoplasm analyzed were within 200 nm of their respective SR pairs.

The measurement of elemental concentrations is based on the fact, pointed out by Hall (1971) that for elements of atomic number $Z \geq 11$ to 20 in concentrations of less than 1 molar/kg, the characteristic peak/X-ray continuum ratio is linearly related to the concentration/dry mass. The peak counts in the unknown peak and its error of measurement are computed together with the counts in the continuum band 1.34-1.64 keV. The basic equation for quantitation is $C_x = \frac{I_x}{I_b} \times W_x$, where C_x is the concentration of element x, I_x is the number of characteristic peak counts for that element, I_b is the number of continuum counts, and W_x is a quantitation parameter relating concentration to peak/background ratio and is obtained from standards. W_x includes the effects of ionization cross sections, fluorescence yield, and detector efficiency for each specific element (Shuman *et al.*, 1976).

The X-ray spectrum obtained in a transmission electron optical column from a thin section includes the characteristic peaks due to elements $Z \geq 11$ present in the specimen and the associated continuum largely arising from the organic matrix. It also contains instrumental peaks (the largest one being the Cu signal K_a = 8.047 keV) from the specimen grid and holder, with an associated extraneous continuum, and low energy noise probably due to electrons reaching the detector. These extraneous signals and the associated continuum can be measured by collecting the spectrum generated with the beam passing through an empty grid hole, and are subtracted by the computer routine as is the low energy noise (Shuman *et al.*, 1976).

For ultrastructural studies of the triad, the muscles were fixed in 2% glutaraldehyde, 4.5% sucrose, 1.2 m*M* $CaCl_2$ in 0.075 *M* cacodylate buffer, stored in 0.1*M* cacodylate buffer with 6% sucrose for three days at 4°C, followed by incubation in 4% tannic acid (Baker 4-0377) in 0.1 *M* cacodylate buffer at pH 7.2 for 4 hr at room temperature. They were cut into small cubes, transferred to 1% w/v $FeCl_3$ in H_2O for 90 min. (Nehls and Schaffner, 1976), dehydrated, and embedded in Spurr's resin.

RESULTS AND DISCUSSION

A. *Frog Fast Twitch Muscle*

Paired analyses of terminal cisternae and adjacent cytoplasm are summarized in Table 1. The probe diameters used for these analyses were approximately 50-70 nm, to minimize the inclusion of cytoplasm in the microvolume of terminal cisternae (TC) analyzed. The mean Ca concentration in these TC in resting frog toe muscles was 66 mmoles/kg dry wt. ± 4.6 S.E.M. (≃16 mmol/l SR), and is within the range of 5-40 mmol/l SR estimated by other techniques (Ebashi and Endo, 1968; Ford and Podolsky, 1972; Peachey and Adrian, 1973; Winegrad, 1968). The greater than cytoplasmic P content of the terminal cisternae could represent the phospholipid in the membranes of the reticulum and/or P associated with Ca.

TABLE 1. *Elemental Concentrations of Terminal Cisternae and Cytoplasm in Normal Frog Toe Muscles*[a]

(mmoles/kg dry wt.) (weighted $\bar{x}$±S.E.M.)

	P	*S*	*Cl*	*K*	*Ca*	*Mg*	*Na*
Terminal cisternae	449±16	208±9.1	54±4.5	587±18.7	66±4.6	40±5.9	44±14.1
Cytoplasm	317±12.3	216±8.8	42±3.8	488±15.5	1±2.3	39±5.2	29±12.6

[a] *50 nm diameter areas of terminal cisternae or adjacent cytoplasm analyzed n=30 pairs. (From Somlyo* et al., *1977b.)*

The Cl content of the normal TC does not resemble the extracellular concentration, contrary to previous suggestions (Conway, 1957; Harris 1963; Keynes and Steinhardt, 1968; Makino and Page, 1975; Rogus and Zierler, 1973). The 54 mmol Cl/kg dry wt. measured over the TC is very significantly less than the ∿360 mmol Cl/kg dry wt. expected if the SR were an extracellular compartment. The error figures for the Na measurements in these experiments are relatively large, due to the low fluorescence yield and absorption of the soft Na X-rays in the beryllium window of the detector. The greater concentration of K in the TC region compared to the cytoplasm could be due to binding of K to calsequestrin (Carvalho and Leo, 1967; MacLennan and Holland, 1975).

B. *Toadfish Swimbladder Muscle*

This fast twitch muscle has a very extensive (approximately 30% of fiber volume, Franzini-Armstrong, personal communication) and well organized (Fawcett and Revel, 1961) sarcoplasmic reticulum (SR) that is regularly arranged with two triads per A band (Fig. 1). A frozen-dried thin section of unfixed muscle is illustrated in Figs. 2 and 3, and the concentrations of Na, P, S, Cl, K, Ca, and Mg in the SR and in the adjacent cytoplasm, determined with 50-100 nm diameter probes, are summarized in Table 2. The mean Ca concentration in the SR was 77 ± 0.9 mmol/kg dry wt. ±S.E.M. The cytoplasmic Ca in the region between the rows of SR was higher than in frog muscle (Table 1, possibly due to obliquely oriented SR in the narrow segments of cytoplasm analyzed. The Cl concentrations, as in frog muscle, are not significantly different in the SR and in the cytoplasm indicating that the SR is not in ionic communication with the extracellular space. Analyses of large areas of fibers (1-6 μm diameter) including SR and cytoplasm also failed to show "excess NaCl," in agreement with the results of small probe analysis. The Na concentrations were measured with better statistics in the swimbladder muscle than in the frog muscle (Tables 1 and 2) and like Cl, show the absence of compartmentalization. The low (similar to cytoplasmic) Na concentration of the SR excludes the possibility that the low Cl concentration in the SR is due to its partial exclusion through a Donnan effect of negatively charged proteins (MacLennan and Holland, 1975). The similar Na and Cl contents of SR and cytoplasm indicate that the T-SR junction is not freely permeable to Na or Cl and that there is not a significant Na or Cl potential across the SR membranes.

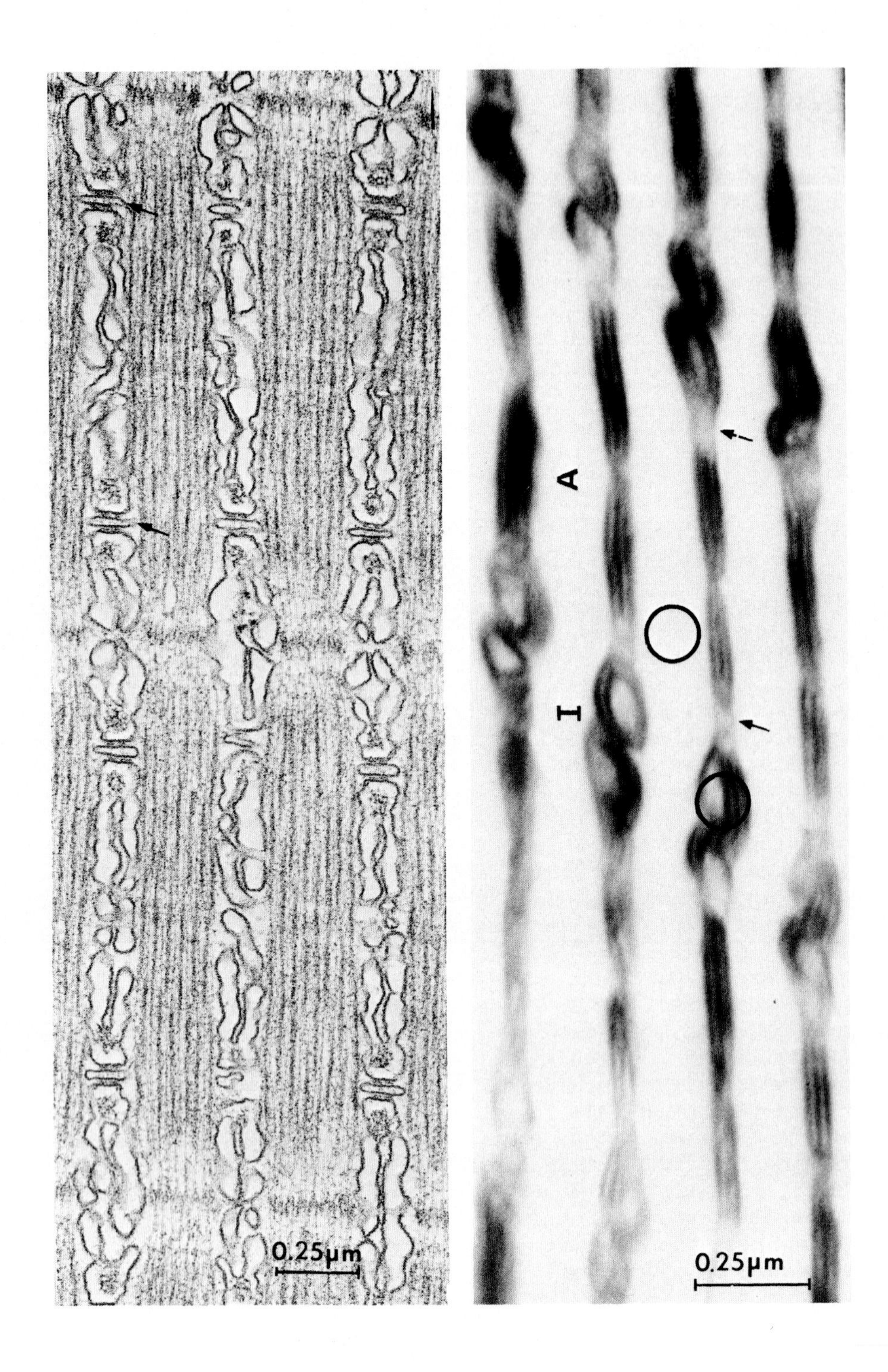
0.25μm
A
I
0.25μm

FIGURE 1. Longitudinal section from a toadfish swim-bladder muscle. Two triads per A band. T-tubules - arrows. Glutaraldehyde, osmium, fixation, plastic embedding. Magnification x46,000. (Reprinted with permission from Nature *(Somlyo* et al., *1977b).)*

FIGURE 2. Longitudinal cryo section from toadfish swim-bladder muscle. Unfixed muscle, frozen in supercooled Freon 22, sectioned at -130°C, picked up dry onto Cu grid with C foil and dried at <10^{-6}Torr. This dried section was stained with osmium vapor under vacuum to enhance contrast. Electron probe X-ray microanalyses were done on similar but unstained material. T-tubule regions indicated by arrows. Circles indicate approximate probe size used for the analyses summarized in Table 2. Magnification x66,000. (Reprinted with permission from Nature *(Somlyo* et al., *1977b).)*

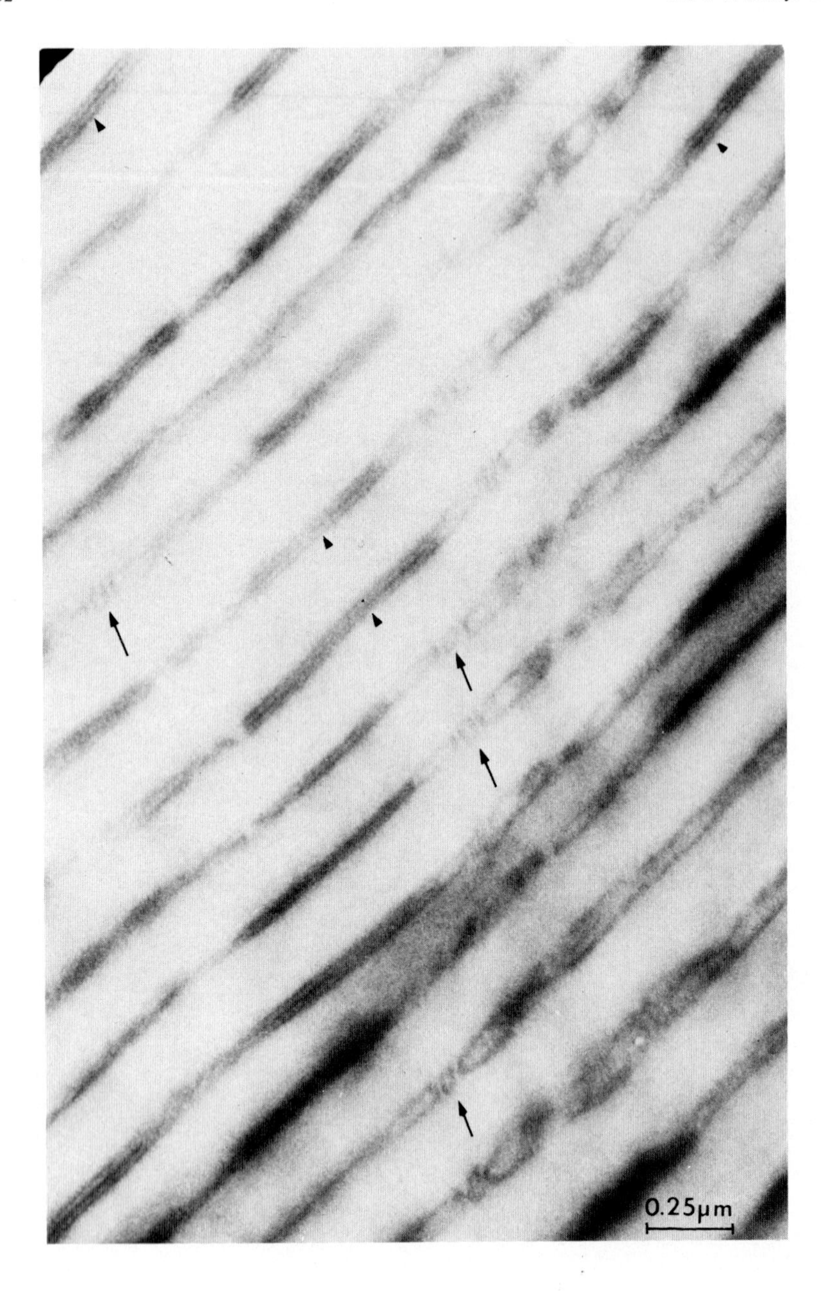

FIGURE 3. Longitudinal cryo section from toadfish swimbladder frozen at a longer sarcomere length than the example in Fig. 2. See Fig. 2 legend for details. T-tubules - arrows; laminated longitudinal reticulum - arrowheads. Magnification x54,000. (Reprinted with permission from Nature *(Somlyo* et al.*, 1977b).)*

C. *Effects of Hypertonicity*

Muscles incubated in 2.2 x hypertonic NaCl, 2.5 x hypertonic sucrose, or 2.2 x hypertonic Na isethionate consistently showed vacuoles along the Z band (Somlyo *et al.*, 1977b). Due to the contortions and poor resolution of the vacuole membranes in the shrunken muscles, we could not determine whether they communicated with the T-tubule or with the SR system. Recently, using freeze-substitution techniques, the vacuoles containing extracellular solutes have been shown to be part of the T-tubule system (Franzini-Armstrong *et al.*, 1978).
The swollen T-tubules often bulged into the glycogen spaces on either side of the two terminal cisternae at the triad, giving rise to the paired vacuoles seen in the unfixed frozen-dried cryo sections.

Occasional granules consisting of Ca,Mg, and P were found in the longitudinal SR in every muscle fiber treated with hypertonic solutions (Somlyo *et al.*, 1977b). Hypertonic solutions produce a transient contracture (Bianchi and Bolton, 1974; Homsher *et al.*, 1974; Howarth, 1958; Lännergen and Noth, 1973), and a slight decrease in total fiber Ca (Bianchi and Bolton, 1974; Homsher *et al.*, 1974). An increase in the slow component of exchangeable ^{45}Ca has been suggested to reflect translocation of Ca by hypertonicity (Homsher *et al.*, 1974). The presence of Ca granules in the longitudinal SR is consistent with this interpretation.

D. *Effect of Fatigue*

Single fibers or small fiber bundles from frog semitendinosus muscles become mechanically refractory to intermittent tetani of 40-50 supramaximal shocks per second during 0.3 sec every second. This fatigue is associated with light microscopically detectable vacuolation that is not blocked by ouabain (Gonzales-Serratos *et al.*, 1973, 1974, 1978). It is not due to failure of the action potential (Eberstein and Sandow, 1963; Grabowski *et al.*, 1972; Mashima *et al.*, 1962) or to the depletion of high energy phosphates (Nassar-Gentina *et al.*, 1977; Vergara *et al.*, 1977). In a recent study, in collaboration with H. Gonzales-Serratos and G. McClellan (Gonzales-Serratos *et al.*, 1978), utilizing electron probe

Table 2. Elemental Concentrations of Sarcoplasmic Reticulum and Cytoplasm of Toadfish Swimbladder determined by Paired "Small Spot" Electron Probe Analysis

mmoles/kg dry wt. ±S.E.M. (n = 25 pairs)

Region analyzed	P	S	Cl	K	Ca	Mg	Na
Sarcoplasmic reticulum	443±2.5	362±2.0	42±1.0	589±2.8	77±0.9	40±1.8	34±4.2
Cytoplasm (between SR)	275±2.0	335±2.0	47±1.1	510±2.7	4±0.8	47±1.9	30±4.3

(From Somlyo et al., 1977a.)

analysis of frozen-dried thin sections and freeze-substitution, we have shown that the vacuoles are a part of the T-tubule system and contain Na and Cl.

The Ca concentration of the terminal cisternae of the fatigued muscles was 102 ± 4 mmol/kg dry wt. ± S.E.M. (Fig. 4) indicating that fatigue is not due to Ca depletion of the

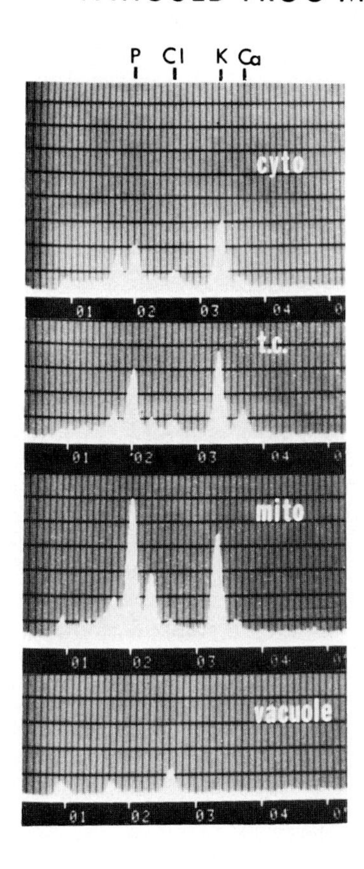

FIGURE 4. Four spectra from analysis of cytoplasm (cyto) terminal cisternae (t.c.), mitochondrion (mito), and a vacuole from a cryo section of a fatigued frog semitendinosus fiber. The number of counts on the ordinate are shown for the energy in keV on the abscissa. The characteristic K shell energy lines for P, Cl, K, and Ca are shown at the top. Note the large Cl peak relative to the K peak in the vacuole, high K over the cyto, and Ca in the t.c. but not the mitochondria. The small peak to the right of the K peak in the cyto and mito spectra is the K_{K_β} *peak and is separated by 100 eV from the* Ca_{K_α}.

terminal cisternae. Compared to the Ca content shown in Table 1, the terminal cisternae of fatigued muscle had a significantly greater than normal Ca content. This could arise from the reduced pH (6.27) in fatigued muscles (Gonzales-Serratos *et al.*, 1978), since the total Ca bound by fragmented SR is increased by 50% with a similar reduction of pH (Nakamaru and Schwartz, 1972). However, the normal muscles were not paired from the same animals used for the fatigue experiments, and the possibility of frog-to-frog variations contributing to the higher SR Ca content in fatigue cannot be ruled out.

Mitochondrial Ca content was relatively low (Fig. 4) indicating that fatigue is not due to uncoupling of mitochondria due to Ca loading as previously considered (Gonzales-Serratos *et al.*, 1973, 1974).

It was concluded (Gonzales-Serratos *et al.*, 1978) that fatigue is not due to the depletion of calcium stores from the TC or to uncoupling of mitochondria due to calcium loading, but may be caused by multiple mechanisms including failure of the T-tubule action potential.

E. *Morphology of the Sarcoplasmic Reticulum*

1) Unfixed frozen muscles: The membranes of the SR could be imaged in the thinnest frozen-dried thin sections obtained from both frog and swimbladder muscles frozen without fixation or cryoprotectants. The cryo sections illustrated in Figs. 2 and 3 were stained *in vacuo* with osmium vapor after drying, to enhance contrast. Osmium stained material was not used for electron probe analysis. Even in unstained sections there was sufficient mass contrast to discern fine structure. The junctional gap between the T-tubule and the TC, which is the site that must be traversed by the signal for Ca release, was present in both the unfixed frog muscle (Somlyo *et al.*, 1977b) and in the swimbladder (Fig. 3). In the frog muscle periodic structures suggestive of foot processes were also observed (Somlyo *et al.*, 1977b). The similarities of the triad structure in the cryo sections and in conventionally fixed material indicate that the observed morphology is not due to the effects of fixation and dehydration.

The volume of the longitudinal reticulum, particularly in the A band region, shows a striking difference between the conventionally fixed swimbladder muscle (Fig. 1) and cryo sections (Figs. 2 and 3). The SR is narrower in the cryo sections and has a laminated appearance, suggesting that SR volume, but not membrane surface area, may be overestimated in the conventionally fixed material.

A collapsed SR has been previously reported in muscles fixed in the presence of polyvalent cations (Wallace and

Sommer, 1975; Howell, 1974) and in freeze-fracture replicas of unfixed muscle (Franzini-Armstrong, *et al.*, 1978). It is not clear from the cryo sections whether the lighter regions of the laminated SR are the lumen of the SR or spaces between elements of SR. The bulbous appearance of the SR in the I band region in Fig. 2 may be due to the shorter sarcomere length as compared to the muscle illustrated in Fig. 3.

The smaller luminal volume of the longitudinal reticulum observed in cryo sections could contribute to the differential distribution of Ca in the A and I bands (Endo, 1977; Winegrad, 1968, 1970).

2) Fixed, tannic acid - $FeCl_3$ treated muscle: In glutaraldehyde fixed muscles exposed to tannic acid and subsequently to iron salts, the sarcoplasmic reticulum and T-tubule (TT) membranes showed fine delineation of the membrane trilaminar structure. In the thinnest sections electron lucent periodicities were seen traversing the junctional gap. At the respective ends of these bridges the cytoplasmic leaflets of the T-tubule and TC membranes diverged at right angles and the electron lucent lamellae of the TC and TT trilaminar membranes were continuous (Fig. 5). These images suggest that the cytoplasmic, but not the luminal leaflets, of the membranes of the TT and TC are continuous, in agreement with electron probe analysis. A similar sharp divergence of the membrane bilayer has been observed in tubular myelin from lamellar bodies in the lung (Sanderson and Vatter, 1977) and during fusion of secretory granules with the cell membrane prior to exocytosis (Palade, 1975). There was enhanced staining in the region of the junctional gap and also of the granular material in the terminal cisternae. It is probable that the bridges observed in this study represent the cores of the foot processes characterized by Franzini-Armstrong (1976).

Tannic acid has been proposed originally as a supplementary fixative (Mizuhira and Futaesaku, 1972), as a mordant between osmicated structures and lead stains or other heavy metals (Simionescu and Simionescu, 1976a,b; Wagner, 1976) resulting in enhanced density and contrast and as a stabilizer of some tissue components against extraction during processing (Simionescu and Simionescu, 1976a,b). Kalina and Pease (1977a,b) attribute the enhanced staining of membranes treated with tannic acid to their demonstration that tannic acid forms complexes with the choline bases of phosphatidyl choline and sphingomyelin. The complex is subsequently stabilized by osmium and is retained during dehydration, unlike conventionally fixed saturated phospholipids. Superior contrast and resolution has been reported with the use of the iron-tannin, compared to the osmium-tannin complex (Nehls and Schaffner, 1976), and was also observed in the present study of the triad.

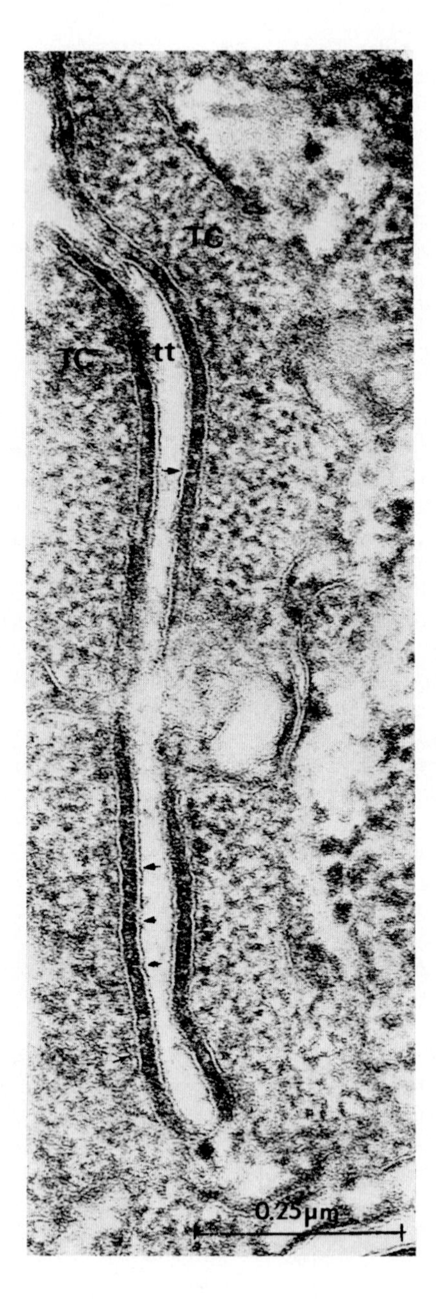

FIGURE 5. Longitudinal section of frog semitendinosus muscle fixed with glutaraldehyde followed by tannic acid and $FeCl_3$ showing an enface view of the triad, dense staining of the junctional gap and electron lucent periodicities (arrows) crossing the junctional gap between the terminal cisternae, TC, and T-tubule, tt. Section lead stained. Magnification x124,000.

CONCLUSIONS

The concentrations of Na and Cl in the sarcoplasmic reticulum of normal resting striated muscle of the frog and toadfish swimbladder do not differ from the cytoplasmic concentrations and do not resemble that of the extracellular space. These findings (albeit on a dry weight basis) do not support the existence of a Cl of Na potential across the SR membrane in resting muscle.

Ca is sequestered in the terminal cisternae in normal resting muscle with a mean Ca concentration in frog muscle of 66 ± 4.6 mmole/kg dry wt. ± S.E.M. and in toadfish swimbladder muscle of 77 ± 0.9.

Hypertonic 2.2 x NaCl, 2.5 x sucrose, or 2.2 x Na isethionate produce vacuoles containing extracellular solutes along the Z line. These vacuoles arise from swelling of the T-tubule system (Franzini-Armstrong *et al.*, 1978). Muscle fatigue is also associated with vacuolation of the T-tubule system. The Ca content of the terminal cisternae is not subnormal. Mitochondria are not swollen and their Ca content is normal. Therefore, fatigue is not due to insufficient Ca for release and may be due to failure of the T-tubule action potential.

The presence of the junctional gap at the triadic junction in cryo sections is similar to that observed in glutaraldehyde-osmium fixed and plastic embedded material. The longitudinal SR is narrower with a laminated appearance in cryo sections suggesting that SR volumes may be overestimated in conventionally fixed muscle.

In glutaraldehyde-tannic acid-$FeCl_3$ fixed muscles electron lucent bridges were observed crossing the junctional gap. These bridges appeared continuous with the cytoplasmic (but not luminal) leaflets of the terminal sac and T-tubule membranes.

ACKNOWLEDGMENTS

This work was supported by Grant HL-15835 to the Pennsylvania Muscle Institute and by Grant GM-00092.

Note Added in Proof: Since this symposium, the study of the membrane bridges has been completed and published. Somlyo A. V. *J. Cell Biol. 80:* 743-750, 1979.

REFERENCES

Bianchi, C. P., and Bolton, T. C. (1974). *J. Pharmacol. Exp. Ther. 188*, 536-552.

Birks, R. I., and Davey, D. F. (1969). *J. Physiol. (Lond.) 202*, 171-188.

Carvalho, A. P., and Leo, B. (1967). *J. Gen. Physiol. 50*, 1327-1352.

Conway, E. J. (1957). *Physiol. Rev. 37*, 84-132.

Ebashi, S. (1976). *Ann. Rev. Physiol. 38*, 293-313.

Ebashi, S., and Endo, M. (1968). *In* "Progress in Biophysics and Molecular Biology" (J.A.V. Butler and D. Noble, eds.), pp. 123-183. Pergamon Press, New York.

Eberstein, A., and Sandow, A. (1963). *In* "The Effect of Use and Disuse on Neuromuscular Function" (Gutmann and Hnik, eds.), pp. 135-216. Elsevier, Amsterdam.

Endo, M. (1977). *Physiol. Rev. 57*, 71-108.

Fawcett, D. W., and Revel, J. P. (1961). *J. Biophys. Biochem. Cytol. 10*, 89-109.

Ford, L. E., and Podolsky, R. J. (1972). *J. Physiol. (Lond.) 223*, 21-33.

Franzini-Armstrong, C. (1976). *Proc. 5th Intern. Conf. Mus. Dys. Assn.*, Durango, Colo., pp. 612-625.

Franzini-Armstrong, C., Heuser, J. E., Reese, T. S., Somlyo, A. P., and Somlyo, A. V. (1978). *J. Physiol. (Lond.), 283*, 133-140.

Fuchs, F. (1974). *Ann. Rev. Physiol. 36*, 461-502.

Gonzales-Serratos, H., Borrero, L. M., and Franzini-Armstrong, C. (1973). *J. Gen. Physiol. 62*, 656.

Gonzales-Serratos, H., Borrero, L. M., and Franzini-Armstrong, C. (1974). *Fed. Proc. 33*, 1401.

Gonzales-Serratos, H., *et al.* (1978). *Proc. Natl. Acad. Sci. 75*, 1-7.

Grabowski, W., Logsiger, E. A., and Lüttgau, H. D. (1972). *Pflügers Arch. ges. Physiol. 334*, 222-239.

Hall, T. A. (1971). *In* "Physical Techniques in Biological Research" (G. Oster, ed.), Vol. Ia. Academic Press, New York.

Harris, E. J. (1963). *J. Physiol. (Lond.) 166*, 87-109.

Homsher, E., Briggs, F., and Wise, R. M. (1974). *Am. J. Physiol. 226*, 855-863.

Howarth, J. V. (1958). *J. Physiol. (Lond.) 144*, 167-175.

Howell, J. N. (1974). *J. Cell Biol. 62*, 242-245.

Kalina, M., and Pease, D. C. (1977a). *J. Cell Biol. 74*, 726-741.

Kalina, M., and Pease, D. C. (1977b). *J. Cell Biol. 74*, 742-746.

Keynes, R. D., and Steinhardt, R. A. (1968). *J. Physiol. (Lond.) 198,* 581-599.

Lännergren, J., and Noth, J. (1973). *J. Gen. Physiol. 61,* 158-175.

MacLennan, D. H., and Holland, P. C. (1975). *Rev. Biophys. Bioeng. 4,* 377-404.

Makino, H., and Page, E. (1975). *Fed. Proc. 34,* 413.

Mashima, H., Matsumura, M., and Nakajama, Y. (1962). *Jap. J. Physiol. 12,* 324.

Mizuhira, V., and Futaesaku, Y. (1972). *Proc. Elect. Micros. Soc. Amer. 29,* 494.

Nakamaru, Y., and Schwartz, A. (1972). *J. Gen. Physiol. 59,* 22-32.

Nassar-Gentina, V., Passonneau, J. V., and Rapoport, S. I. (1977). *Biophys. J. 17,* 173a.

Nehls, R., and Schaffner, G. (1976). *Cytobiologie 13,* 285-290.

Palade, G. (1975). *Science 189,* 347-358.

Peachey, L. D., and Adrian, R. H. (1973). *In* "The Structure and Function of Muscle" (G. H. Bourne, ed.), Vol. III, pp. 1-29. Academic Press, New York.

Rogus, E., and Zierler, K. L. (1973). *J. Physiol. (Lond.) 233,* 227-270.

Sanderson, F. J., and Vatter, A. E. (1977). *J. Cell Biol. 74,* 1027-1031.

Shuman, H., Somlyo, A. V., and Somlyo, A. P. (1976). *Ultramicroscopy 1,* 317-339.

Simionescu, N., and Simionescu, M. (1976a). *J. Cell Biol. 70,* 608-621.

Simionescu, N., and Simionescu, M. (1976b). *J. Cell Biol. 70,* 622-633.

Somlyo, A. V., Shuman, H., and Somlyo, A. P. (1977a). *Nature 268,* 556-558.

Somlyo, A. V., Shuman, H., and Somlyo, A. P. (1977b). *J. Cell Biol. 74,* 828-857.

Vergara, J. L., Rapoport, S. I., and Nassar-Gentina, V. (1977). *Am. J. Physiol. 232,* C185-C190.

Wagner, R. C. (1976). *J. Ultrastruct. Res. 57,* 132-139.

Wallace, N., and Sommer, J. R. (1975). *Proc. Electron Micros. Soc. Am. 33rd A. Mtg.,* 500.

Winegrad, S. (1968). *J. Gen. Physiol. 51,* 65-83.

Motility in Cell Function
Proceedings of the First John M. Marshall Symposium in Cell Biology

MEMBRANE CONTROL OF CARDIAC CONTRACTILITY

Saul Winegrad and George B. McClellan

School of Medicine
University of Pennsylvania
Department of Physiology
Philadelphia, Pennsylvania

By appropriate interventions the contraction of individual cardiac cells or small groups of cells can be modified in the amount of force generated, the duration of the contraction, the rate of force development, or the rate of relaxation. Variation of the voltage or the duration of the action potential changes the contraction, presumably as a result of alterations in the transmembrane movements of Ca and the intensity of the signal to intracellular Ca stores (Morad and Orkand, 1971). Phosphorylation of the sarcoplasmic reticulum increases the rate of accumulation of Ca by the sarcoplasmic reticulum (Tada *et al.*, 1974). There is, however, a paucity of information regarding the way in which changes in the contractile proteins themselves might influence the contraction. Several phosphorylation sites exist on the contractile proteins, including two cAMP-dependent sites on troponin-I (TN-I) and a Ca-dependent site on the regulatory light chain of myosin (Perry, 1975). In smooth muscle and certain nonmuscle forms, phosphorylation of one of the light chains of myosin appears to be essential for the activation of the contraction. Because of the increase in the concentration of cAMP in the cardiac cell during the inotropic response to epinephrine, attempts have been made to associate the enhanced contractility with phosphorylation, but as yet no convincing data have been produced (Tsien, 1977).

For studying the ability of the contractile proteins of cardiac muscle to undergo changes in performance, direct activation from Ca in bathing solutions must be possible in order to eliminate the electromechanical coupling system as a cause for modified function, but the regulatory system must not be lost in creating the accessibility of the contractile proteins to Ca in the bathing medium. The "hyperpermeable" cardiac

ISBN 0-12-551750-5

fiber, which is produced by a prolonged exposure of cardiac muscle to EGTA at low temperature, is the closest model to the intact cell in which access to the contractile proteins for small extracellular ions is available without the interference of a major diffusion barrier (Winegrad, 1971; McClellan and Winegrad, 1977). Its surface membrane is highly permeable to small molecules and ions such as Ca, but it still remains a diffusion barrier for proteins and other large molecules. The degree of activation of the preparation can be controlled by the Ca concentration in the bathing fluid, which consists of 5 m*M* ATP, 7 m*M* $MgCl_2$, 120 m*M* KCl, 25 m*M* imidazole buffer at pH 7.0, 15 m*M* creatine phosphate, 1 mg per 1 ml creatine phosphokinase, 3 m*M* EGTA and the necessary amount of $CaCl_2$ to give the desired pCa.

In the following studies, small bundles of hyperpermeable fibers prepared from trabeculae from the right ventricle of a rat have been used to study two specific properties of the contractile system, the range of Ca concentration over which activation occurs, referred to as Ca sensitivity, and the maximum Ca-activated force. Emphasis has been placed on cyclic nucleotide-regulated control mechanisms because of the several phosphorylation sites on the contractile proteins and the large body of evidence implicating phosphorylation in the regulation of cellular reactions.

REGULATION OF Ca SENSITIVITY

The first set of experiments was conducted on tissue that had been removed and rapidly cooled to approximately $0^{o}C$ immediately after decapitation of the rat. Since the plasma catecholamine concentration, which varies according to the handling of the animal prior to sacrifice, is high in decapitated rats, these tissues had a high level of catecholamine stimulation at the time they were chilled. In hyperpermeable bundles prepared this way, a pCa($-\log[Ca^{++}]$) of about 4.5 is required for maximum activation and about 5.0 for 50% activation (Fig. 1). The relation between force and Ca concentration is unchanged by the addition of either cAMP or cGMP over the concentration range of 10^{-9} to 10^{-5} *M*, but either theophylline or Squibb 20009, which are potent phosphodiesterase inhibitors, produces a significant and completely reversible increase in Ca sensitivity. The pCa required for 50% and maximum Ca-activated force are increased to 5.6 and 5.2, respectively. The addition of 10^{-7} *M* cGMP to the bathing medium already containing the phosphodiesterase inhibitor produces a small but significant further increase in Ca sensitivity (Fig. 2), while 10^{-6} *M* cAMP, on the other hand, reduces the extent of the

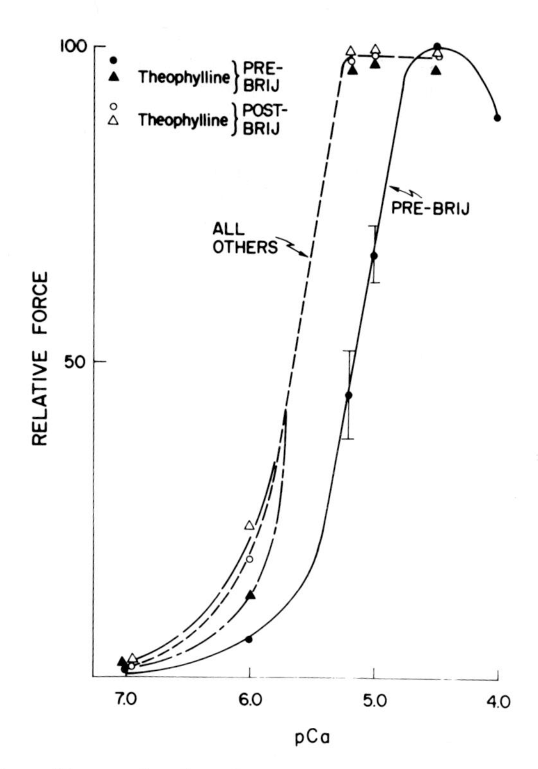

FIGURE 1. The relation between pCa and force in hyperpermeable rat ventricular bundles. Treatment with theophylline increases Ca sensitivity to the same degree as treatment with the detergent Brij, but theophylline produces no further increase in preparations that have already been treated with Brij.

increase from the inhibition of phosphodiesterase by a modest but significant amount. All of these changes are completely reversed by removing the drugs.

These experiments indicate that the preparation contains an active phosphodiesterase that must be inhibited for the cyclic nucleotides from the bathing medium to be effective, but when the hydrolysis of the cyclic nucleotides has been prevented, cGMP increases and cAMP decreases Ca sensitivity. Theophylline by itself also increases Ca sensitivity. The hyperpermeable cardiac cells must therefore be actively synthesizing cyclic nucleotides, and inhibition of phosphodiesterase, which hydrolyzes both cAMP and cGMP, should raise the concentration of both nucleotides. In the presence of elevated concentrations of both cyclic nucleotides produced by the inhibitor, the effect of cGMP predominates.

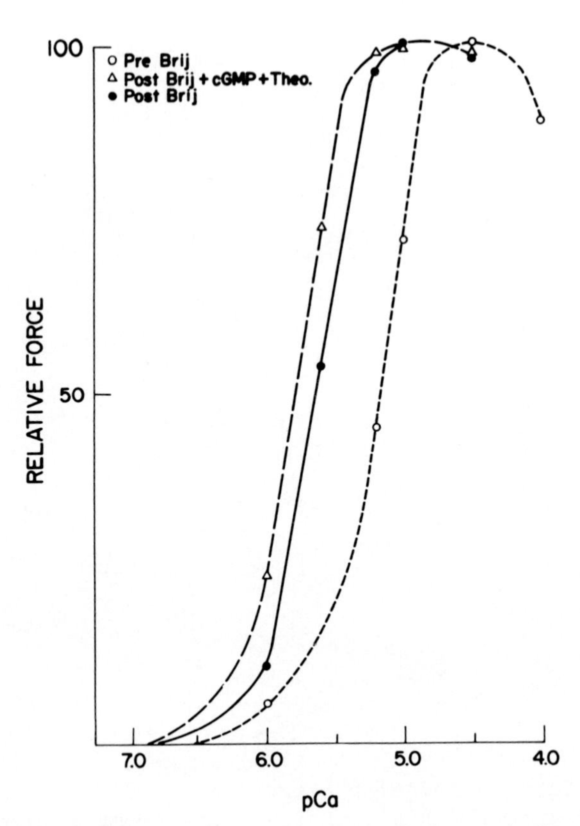

FIGURE 2. Relation between pCa and force before and after Brij treatment. Cyclic GMP plus theophylline increase Ca sensitivity even after Brij treatment.

Although most tissues prepared from hearts chilled immediately after removal from the decapitated rat have a relatively low sensitivity to Ca, occasionally there is a high sensitivity. This is more common with myocardial specimens from older rats. A net decrease in Ca sensitivity in these tissues is produced by the addition of cAMP with theophylline to the bathing medium instead of merely a reduction of the increase that theophylline alone produces in preparations with an initially low Ca sensitivity (Fig. 3).

The Ca sensitivity of hyperpermeable rat ventricular fibers can also be increased by treatment with a bathing solution containing 0.5 to 1.0% of a nonionic detergent such as Brij 58 or Triton X-100 for 30 min. The shift in Ca sensitivity is the same as with theophylline (Fig. 1) except for the irreversible nature of the change with the detergent. A small but significant further increase in Ca sensitivity can usually

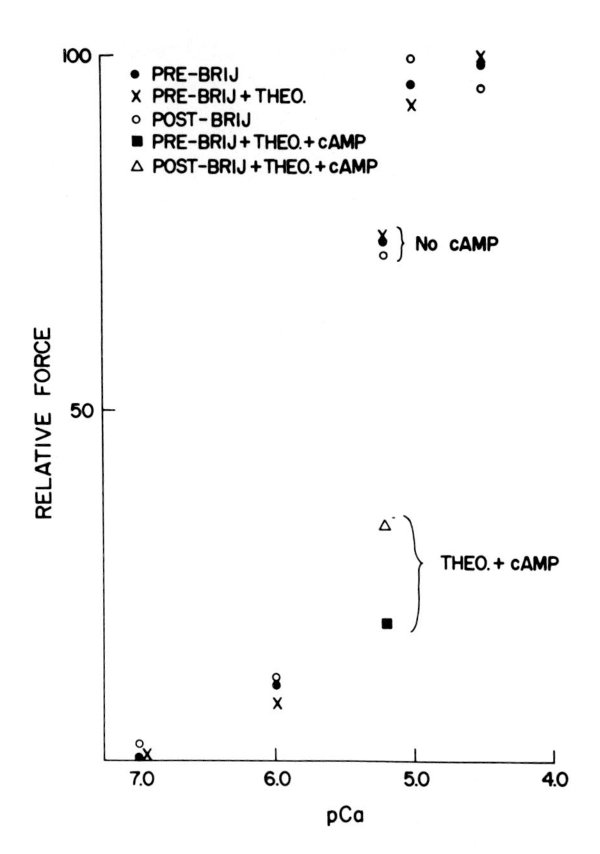

FIGURE 3. The effect of cAMP plus theophylline on the Ca sensitivity of hyperpermeable bundles that initially had a relatively high sensitivity to Ca. Note particularly the decrease in force at pCa 5.2 that is produced by cAMP.

be produced by cGMP and theophylline after exposure to the detergent (Fig. 2) and a small decrease with cAMP plus theophylline. The Ca sensitivity in the presence of cGMP and theophylline is the same, however, before and after detergent, indicating that the detergent-soluble parts of the tissue, presumably the membranes, are not important for this cyclic nucleotide to produce its effect. On the other hand, the effect of cAMP with theophylline is never as great after treatment with detergent as before even when the concentration of the cAMP is increased tenfold. Something that enhances the effect of cAMP is either inactivated or removed by the detergent.

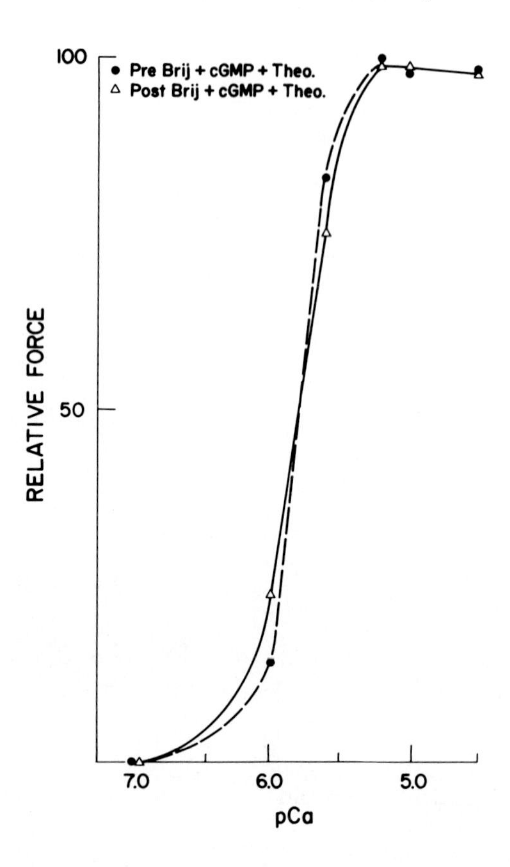

FIGURE 4. The effect of cGMP plus theophylline on Ca sensitivity before and after Brij treatment. The relation is the same, indicating that the sarcolemma is not required for the Brij effect.

Some experiments that have been performed on a mechanically skinned rat ventricular fiber are very helpful in localizing the site of the reactions involved in regulating Ca sensitivity (Fabiato and Fabiato, 1975). The Ca sensitivity of this preparation, which has no sarcolemma, and the detergent-treated hyperpermeable fibers are the same, and no alteration in mechanically skinned fiber is produced by detergents. Since the skinned fiber has an actively functioning sarcoplasmic reticulum that is inhibited by Brij 58, the surface membrane must be the site of action of the detergent in the hyperpermeable fiber.

One can conclude, therefore, that the mechanism for regulation of Ca sensitivity of the contractile proteins involves one reaction controlled primarily by the sarcolemma that decreases sensitivity and a second cytosolic reaction that increases the sensitivity. The first involves cAMP and the second, which can prevent the effects of the first, is regulated by cGMP. The basic reaction in the most obvious model that one can infer from these results is a cAMP-dependent phosphorylation of the contractile proteins that decreases Ca sensitivity. It is entirely consistent with observations on isolated proteins that a cAMP-dependent phosphorylation of cardiac TN-I decreases the Ca sensitivity of the troponin-regulated cardiac contractile protein system (Ray and England, 1976). This model can be tested by replacing ATP with cytosine triphosphate (CTP) in the bathing solution, for CTP is an excellent substitute for ATP in the force-generation reaction between myosin and actin but a poor substitute in phosphorylating reactions. Total replacement of CTP for ATP, which should produce dephosphorylation of the contractile proteins, causes an increase in the Ca sensitivity of the contractile system to the same level as treatment with detergent (Fig. 5). The nonphosphorylated contractile proteins have the high Ca sensitivity, and regulation occurs by a cAMP-dependent phosphorylation that reduces sensitivity.

The initial Ca sensitivity of the hyperpermeable fibers depends on the treatment of the isolated heart before it is placed in the EGTA solution (Solaro *et al.*, 1976; Ray and England, 1976; Depocas and Behrens, 1977). If, instead of being immediately cooled to 0°C. the isolated heart is first perfused through its circulation with a modified Kreb's solution at room temperature for about 10 min, the large amount of catecholamines that is present as a result of the terminal stress of sacrifice to the animal is washed out. Hyperpermeable fibers made from these ventricles are more sensitive to Ca ions and in fact, their Ca sensitivity is almost as great as hyperpermeable fibers that have been treated with detergent. The effect of perfusion with Kreb's on the Ca sensitivity can be partially reversed if 2 μM epinephrine is added to the Kreb's after several minutes of perfusion. The Ca sensitivity of the hyperpermeable fibers from all three kinds of hearts, that is, nonperfused, Kreb's-perfused and Kreb's-then-epinephrine-perfused, is the same after treatment with detergent. Epinephrine, therefore, acting at the same cell surface presumably on adenylate cyclase in the membrane, decreases Ca sensitivity by promoting a cAMP-dependent phosphorylation. These observations complement those of Solaro *et al.* (1976), who have shown a similar correlation between phosphorylation of TN-I and the presumed epinephrine concentration at the time the heart was frozen for isolation of the contractile proteins. The inhibitory reaction in the membrane is controlled by circulating catecholamines, but a second reaction in the cytosol can oppose the commands of the membrane.

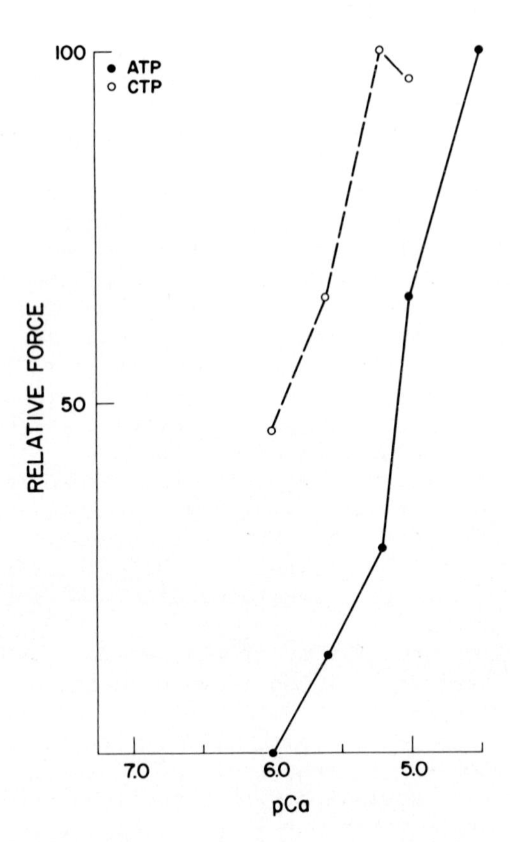

FIGURE 5. The effect of replacing ATP with CTP on the tension-pCa relation. The substitution produces an increase in Ca sensitivity of about the same degree or Brij as theophylline.

This type of regulatory system, which permits input from both the organism and cell metabolism, makes Ca sensitivity a function of both the organism's requirements and the cell's metabolic state.

REGULATION OF THE MAGNITUDE OF FORCE

In all of the experiments described so far, exposure of hyperpermeable fibers to cyclic nucleotides, phosphodiesterase inhibitors, and detergents produced little if any change in the magnitude of the maximum Ca-activated contraction. Although treatment with detergent decreased force about 10%, the addition of cAMP or cGMP in concentrations up to 10^{-5} *M* with or without a phosphodiesterase inhibitor had no effect on the maximum Ca-activated force. Clearly, the protein kinase involved in regulating Ca sensitivity was either inactive or insufficient to produce any changes in the capacity for generating force.

When the cyclic nucleotides were added in the presence of detergent, the results were strikingly different (Table 1). Addition of the combination of 10^{-6} *M* cAMP plus 1% Triton X-100 to the normal bathing solution for 30 min increased the maximum Ca-activated force by 39%, while 10^{-6} *M* cGMP plus Triton produced a 100% increase. The contractility was stable at the high level after either of these treatments, and a second exposure to the detergent with or without cyclic nucleotides did not change the Ca-activated force. The enhanced contractility after the cyclic nucleotide in detergent was prevented by a prior exposure to detergent. The detergent in some way seems to facilitate the access of the cyclic nucleotide to the involved protein kinase but at the same time slowly inactivates or removes the protein kinase as well as the phosphatase from the preparation. Solubilization of a membrane-bound cyclase by detergents has already been observed in isolated membrane preparations (Hofmann *et al.*, 1975; Corbin *et al.*, 1977).

From these data it is clear that the protein kinase regulating contractility is different from the one controlling Ca sensitivity, and there is a reasonable argument for inferring its location in the sarcolemma. Since the involved protein kinase would have to be accessible to the cyclic nucleotides under normal physiological conditions, in the hyperpermeable fiber it should be located in a lipid phase that has been altered by the soak in EGTA. The organelle changed the most by the EGTA is the sarcolemma as intracellular membranes are normally exposed to a very low concentration of Ca in resting cardiac cells. The most likely site of the protein kinase is, therefore, the sarcolemma. The failure of detergents to

TABLE 1. Effect of Drugs on Response to Detergent[a]

Drug	*Catecholamine Stimulated*	*Noncatecholamine Stimulated*
None	*0.85 ± 0.02*	*1.46 ± 0.16*
cAMP	*1.39 ± 0.10*	*1.30 ± 0.04*
cGMP	*2.04 ± 0.10*	*1.11 ± 0.12*

[a]*Expressed as* $\dfrac{\textit{maximum Ca-activated force after detergent}}{\textit{maximum Ca-activated force before detergent}}$.

produce any change in contractility of the mechanically skinned cardiac cell, which has no sarcolemma, is consistant with this conclusion.

The simplest explanation for these observations is that in the presence of a detergent, cAMP releases the catalytic subunit from the holoenzyme in the membrane and allows it to diffuse to the myofibril. This type of mechanism has recently been described for protein kinase in isolated mammalian cardiac membranes (Corbin *et al.*, 1977). The site of action of cGMP is less clear, although it too should be in the membrane since detergent is necessary for its action. It is well known that GTP and GMP interact with a protein to greatly enhance the activation of membrane adenylate cyclase when catecholamines have combined with the β receptors (Rodbell *et al.*, 1971; Pfeuffer, 1977; Lefkowitz, 1975). In hyperpermeable fibers that have been prepared from catecholamine-stimulated hearts, as is the case in these studies, cGMP may exert its positive inotropic effect by facilitating the synthesis of cAMP. Preliminary studies of hyperpermeable fibers prepared from hearts that have been perfused to wash out the catecholamines support this interpretation. With these cells, cAMP plus detergent increases contractility but cGMP plus detergent does not.

SUMMARY

A hyperpermeable bundle of rat ventricular fibers has been used to study the regulation of Ca sensitivity and force generation by the contractile proteins. The advantage of this preparation is that its permeable membrane allows Ca and other small ions or molecules to pass readily into the cytoplasm, but it retains soluble proteins in the cytosol and active enzymes in the membrane. Either catecholamine-activated or nonactivated preparations can be created by varying the condition of the heart at the time it is put into chilled EGTA to increase membrane permeability.

The Ca sensitivity of the contractile proteins can be varied over a range of up to sixfold by the balance of two opposing reactions, one cAMP-dependent decreasing sensitivity and the second cGMP-dependent increasing sensitivity. The nonphosphorylated contractile proteins have a high Ca sensitivity, and modulation of this property occurs by phosphorylation of TN-I. Epinephrine, through its stimulation of adenylate cyclase, decreases Ca sensitivity.

Force production is regulated by another protein kinase, which is located in a lipid portion of the cell, probably the sarcolemma. This conclusion has been inferred from the need for a detergent in order for cyclic nucleotides to alter force production. Both cAMP and cGMP enhance maximum Ca-activated force in hyperpermeable cells from catecholamine-stimulated hearts, but only cAMP is effective in the unstimulated heart. These observations can be explained by a protein kinase in the membrane that is activated by cAMP and an adenylate cyclase that requires a guanosine nucleotide-binding protein for maximal activation.

ACKNOWLEDGMENT

This work was supported by USPHS Grant HL 16010.

REFERENCES

Corbin, J. D., Sugden, P., Lincoln, T. M., and Keely, S. L. (1977). *J. Biol. Chem. 252,* 3854.

Depocas, F., and Behrens, W. A. (1977). *Canad. J. Physiol. Pharmacol. 55,* 212.

Fabiato, A., and Fabiato, F., (1975). *J. Physiol. 249,* 497.

Hofmann, F., Beavo, J., Bechtel, P., and Krebs, E. (1975). *J. Biol. Chem. 250,* 7798.

Lefkowitz, R. (1975). *J. Mol. Cell Cardiol. 7,* 237.

McClellan, G. B., and Winegrad, S. (1977). *Nature 268,* 261.

Morad, M., and Orkand, R. (1971). *J. Physiol. 219,* 167-189.

Perry, S. V. (1975). The contractile and regulatory proteins of the myocardium, *in* "Contraction and Relaxation in the Myocardium" (W. Nayler, ed.). Academic Press, London.

Pfeuffer, T. (1977). *J. Biol. Chem. 252,* 7224.

Rey, K., and England, P. (1976). *FEBS Lett. 70,* 11.

Rodbell, M., Krams, H. M., Pohl, S. L., and Birnbaumer, L. (1971). *J. Biol. Chem. 246,* 1872.

Solaro, J., Moir, A., and Perry, S. V. (1976). *Nature 262,* 615.

Tada, M., Kirchberger, M., Repke, D., and Katz, A. M. (1974). *J. Biol. Chem. 249,* 6174.

Tsien, R. (1977). *In* "Advances in Cyclic Nucleotide Research" (P. Greengard and G. A. Robison, eds.). Vol 8.

Winegrad, S. (1971). *J. Gen. Physiol. 58,* 71.

Motility in Cell Function
Proceedings of the First John M. Marshall Symposium in Cell Biology

METABOLISM OF ACTIVE SMOOTH MUSCLE

Robert E. Davies

Department of Animal Biology
School of Veterinary Medicine
University of Pennsylvania
Philadelphia, Pennsylvania

INTRODUCTION

When I arrived here in 1956 from Sheffield via Oxford there was only a small fraction of the interest in all aspects of muscle that there is now. Britton Chance and his colleagues in the Johnson Foundation were measuring tiny rapid optical changes in contracting muscle; Victor Heilbrunn in Biology was studying the role of calcium in muscle and nearly everything else; and John Marshall in Anatomy was interested in structure and function.

I was trying to solve A. V. Hill's "Challenge to Biochemists" and thus to find out whether adenosinetriphosphate (ATP) has a direct role in supplying the energy for contraction in living muscle (Hill, 1950). This was a discouraging time since almost all my ideas proved useless and it was impossible to show any chemical change in anything during a single contraction of an isolated muscle (Davies, 1965).

During this time I spent many hours with John Marshall talking over our problems. One of them was that a sink in my laboratory occasionally leaked water down through the floor and slowed up his and Vivianne Nachmias' researches by ruining their photomicrographs. That was easy to solve, but the other discussions sustained me during the latter part of the 10 years of time and about 30 person years of effort and continual failure associated with that project until success was finally achieved.

That was with Dennis Cain, then a graduate student, when it first became technically possible to show changes, first in inorganic phosphate in the retractor penis of the turtle, then

ISBN 0-12-551750-5

in phosphorylcreatine (PCr), then in creatine, and finally in ATP itself during a single working contraction of isolated frog striated muscle (Cain and Davies, 1962). The keys were the use of new and improved methods of analysis and of 2:4-dinitrofluorobenzene (DNFB), the reagent Sanger had used to uncover the primary structure of insulin, and that Kuby and Mahowald (1959) had showed was a powerful inhibitor of creatine phosphokinase.

We were able to define conditions at $0^{o}C$, where for about 20 min this aggressive reagent had blocked that enzyme but had not yet damaged the rest of the contractile machinery. We used it to show many things. The chemical energy from the splitting of ATP is not used in a constant amount during an initial burst, as occurs when a cartridge propels a bullet from a gun or an arrow is shot from a bow. The ATP is mainly used during the working, not the relaxation, phase of a contraction. The thermodynamic efficiency varies with velocity and is remarkably high. The rate of ATP usage during isotonic contractions depends on the load, can be varied after electrical stimulation and is not predetermined by the stimulus.

ATP is used continuously during the maintenance of a force under isometric conditions, but usually at a rate less than that needed to develop the force despite uniform continuous stimulation (Kushmerick and Davies, 1969). The isometric rates are high in frog sartorius, lower in rectus abdominus, and almost zero in the anterior byssus retractor muscle during a "catch" (Nauss and Davies, 1966b). Thus, the energy turnover depends very much on the type of muscle. It also depends on the length, being greater at P_0 and less at lengths longer or shorter than ℓ_0 (Infante *et al.*, 1964a).

A. V. Hill's "Further Challenge to Biochemists" was met by following a protocol defined by him and comparing the ATP used in an isometric contraction with one in which the muscle contracted under a light load (Hill, 1966). Other muscles shortened various distances under various loads. The results clearly showed that "shortening heat" is not degraded free energy from ATP splitting. The ATP is used to do the external work, not for shortening per se (Davies *et al.*, 1967). The source of shortening heat is still a mystery.

The rate of ATP usage is dramatically reduced if the muscle is being slowly stretched during stimulation, even though the developed force is much greater than P_0 (Infante *et al.*, 1964b; Curtin and Davies, 1973). No measurable ATP splitting is associated with the release of bound calcium but some is needed to pump this calcium back again to its storage sites (Infante *et al.*, 1964a). Far less ATP is needed to restore the sodium and potassium ions that move during the passage of an action potential before a twitch or during a tetanus (Kushmerick *et al.*, 1969).

Although the stiffness of striated muscle in rigor mortis is maintained without ATP usage and, in fact, in its absence, the extensive shortening often observed as muscles go into rigor certainly depends on the liberation of calcium and the splitting of ATP (Nauss and Davies, 1966a). This is especially clear in, for example, thaw rigor when a living muscle is frozen and then thawed. The muscle can shorten and do work before it stiffens and becomes immobile (Kushmerick and Davies, 1968).

SMOOTH MUSCLE

John Marshall had often talked about the need to work with smooth muscle, and soon after the formation of the Pennsylvania Muscle Institute and with advice from Andrew and Avril Somlyo, and with Marion Siegman, Susan Mooers, Tom Butler, and Adelaide Delluva, I turned my attention mainly to smooth muscle.

We had much to learn. Many smooth muscles such as rabbit taenia coli have spontaneous rythmic activity (tonus) at 38°C and do not respond to electrical stimulation at 0°C. At 18°C, however, they remain relaxed and can be maximally stimulated electrically.

Early problems were stress relaxation (Siegman *et al.*, 1976) and rigor mortis (Butler *et al.*, 1976). Although the muscle remains floppy after "death" and the loss of ATP, i.e., it bends easily, its resistance to stretch is high but quite similar to the situation during life. Removal of external calcium ions, however, reduced the high initial resistance to stretch of the living muscle, but had no effect on the muscle in rigor mortis. Stress relaxation, a dramatic property of normal resting smooth muscle, virtually disappeared in the absence of external calcium ions. Thus, except at long muscle lengths where passive tension due to connective tissue elements was high, a stepwise increase in the length of the muscle led to a stepwise increase in tension instead of to a large transient increase in the presence of calcium ions that then decayed to a much lower steady level.

A detailed investigation of the length-tension relationship of the resting muscle and of samples of collagen and connective tissue made it appear that in smooth muscle at rest the cross bridges between the myosin and actin filaments are completed without any steady-rate force maintenance. They can transiently resist stretching then reform, in the numbers appropriate to the filament overlap, at the new muscle length.

Most of the stress relaxation of resting muscle is calcium dependent. Thus, at very low external calcium concentrations the muscle is atonic, does not show stress relaxation, and does not respond to electrical stimulation. At normal external calcium concentrations it is also atonic, exhibits stress relaxation, is as stiff in its initial resistance to stretch as it is during rigor mortis, and can contract if stimulated at 18°C but not at 0°C. It contracts spontaneously and rythmically at 25°C and 37°C.

Early experiments on the energetics of contraction showed that this muscle is so slow mechanically that the metabolic recovery processes have significant effects even during a single tetanus. It was, therefore, necessary to inhibit respiration and glycolysis.

Unfortunately, DNFB is not suitable as an inhibitor for use at 18°C since it changes the mechanical behavior of the muscle.

Pretreatment at 0°C with iodoacetate and fluoracetate gave a preparation which, anaerobically, behaved mechanically in an identical manner with untreated muscles, but neither respired nor glycolyzed. This preparation uses ATP and PCr as its energy source and two or three samples showing no significant variation in chemical content could be obtained from each muscle. The method involved the following steps (Butler *et al.*, 1978):

> equilibration of the rabbit taenia coli (15 mgm, slack length 12 mm) in Krebs bicarbonate solution at 20°C under 2 gm-wt for at least 2 hr; adjustment of the muscle length to the desired length relative to ℓ_0; anaerobic treatment with 0.5 m*M* iodoacetate, 5.0 m*M* fluoracetate, (0 m*M* glucose) Krebs solution for 30 min at 5°C followed by rewarming to 18°C for 3-5 min; electrical stimulation (60 H_z AC, 10 V rms) according to the experimental design; rapid freezing in a pair of modified Wallenberger tongs precooled in liquid nitrogen; chemical extraction and analysis for PCr, total creatine (Ct) and adenine nucleotides. This preparation has a very low content of AMP and the highest ratio of PCr to Ct ever recorded. The ratio of PCr to ATP is much lower than in normal skeletal muscles. Even after a 60 sec tetanus and a 15 min additional incubation period the preparation made no significant amount of lactate.

This muscle is very properly called slow. At 18°C it uses energy 45 times slower than the frog sartorius muscle does at 0°C. It takes over 20 sec to develop its maximum tension. Taking a Q_{10} for frog sartorius of 2-5 this muscle is thus about 250 times faster in its usage of high-energy phosphate (~P) than the rabbit taenia coli at the same temperature (Butler *et al.*, 1977). This fits with the finding that there is about 7 times less myosin in the smooth muscle and its myosin ATPase is about 30-40 times slower. Interestingly, they both use about 0.35 μmoles ~P/g extra energy during the initial force development. The internal work against elastic elements is about 0.3 mcal/g, giving a thermodynamic efficiency of less than 10% for this work in both cases, based on 10,000 cal/mole ~P. For isotonic work sartorius can reach over 70% (Kushmerick and Davies, 1969).

Even more interestingly, in rabbit taenia coli but not in frog sartorius there is a net ATP usage at 10 sec. There is no significant change in PCr at this time, but the rate of PCr breakdown increases when there is net ATP resynthesis by 20-25 sec. The precise cause of this remains unexplained and is being investigated.

The maintenance of force requires three times less chemical energy than the development of that force, and, if stimulation is stopped the relaxation of that developed force is very slow (Butler *et al.*, 1978).

More force-times-time, or impulse, is maintained per ATP split during relaxation than during stimulation in a smooth tetanus where crossbridge cycling processes occur driven by ATP. This probably means that, during relaxation, active cross bridge cycling slows at a rate that exceeds the rate of increase of force maintained by the muscle. The active state, or ability to redevelop tension after a quick release also disappears quicker than the active force. This suggests that during relaxation crossbridges can transiently maintain force without ATP usage. Such an attached state is similar to that which is believed to exist in the resting muscle (Butler *et al.*, 1976).

Thus once again, the muscle without significant extra energy turnover, can maintain tension in cross bridges which slowly detach and reform until the muscle reaches the resting tension, appropriate to its new length, which is maintained by the connective tissue and elastic elements. However, stiffness remains in the contractile apparatus that is both length and calcium dependent.

Smooth muscle is full of surprises. John Marshall would have been fascinated. I am sorry he is not here, but then, if he were, we would be holding this symposium as an honor to him rather than as a memorial.

REFERENCES

Butler, T. M., Siegman, M. J., and Davies, R. E. (1976). *Am. J. Physiol. 231*, 1509-1514.

Butler, T. M., Siegman, M. J., Mooers, S. U., and Davies, R. E. (1977). *In* "Exitation-Contraction Coupling in Smooth Muscle" (R. Casteels, T. Godfraind, and J. C. Rüegg, eds.) pp. 463-469. Elsevier/North Holland Biomedical Press, Amsterdam.

Butler, T. M., Siegman, M. J., Mooers, S. U., and Davies, R. E. (1978). *Am. J. Physiol*, in press.

Cain, D. F., and Davies, R. E. (1962) *Biochem. Biophys. Res. Commun. 8*, 361-366.

Curtin, N. A., and Davies, R. E. (1973). *Cold Spring Harbor Symp. Quant. Biol. 37*, 619-626.

Davies, R. E. (1965). *In* "Essays in Biochemistry" (P. N. Campbell and G. D. Greville, eds.), Vol 1, pp. 29-55.

Davies, R. E., Kushmerick, M. J., and Larson, R. E. (1967). *Nature 214*, 148-151.

Hill, A. V. (1950). *Biochim. Biophys. Acta 4*, 4-11.

Hill, A. V. (1966). *Biochem. Z. 345*, 1-8.

Infante, A. A., Klaupiks, D., and Davies, R. E. (1964a). *Nature 201*, 620.

Infante, A. A., Klaupiks, D., and Davies, R. E. (1964b). *Science 144*, 1577-1578.

Kuby, S. A., and Mahowald, T. A. (1959). *Fed. Proc. 18*, 267.

Kushmerick, M. J., and Davies, R. E. (1968). *Biochim. Biophys. Acta*, 279-287.

Kushmerick, M. J., and Davies, R. E. (1959). *Proc. Roy. Soc. B 174*, 315-353.

Kushmerick, M. J., Larson, R. E., and Davies, R. E. (1969). *Proc. Roy. Soc. B 174*, 293-313.

Nauss, K. M., and Davies, R. E. (1966a). *J. Biol. Chem. 241*, 2918-2922.

Nauss, K. M., and Davies, R. E. (1966b). *Biochem. Z. 345*, 173-187.

Siegman, M. J., Butler, T. M., Mooers, S. U., and Davies, R. E. (1976). *Am. J. Physiol. 231*, 1501-1508.

Motility in Cell Function
Proceedings of the First John M. Marshall Symposium in Cell Biology

THE NATURE OF CATCH AND ITS CONTROL

Betty M. Twarog

Department of Anatomical Sciences
Health Sciences Center
State University of New York at Stony Brook
Stony Brook, New York

INTRODUCTION

This paper presents a brief review of recent studies on catch muscles, and attempts to put in perspective and summarize the general implications of these studies; it will also propose a new hypothesis on the nature of catch and of the interactions that control the turning-on and turning-off of the catch state by neural stimuli. Some of these matters have been discussed in two recent reviews (Twarog, 1976, 1977). Other work is contained in publications by Marchand-Dumont and Baguet (1975), Köhler *et al.* (1976), Köhler and Lindl (1977), Achazi (1974, 1976, 1977), Achazi *et al.* (1974), Muneoka *et al.* (1977), Hidaka *et al.* (1977), and Twarog *et al.* (1977).

THE NATURE OF CATCH

Evidence has been accumulating that stiffness (resistance to stretch) in an apparent resting state in smooth muscles may be maintained by attached cross bridges that do not cycle (Butler *et al.*, 1976; Siegman *et al.*, 1977). In molluscan catch muscles, stiffness is sustained for long periods after the active state has ceased. This stiffness is termed catch, and is measured as persistent tension after excitation has terminated. The evidence is strong that in catch the actin-myosin cross bridges remain attached and cycle very slowly, if at all. This explanation of the molecular basis of catch was first suggested by Lowy *et al.* (1964). Catch muscles are pos-

ISBN 0-12-551750-5

sibly displaying a property seen less dramatically in other smooth muscles and in striated muscles: that of resting cross-bridge attachment.

A unique feature of catch muscles is the existence of a specialized control system that turns off catch without inhibiting contraction; i.e., relaxing nerves release serotonin to terminate catch tension, but do not inhibit the development of tension during the active state. Bridge attachment, i.e., actin-myosin interaction, can be controlled independently of cycling and of actomyosin ATPase activity.

Intracellular Ca^{2+} concentration controls activation and tension development, but catch is sustained when intracellular Ca^{2+} returns to resting levels after excitation (Atsumi and Sugi, 1976). Catch is independent of Ca^{2+} in the external bathing medium.

Paramyosin is now known to be found in virtually all thick filaments in varying amounts. (Vertebrate smooth and striated muscles, however, lack this protein.) Muscles that have the property of catch also have very high proportions of paramyosin (Rüegg, 1961; Szent-Györgyi *et al.*, 1971). In molluscan catch muscle, it forms a core on which myosin molecules are arrayed to form thick filaments (Szent-Györgyi *et al.*, 1971). The functional role of paramyosin has not been established. Clearly, it is widely distributed among muscles, only a few of which display the phenomenon of catch (Elfvin *et al.*, 1976; Levine *et al.*, 1976).

Several functions have been hypothesized for paramyosin. Two groups of workers conclude that interaction of paramyosin in the cores of adjacent thick filaments maintains catch tension, independent of the actin-myosin interaction by which tension is developed during the active state (Rüegg, 1971; Johnson, 1962). That paramyosin interaction is the mechanical basis of catch tension has not been completely ruled out, but it is very improbable on several grounds, including the molecular structure of the thick filaments (Szent-Györgyi *et al.*, 1971; Nonomura, 1974) and the physiological properties of the muscle (Sugi and Yamaguchi, 1976). Others have suggested that the role of paramyosin in catch muscles is structural, and thus makes possible the very long, thick filaments characteristic of mollusc muscles (30 μm in length and 40-60 nm in diameter). The length of the filaments accounts for the exceptionally large tensions developed by catch muscles (up to 15 kg/cm^2). A third possible function of paramyosin is in the control system that turns catch on and off. Several workers (Szent-Györgyi *et al.*, 1971; Epstein *et al.*, 1975; Achazi, 1977) have made observations that strongly suggest a regulatory role for paramyosin in the relaxation of catch. Nonomura (1974), working with paramyosin in formed filaments rather than in solution, has obtained different results, but his observations are not inconsistent with a role of paramyosin in catch control mechanisms.

STIMULI THAT DETERMINE WHETHER CONTRACTION IS PHASIC OR CATCH

What is known of the stimuli that determine whether phasic contraction, catch contraction, or relaxation will occur is summarized in Fig. 1. When the nerve innervating the byssus retractor muscle of *Mytilus* (a molluscan catch muscle) is stimulated, phasic contraction results. This nerve is "mixed," i.e., it contains excitatory and relaxing nerve fibers. If an antagonist is present to block the relaxing mediator serotonin, neural stimulation will result in a catch contraction. The excitatory mediator acetylcholine, when applied alone, results in a catch contraction; however, when serotonin is simultaneously present, a phasic contraction results. Prolonged pulses of direct current (dc) will also result in a catch con-

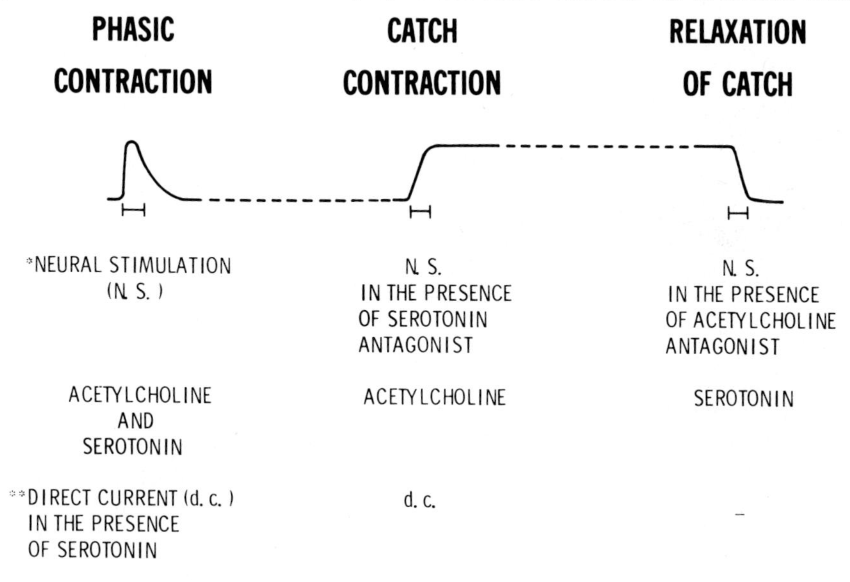

*STIMULATION OF ABRM NERVE (MIXED EXCITATORY AND RELAXING NERVE FIBERS) VIA SUCTION ELECTRODE, OR STIMULATION OF INTRAMUSCULAR, MIXED NERVE FIBERS BY REPETITIVE PULSES DIRECTLY APPLIED TO MUSCLE (1 TO 5 msec DURATION; 10/sec)

**DIRECT CURRENT (d. c.): CATHODAL PULSES 10 sec OR LONGER IN DURATION

FIGURE 1. The effects of various stimuli on triggering of phasic contraction, catch contraction, and relaxation.

traction, unless serotonin is present. These pulses presumably act directly on the muscle membrane, depolarizing it. Apparently, all excitatory stimuli applied to catch muscle result in a catch contraction unless the relaxing nerve fibers are simultaneously stimulated, or unless the relaxing mediator serotonin is present at the time of the excitatory stimulation (Twarog *et al.*, 1977).

EXCITATION-CONTRACTION COUPLING IN CATCH AND PHASIC CONTRACTION

It can now be stated that in all instances where excitatory stimuli produce a catch contraction, the activating calcium is released from internal stores. What is now known concerning excitation-contraction coupling in catch muscle is summarized in Fig. 2. Acetylcholine acts on a receptor that can be blocked by d-tubocurarine or propantheline (Twarog, 1959). When acetylcholine activates the receptors, the membrane conductance

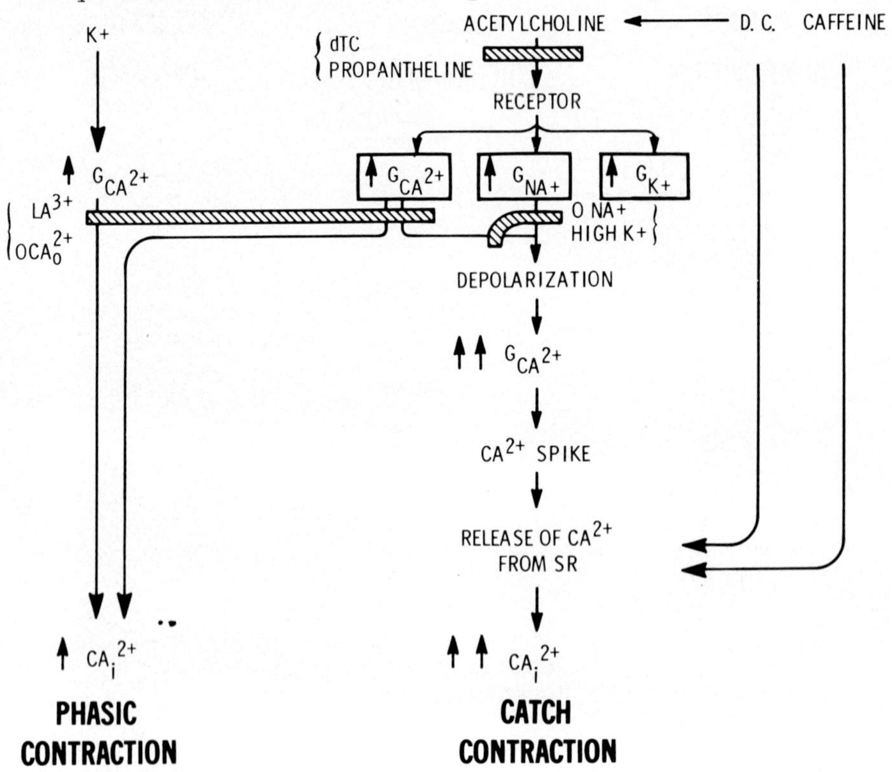

FIGURE 2. Mechanisms of excitation-contraction coupling in catch muscle. CA_o^{2+}: Calcium ion in external bathing medium; Ca_i^{2+}: Intracellular calcium ion; [hatched bar]: Block

is increased to Ca^{2+}, Na^{+}, and K^{+}. A depolarization results (Twarog, 1954; Muneoka, 1966; Hidaka and Twarog, 1977), which, when it reaches a critical level, results in further increase in Ca^{2+} conductance and the triggering of a Ca^{2+}-dependent action potential (Twarog and Muneoka, 1973). The firing of the spike apparently releases further Ca^{2+} from internal stores (Sugi and Yamaguchi, 1976), which probably include elements of the sarcoplasmic reticulum as well as other intracellular organelles. Catch then ensues. Two other forms of stimulation result in catch contraction. At the onset of cathodal dc pulses, mediators are transiently released from intramuscular nerve endings, but the primary effect of dc pulses is on the muscle fiber membrane, at which site they cause depolarization and consequent release of Ca^{2+} from internal stores. Caffeine, at concentrations of 5 m*M* and higher, produces a catch contraction by stimulating release of Ca^{2+} from internal sites.

Catch tension is essentially independent of Ca^{2+} concentration in the external medium, although in some cases calcium must be present in the external bathing solution to permit the initial depolarization that triggers contraction. In all cases, however, the tension maximum of the catch contraction is not proportional to Ca^{2+} in the external bathing medium, and relaxation does not ensue when Ca^{2+} is removed from this medium, even in the presence of 10^{-3} *M* EGTA, a calcium-chelating agent (Twarog, 1967). As mentioned above, Atsumi and Sugi (1976) have shown that in catch muscle, Ca^{2+} distribution appears identical in the resting and catch states.

Phasic contractions under all circumstances result from stimuli that trigger Ca^{2+} entry across the outer cell membrane (see Fig. 2). Physiological solution high in potassium stimulates a phasic contraction, not catch. Hagiwara and Nagai (1970) have demonstrated net influx of Ca^{2+} during K^{+} contraction, and studies with La^{3+} show that this contraction requires Ca^{2+} entry across the muscle membrane (Muneoka and Twarog, 1977). If acetylcholine is applied to a muscle under conditions where depolarization is prevented, e.g., when Na^{+} is omitted from the bathing solution, or if it is high in K^{+}, then addition of Ca^{2+} to the external medium triggers phasic contraction. The contraction tension is proportional to the Ca^{2+} concentration and relaxes when the Ca^{2+} is removed. Thus, phasic tension, unlike catch tension, depends on the presence of Ca^{2+} in the external bathing solution; the tension is proportional to the log of such Ca^{2+}. When Ca^{2+} in the external solution is removed, the tension falls. Quick-release studies have shown that the muscle undergoing phasic contraction, unlike catch, is in the active state (Jewell, 1959; Johnson and Twarog, 1960), and this active-state phasic contraction, in contrast to catch, apparently depends on Ca^{2+} which enters through the membrane.

What ultimately determines whether a contraction will be catch or phasic? A diagram showing how catch may be established is presented in Fig. 3. The threshold for activation of either type of contraction is at pCa 7 and activation is maximum at pCa 6 (Baguet, 1973; Baguet and Marchand-Dumont, 1975; Nichols, 1975). The level of internal Ca^{2+} at the time of excitation may control whether catch will be established. According to this idea, when internal Ca^{2+} levels are higher than those required for maximum activation, catch is established. The upper half of Fig. 3 shows that there may be a Ca^{2+} threshold *below* which the activating Ca^{2+} turns on *phasic* contraction, and *above* which the Ca^{2+} released stimulates *catch*. Calcium ion levels would remain below this threshold when Ca^{2+} enters from the external medium, but would reach catch threshold when Ca^{2+} is released from internal stores.

Several lines of evidence suggest that the high levels of internal Ca^{2+} that establish catch may do so by inhibiting relaxation. First, there is the observation, mentioned above, that all excitatory stimuli cause catch in the absence of simultaneous activation of the relaxing system (Twarog *et al.*,

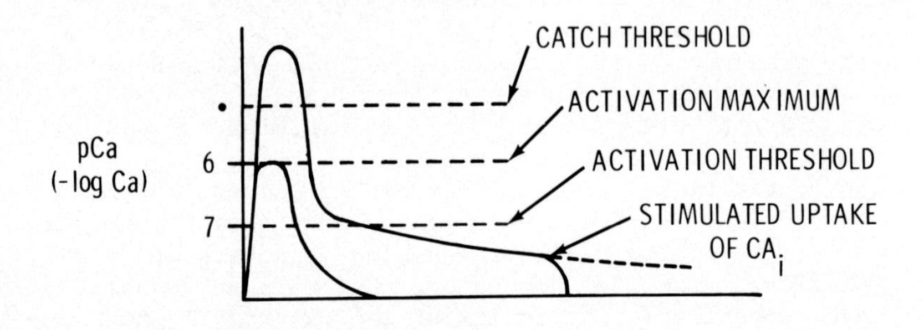

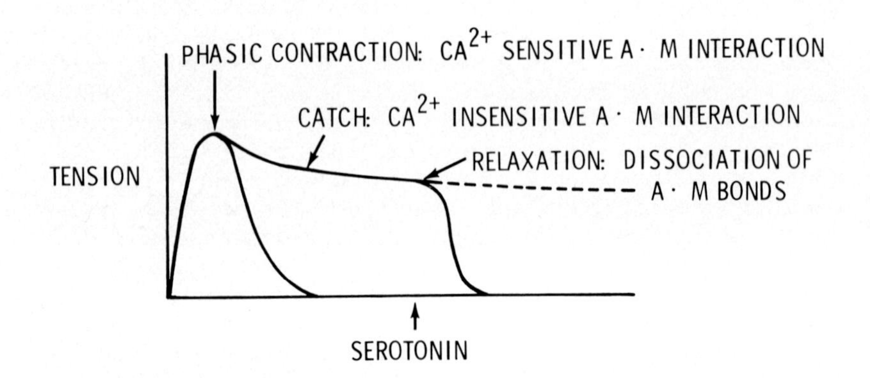

FIGURE 3. Phasic or catch contraction: A function of the intracellular calcium ion concentration during activation?

1977). Second, experiments with lanthanum (La^{3+}) demonstrate that La^{3+} (a Ca^{2+} antagonist) blocks relaxation *only* when it penetrates beyond superficial sites in the muscle. The blockage of relaxation by penetration of La^{3+} to sites beyond the superficial takes place quite independently of its combination with sites on the fiber membrane and consequent block of Ca^{2+} influx from the external medium.

THE RELAXING SYSTEM: CYCLIC-AMP, PARAMYOSIN PHOSPHORYLATION, AND CALCIUM ION

Pharmacological and biochemical studies suggest that relaxation by serotonin involves a second messenger, c-AMP (Cole and Twarog, 1972; Achazi *et al.*, 1974; Köhler and Lindl, 1977). Both Achazi and Köhler have shown a significant increase in c-AMP preceding relaxation in response to serotonin. However, Köhler has pointed out that the time course of the increase in c-AMP needs more precise study and that its pharmacology must be more carefully related to the corresponding pharmacology of relaxation by serotonin. Achazi found that the serotonin-induced increase in c-AMP appears to be inhibited by Ca^{2+}; the formation of c-AMP is greatest when internal Ca^{2+} levels are low.

In an interesting series of experiments that complement those of Köhler and Achazi, Marchand-Dumont and Baguet (1975) reported observations that in chemically-skinned catch muscle bundles, both serotonin and c-AMP caused relaxation of sustained tension, whereas in the intact fiber, only serotonin effected relaxation. They concluded that in the skinned muscle, the c-AMP reaches sites that are inaccessible when membrane permeability is normal. Neither serotonin nor c-AMP affect tension in glycerinated bundles; this fact implies that their relaxing properties do not directly affect the contractile elements.

Phosphorylation in various fractions of catch muscle was sought following relaxation by serotonin. Köhler and associates (1976) examined the microsomal fraction, because it has frequently been suggested that catch may depend on the uptake of intracellular calcium following activation. They found evidence for phosphorylation of an unidentified protein with a high molecular weight. Achazi (1976,1977), on the other hand, examined the contractile protein fraction and found that two proteins are phosphorylated during serotonin-induced relaxation, one having a weight of 300,000 D, the other of 106,000 D. He identified the lighter protein as paramyosin and went on to show that its phosphorylation is effected by a c-AMP-dependent kinase. Szent-Györgyi *et al.* (1971) and Epstein *et al.* (1975) had previously investigated inhibition by paramyosin of actomyosin ATPase activity; Achazi determined the effect of paramyosin phosphorylation on this interaction. The former workers had observed that paramyosin inhibited actomyosin ATPase in proportion to the paramyosin:myosin ratio. Achazi found that the phosphorylated paramyosin more powerfully inhibited the actomyosin ATPase than did its unphosphorylated counterpart. On the basis of these results, Achazi proposed that relaxation by serotonin may be due to paramyosin phosphorylation and the consequent inhibition of actin-myosin interaction. He carefully pointed out the difficulties inherent in this suggestion.

It is useful, for the purposes of this paper, to speculate further on the mechanism by which the phosphorylation of paramyosin could result in relaxation of catch. While the experiments of Szent-Györgyi *et al.*, Epstein *et al.*, and Achazi showed that addition of paramyosin to myosin in solution can inhibit actomyosin ATPase, this demonstration cannot be taken directly as a model of the relaxing action of paramyosin because in catch there is no cycling and the actomyosin ATPase activity is low, similar to that in the resting muscle. Rather, the utility of these observations as a model lies in their demonstration that the influence of paramyosin on ATPase activity occurs by way of its interference with actin-myosin interaction. Marston and Taylor (1977) presented some results of studies on chicken gizzard smooth muscle. They demonstrated that the lifetime of attached bridges is very long in this muscle, and proposed that the rate-limiting step in contraction may well be the dissociation of actin and myosin.

Similarly, a scheme could be formulated for catch muscles in which the rate-limiting actin-myosin dissociation is affected by the phosphorylation of paramyosin. Since the myosin filaments in catch muscle have a paramyosin core, it is within the realm of possibility that rearrangements of the core could affect the conformation of the myosin molecules on the surface in a way that could regulate myosin-actin interaction as first suggested by Szent-Györgyi *et al.* (1971).

CONTROL SYSTEMS OF CATCH AND RELAXATION

A scheme can be pictured wherein the primary relaxing effect of serotonin is its action on the contractile elements--mediated by c-AMP--via the control function of phosphorylated paramyosin; in this scheme Ca^{2+} serves to modulate the relaxing effect. The scheme is incorporated in Fig. 4. Stimulation of the excitatory nerve causes a greatly-increased level of intracellular Ca^{2+}, which activates the contractile elements and simultaneously inhibits the adenylate cyclase activity that generates c-AMP. Thus, the activating Ca^{2+}

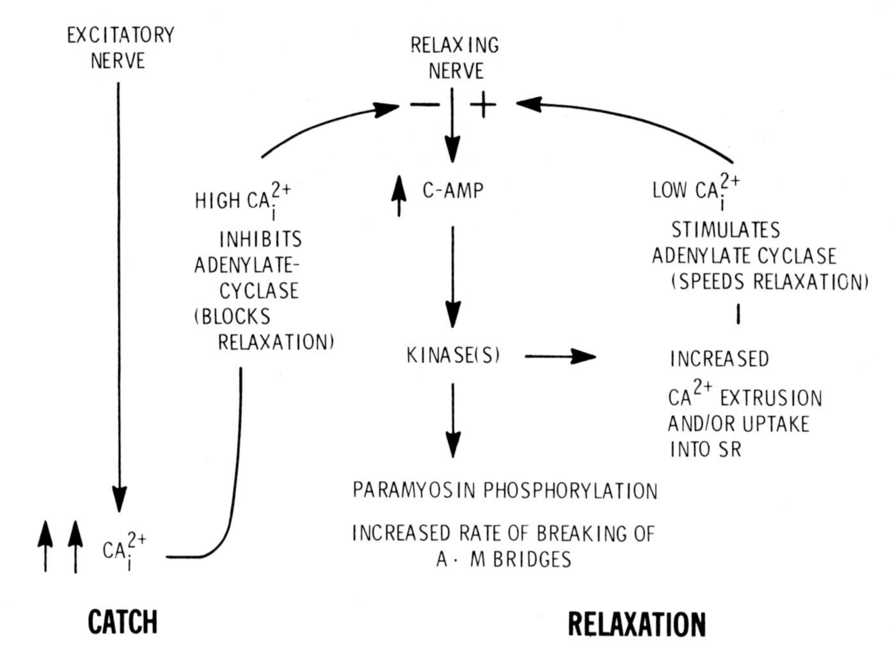

FIGURE 4. Activation of catch and control systems of its relaxation. Ca_i^{2+}: Intracellular calcium ion; A·M: Actomyosin; SR: Sarcoplasmic reticulum.

blocks the relaxing system and actin-myosin interaction persists. When excitation ceases, the Ca^{2+} concentration falls to resting levels. There is no further cycling of ATPase activity, but the actin-myosin interaction still persists. This state is catch. When the relaxing nerve is stimulated during catch, the Ca^{2+} levels have already returned to the resting level and the released serotonin triggers a significant increase in c-AMP. The c-AMP-dependent kinase activity demonstrated by Achazi then leads to phosphorylation of paramyosin and--by a mechanism that will probably have to be elucidated by structural methods--causes an increased rate of breaking of bridges. Other kinases, which lead to phosphorylation of proteins in the microsomal fraction, may cause increased Ca^{2+} uptake into internal stores or increased Ca^{2+} extrusion. The lowered intracellular Ca^{2+} concentrations that would result are not a primary control in relaxation of catch, but this system provides positive feedback: the action of adenylate cyclase in increasing c-AMP is enhanced by c-AMP stimulation of Ca^{2+} uptake.

ACKNOWLEDGMENTS

The author's research has been supported by a grant from the U.S.P.H.S. (NS 12857). Thanks are due to Dr. R. K. Achazi for providing a copy of his Habilitationsschrift, and for discussions of his observations.

REFERENCES

Achazi, R. K. (1974). *Z. Naturforsch. 29,* 451-452.
Achazi, R. K. (1976). *Verh. Dtsch Zool. Ges. 1976,* 90.
Achazi, R. K. (1977). Zür 5-HT-induzierten Erschlaffung von glatten Mollusken-muskeln. Habilitationsschrift, Westfälischen Wilhelms-Universität, Münster.
Achazi, R. K., Dolling, B., and Haakshorst, R. (1974). *Pflügers. Arch. 349,* 19-27.
Atsumi, S., and Sugi, H. (1976). *J. Physiol. 257,* 549-560.
Baguet, F. (1973). *Pflügers Arch. 340,* 19-34.
Baguet, F., and Marchand-Dumont, G. (1975). *Pflügers Arch. 354,* 75-85.
Butler, T. M., Siegman, M. J., and Davies, R. E. (1976). *Am. J. Physiol. 231,* 1509-1514.
Cole, R. A., and Twarog, B. M. (1972). *Comp. Biochem. Physiol. 42A,* 321-330.
Elfvin, M., Levine, R. J. C., and Dewey, M. M. (1976). *J. Cell Biol. 71,* 261-272.
Epstein, H. F., Aronow, B. U., and Harris, H. E. (1975). *J. Supramolecular Structure 3,* 354-360.
Hagiwara, E., and Nagai, T. (1970). *Japan J. Physiol. 20,* 72-83.
Hidaka, T., and Twarog, B. M. (1977). *Gen. Pharmacol. 8,* 83-86.
Hidaka, T., Yamaguchi, H., Twarog, B. M., and Muneoka, Y. (1977). *Gen. Pharmacol. 8,* 87-91.
Jewell, B. R. (1959). *J. Physiol. 149,* 154-177.
Johnson, W. H. (1962). *Physiol. Rev. 42 (Suppl. 5),* 113-143.
Johnson, W. H., and Twarog, B. M. (1960). *J. Gen. Physiol. 43,* 941-960.
Köhler, G., and Lindl, T. (1977). *Proc. I.U.P.S. 13,* Abst. 1170.
Köhler, G., Heilmann, C., Nickel, E., and Florey, E. (1976). *Pflügers Arch. 365,* Abst.
Levine, R. J. C., Elfvin, M., Dewey, M. M., and Walcott, B. (1976). *J. Cell Biol. 71,* 273-279.
Lowy, J., Millman, B. M., and Hanson, J. (1964). *Proc. Roy. Soc. B. 160,* 525-536.

Marchand-Dumont, G., and Baguet, F. (1975). *Pflügers Arch. 354,* 87-100.
Marston, S. B., and Taylor, E. W. (1978). *F.E.B.S. Letters 86,* 167-168.
Muneoka, Y. (1966). *Bull. Biol. Soc. Hiroshima Univ. 33,* 25-33.
Muneoka, Y., and Twarog, B. M. (1977). *J. Pharmacol. Exptl. Therap. 202,* 601-609.
Muneoka, Y., Cottrell, G. A., and Twarog, B. M. (1977). *Gen. Pharmacol. 8,* 93-96.
Nichols, M. R. (1975). Calcium and contraction in a molluscan smooth muscle. PhD. dissertation, Tufts University, Medford, Massachusetts.
Nonomura, Y. (1974). *J. Mol. Biol. 88,* 445-455.
Ruegg, J. C. (1961). *Proc. Roy. Soc., B, 154,* 224-249.
Ruegg, J. C. (1971). *Physiol. Rev. 51,* 201-248.
Siegman, M. J., Butler, T. M., Mooers, S. U., and Davies, R. E. (1977). *In* "Excitation-Contraction Coupling in Smooth Muscle" (R. Casteels *et al.*, eds.), pp. 449-453. Elsevier/North-Holland Press, Amsterdam.
Sugi, H., and Yamaguchi, T. (1976). *J. Physiol. 257,* 531-548.
Szent-Györgyi, A. G., Cohen, C., and Kendrick-Jones, J. (1971). *J. Mol. Biol. 56,* 239-258.
Twarog, B. M. (1954). *J. Cell. Comp. Physiol. 44,* 141-163.
Twarog, B. M. (1959). *Brit. J. Pharmacol. 14,* 555-558.
Twarog, B. M. (1967). *J. Physiol. 192,* 847-856.
Twarog, B. M. (1976). *Physiol. Rev. 56,* 829-838.
Twarog, B. M. (1977). *In* "Excitation-Contraction Coupling in Smooth Muscle" (R. Casteels *et al.*, eds.), pp. 261-271. Elsevier/North-Holland Press, Amsterdam.
Twarog, B. M., and Muneoka, Y. (1973). *Cold Spring Harbor Symp. Quant. Biol. 37,* 489-504.
Twarog, B. M., Muneoka, Y., and Ledgere, M. (1977). *J. Pharmacol. Exptl. Therap. 201,* 350-356.

Motility in Cell Function
Proceedings of the First John M. Marshall Symposium in Cell Biology

OBSERVATIONS ON INTERMEDIATE-SIZED, 100Å FILAMENTS

G. S. Bennett
J. M. Croop
J. J. Otto[*]
S. A. Fellini
Y. Toyama
and
H. Holtzer

Department of Anatomy, School of Medicine,
and Department of Biology,[*]
University of Pennsylvania
Philadelphia, Pennsylvania

INTRODUCTION

Ishikawa *et al.* (1968,1969) described two distinct sets of filaments present in surprisingly large numbers in cells as diverse as myoblasts, chondroblasts, neuroblasts, skin cells, and intestinal epithelial cells. One set, approximately 60 Å in diameter, subtended the cell membrane; the other, approximately 100 Å in diameter, was distributed throughout the cytoplasm.

The finer filaments, often difficult to resolve in electron micrographs of untreated cells, were rendered most conspicuous as arrow-head complexes following decoration with heavy meromyosin. These filaments were termed actinlike microfilaments. Many investigators have attributed a contractile function to these microfilaments (Goldman *et al.*, 1976). Nevertheless, details remain obscure regarding their composition, assembly, disassembly, and role in cell motility. For example, Cytochalasin-B (CB), while fragmenting the microfilaments, promptly induced contraction and arborization in nonmyogenic and precursor myogenic cells. In contrast, CB had no detectable effect on the stability or contractile activity of the actin filaments in myoblasts or myotubes (Sanger *et al.*, 1971,

ISBN 0-12-551750-5

Croop and Holtzer, 1975, Holtzer *et al.*, 1975). These and related observations (Weber *et al.*, 1976) are consistent with the notion that the changes in cell shape induced by CB result from dismantling the microfilaments, which constitute a skeletal, rather than a contractile, system.

It is, however, the set of filaments termed intermediate-sized filaments that will be reviewed here (Ishikawa *et al.*, 1968). These 100 Å filaments have been described in many types of normal embryonic and mature cells. They vary in number and intracellular distribution depending on cell type, and whether the cell had been exposed to Colcemid (Col). In mature smooth muscle (Uehara *et al.*, 1971, Cooke, 1976, Small and Sobieszek, 1977), nerve processes (Schlaepfer, 1977), astrocytes (Peters and Vaughn, 1967), melanocytes (Moellmann and McGuire, 1975), bone cells (Holtrup, 1974), hemopoetic cells (DePetris *et al.*, 1962), and endothelial cells (Blose *et al.*, 1977), 100 Å filaments give the impression of being relatively stable and ordered structures. Although the degree of clustering and center-to-center spacing varies enormously depending on cell type, nevertheless within each cell type they assume a characteristic pattern. In the above cell types the 100 Å filaments course parallel to one another and appear to maintain some constant relationship to the overall geometry of the cell. In contrast, however, the number and arrangement of filaments varies considerably in cells whose shape changes, such as cultured fibroblasts, chondroblasts, and smooth muscle cells. The filaments in these cell types are often isolated, randomly oriented, and vary greatly in number (Holtzer *et al.*, 1970). This great variability in total number and lack of obvious correlation with cell geometry suggests that under some conditions the 100 Å filaments may be labile and that their polymerization and depolymerization may depend on the physiological state of the cell.

CYTOCHALASIN-B AND COLCEMID AND THE INDUCTION OF CABLE OF 100 Å FILAMENTS

Evidence for possible lability of the 100 Å filaments was first shown by exposing myotubes (Ishikawa *et al.*, 1968) and nerve cells (Wisniewski *et al.*, 1968) to colcemid or vinblastine. Both types of cells responded by augmenting the numbers of intermediate-sized filaments readily detectable by electron microscopy. Goldman and Knipe (1973) and Blose *et al.* (1977) have also reported that treating BHK or endothelial cells with Col induces the formation of justa-nuclear caps of 100 Å filaments, and that as rounded BHK cells settle and spread, the filaments of the juxta-nuclear cap were redistributed throughout the cytoplasm.

A fuller appreciation of the 1) ubiquity, 2) quantity, and 3) lability of the 100 Å filaments has recently been provided by observing living cells sequentially treated with Cytochalsin-B (CB) and then Col (Croop and Holtzer, 1975). Phase-lucent, convoluted bands of the kind illustrated in Fig. 1 have been induced in many kinds of embryonic cells and in RSV-transformed myogenic, chondrogenic and melanogenic cells. Figures 2-4 are micrographs of CB-Col treated chick embryonic fibroblasts observed under the scanning and transmission microscopes. The phase-lucent bands (Fig. 1) consist of enormous numbers of aggregated 100 Å filaments (Holtzer *et al.*, 1975, 1976). The filaments in these cables may merge with the skein of microfilaments subtending the plasma membrane; in other places they appear to insert directly into the plasma membrane (Fig. 4). A rough, but conservative, estimate suggests that following CB-Col 20% of the total volume of the cell may be occupied by these cables (Croop and Holtzer, 1975). Cable formation is completely reversible by returning the cells to normal medium.

The respective role of CB and Col in inducing the cables is unclear. Col alone induces modest-sized cables. However, pretreatment with CB gives the impression of inducing them more rapidly, and of enhancing their number, though this is

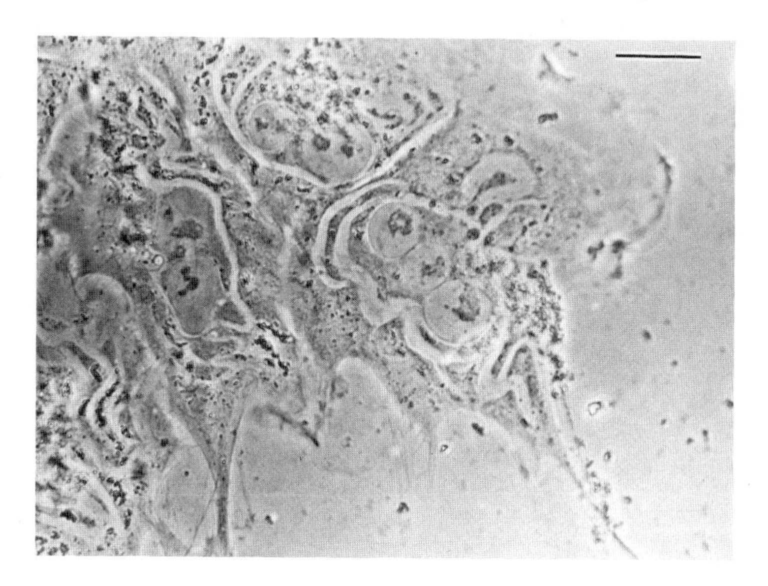

FIGURE 1. Phase micrograph of living fibroblasts after 2 days exposure to 5μg/ml Cytochalsin-B followed by 2 days in 10^{-6}M Colcemid. Observe the twisting phase-lucent cables. Bar = 25 μM.

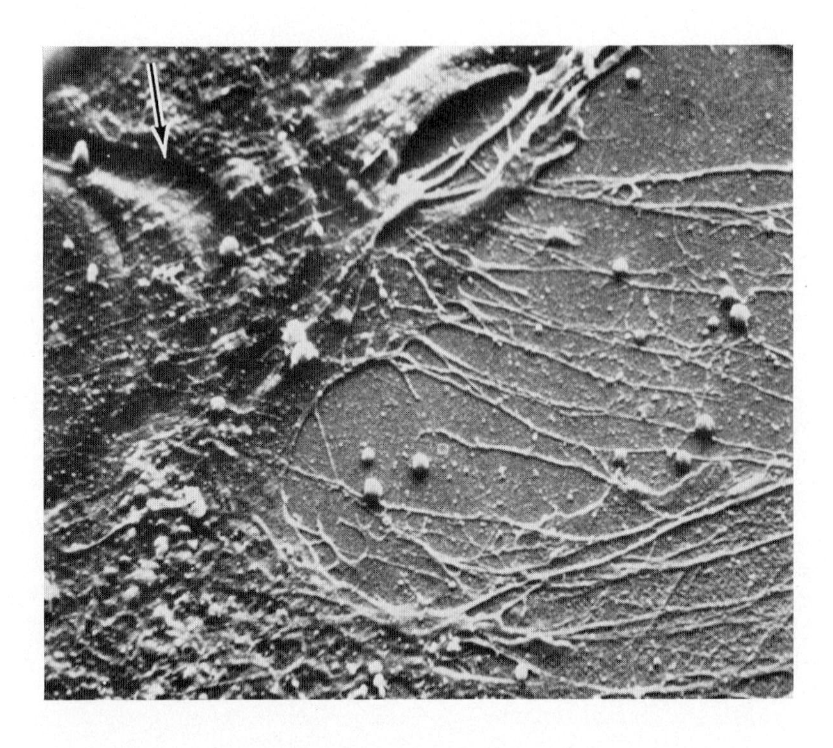

FIGURE 2. Scanning electron micrograph of a cell similar to that in Fig. 1. The location of a portion of the cable is indicated by the arrow. Long processes and microspikes extend from the cell (Gershon and Holtzer, unpublished data).

yet to be proven. The assembly of the cables is not significantly retarded by inhibitors of phosphorylation, nor by inhibitors of protein synthesis, when present during exposure to Col.

The cables in living cells have been observed to shift their position, but whether this is due to passive displacement, or to inherent motility is still unclear. It is also not yet clear whether Col only induces the *aggregation* of already existing 100 Å filaments, or stimulates the *assembly* from a pool of free subunits into filaments as well. Gross aggregation into cables, as well as assembly and disassembly, could generate considerable local currents on a molecular level. To the extent that these phenomena occur, though to a lesser degree, in untreated cells, they may be significant in changing or maintaining cell shape.

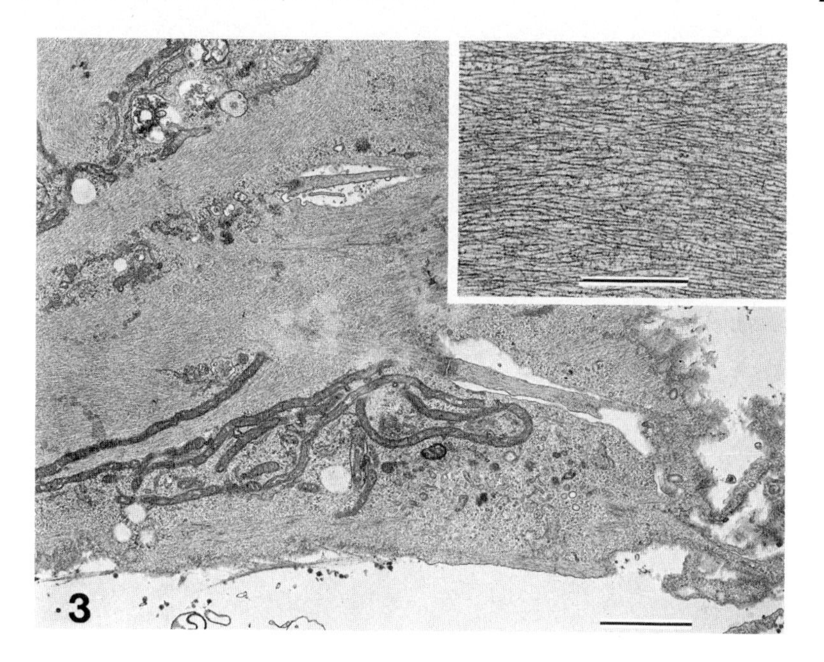

FIGURE 3. Electron micrograph of a CB-Col treated cell. Much of the "cytoplasm" of the cell consists largely of bands of closely packed 100 Å filaments. Cell organelles such as mitochondria, polysomes, lysosomes, etc., are excluded from the domains of these cables; such basic cell organelles are segregated into islands surrounded by the intermediate-sized filaments. The inset, at higher magnification, illustrates the arrangement of the filaments within the cables. Bar = 2 µM; inset: Bar = 0.5 µM.

The mechanism by which colcemid induces cable formation is open to speculation. It could involve either a direct action of Col on 100 Å filaments or be a consequence of the disassembly of microtubules. That these cables become most conspicuous after Col treatment suggests the presence in normal cells of an inhibitor to cable formation. One possibility is that a threshold value of tubulin subunits is required to inactivate the inhibitor. In untreated cells the concentration of tubulin subunits would be low owing to their sequestration in microtubules. Exposure to Col would augment the local concentration of subunits, thus relieving the block to aggregation into cables.

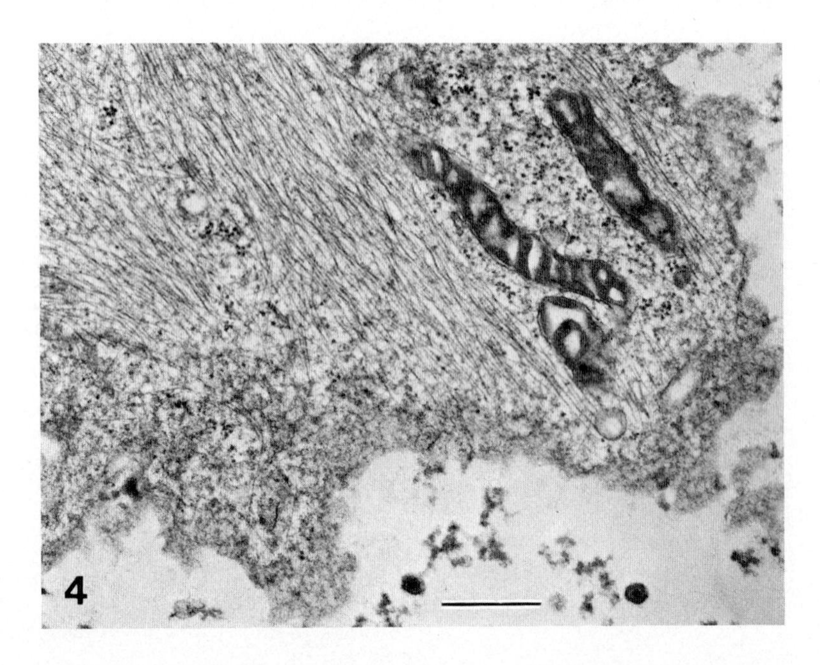

FIGURE 4. A tangential section through a peripheral portion of a CB-Col treated cell. Note that the subplasmalemmal region is devoid of 50-70 Å microfilaments, and that the 100 Å filaments closely approach the cell membrane. Bar = 0.5 μM.

LOCALIZATION OF ANTIBODIES TO A SUBUNIT OF FIBROBLAST 100 Å FILAMENTS

Cooke (1976) has reported that the major subunit of the 100 Å filaments from gizzard smooth muscle is a high salt insoluble protein with a molecular weight of approximately 55,000 (see, also, Holtzer *et al.*, 1976). We have isolated a comparable protein from cultured fibroblasts, using a procedure based on that of Brown *et al.* (1976) and have prepared antibodies against it. The protein, which proved to have a molecular weight of approximately 58,000, was cut from an SDS-polyacrylamide gel and injected into rabbits. These antibodies, hereafter referred to as "anti-58K," were used in the indirect immunofluorescence technique. Figure 5 is a micrograph of an untreated fibroblast "stained" with anti-58K IgG. One hundred percent of the cells exhibited numerous, gently curved, slightly irregular filaments averaging between 0.4 and 0.8 μm in diameter, but often over 20 μm in length. Similar filaments, but varying in numbers and in organization have been observed in cultured chondroblasts, myoblasts, gizzard

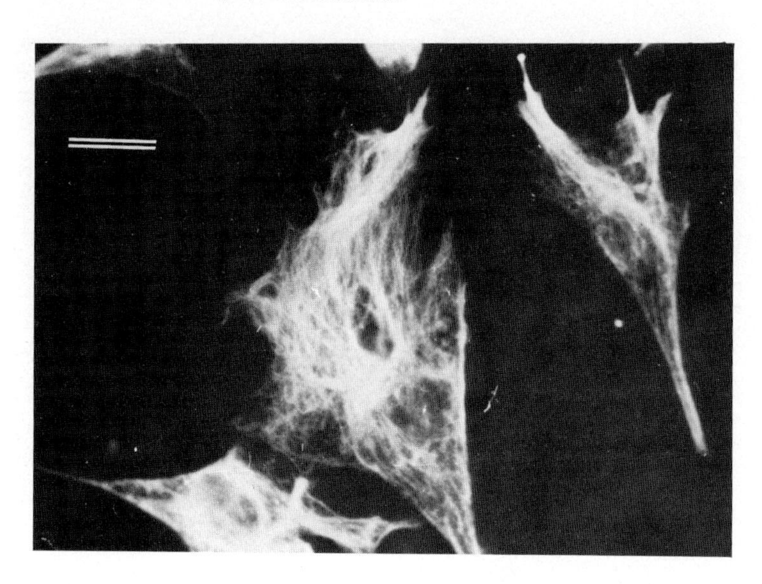

FIGURE 5. Fluorescence micrograph of untreated fibroblasts fixed in 2% formaldehyde followed by acetone at -20°C and incubated with anti-58K IgG (1 mg/ml) followed by fluorescein-labelled goat anti-rabbit IgG (1 mg/ml). Filamentous structures of varying thickness are unevenly distributed throughout the cytoplasm. Bar = 20 μM.

cells, pigment cells, and replicating "brain" cells.

Many different batches of control sera have been used: from unimmunized animals and from animals immunized with gizzard tropomyosin, skeletal muscle myosin, bovine serum albumin and actin (the last, prepared from chick lens cells, kindly supplied by Dr. H. Maisel, Wayne State University). In no instance did the control sera or IgG fraction stain the set of filaments illustrated in Fig. 5. On the other hand, different batches of control sera varied in their binding to nuclei and nucleoli, tended to stain the perinuclear region, and often bound to the "stress fibers" of stretched cells.

When cells that had been treated with CB-Col were fixed and treated with anti-58K all of them displayed the tortuous cables illustrated in Figs. 6 and 7. These twisting cables varied in pattern and displayed no particular relationship with the nucleus. The cytoplasm exhibited either a feltwork or diffuse fluorescence. The brightly fluorescent cables could not be resolved into individual filaments, but compared to the phase contrast visualization, the antibody staining demonstrated a more elaborate structure of thick cables merging with much finer ones.

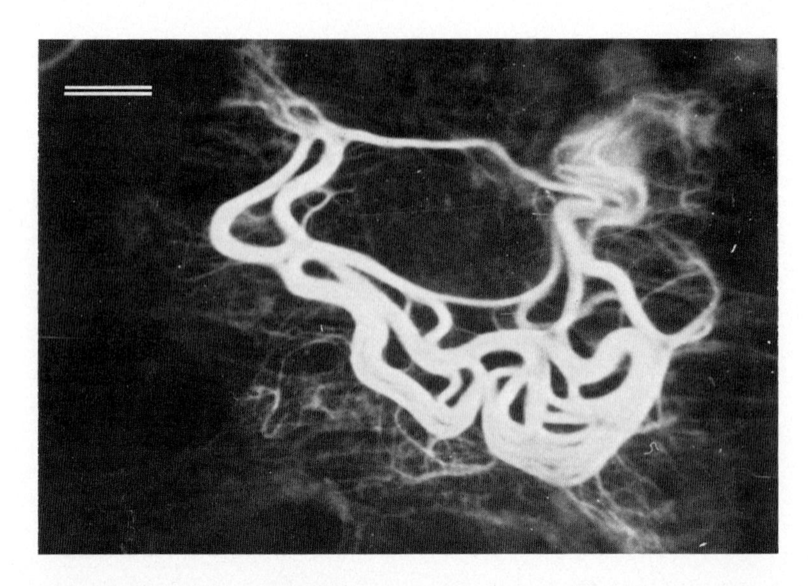

FIGURE 6. Fluorescence micrograph of a CB-Col treated cell fixed and incubated with anti-58K IgG as in Fig. 5. Note the extent of the brightly fluorescent cables. Bar = 20 μM.

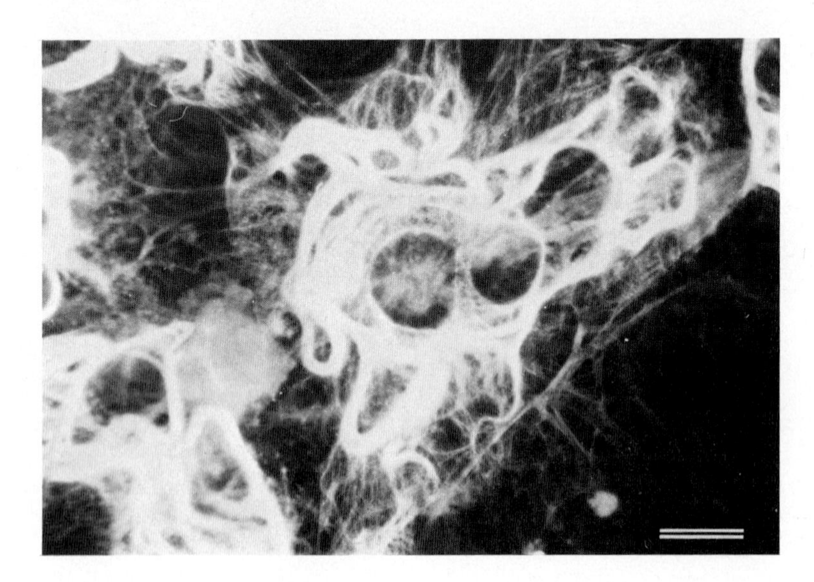

FIGURE 7. A group of cells treated as in Fig. 6, demonstrating the great variability in configuration of the cables as well as the positively stained microspikes and extracellular processes. Cell boundaries are difficult to discern. Bar = 20 μM.

The processes and microspikes bridging cells and remaining on the substrate severed from the cell body (footprints) (Fig. 7) are stained by a contaminating component in anti-58K, which is apparently directed against LETS protein (for a description of this cell surface protein, see Mautner and Hynes, 1977). This minor component developed after immunizing rabbits with 58K protein four or more times, despite the consistent use of electrophoretically purified 58K. A 58K proteolytic fragment of the ∿230K dalton LETS protein is probably responsible. Antisera obtained after less than 3 injections of 58K subunits generally did not contain the anti-LETS component, but the anti-58K titer was lower than that obtained after further immunization. Staining due to anti-58K and anti-LETS antibodies could be distinguished by fixation with formaldehyde alone, without the use of organic solvents. It has been shown that antibody cannot penetrate cells fixed in this manner (Stenman *et al.*, 1977). In this case, the cables remained unstained, and the footprints and processes were stained only by those batches of anti-58K that contained the anti-LETS component. In addition, the staining of intracellular filaments and cables was abolished by absorption of anti-58K with purified 55K 100 Å filament subunits from smooth muscle (Bennett *et al.*, 1978a) while the staining of processes and microspikes remained.

ANTIGENIC DIFFERENCES AMONG 100 Å FILAMENT SUBUNITS IN DIFFERENT CELL-TYPES

The finding that myosin heavy and light chains and tropomyosin in postmitotic myoblasts are products of different structural genes from those regulating the synthesis of the myosins in nonmyogenic cells (Holtzer *et al.*, 1973; Chi *et al.*, 1975a,b) encouraged us to ask whether the 100 Å filaments from different cell types were in fact identical or even related molecules. We have therefore compared the binding to several different cultured chick cell-types of anti-58K and of two other antisera we have prepared: 1) antiserum to the 55K dalton subunit of chick smooth muscle (gizzard) 100 Å filaments, "anti-55K" (Fellini *et al.*, 1978a,b; Bennett *et al.*, 1978b), and 2) antiserum to subunits of brain filaments, "anti-BF" (Bennett *et al.*, 1978b).

Anti-58K stained a filament network in a variety of different cells in addition to fibroblasts: chondroblasts, pigment cells, smooth muscle, cardiac cells, replicating presumptive skeletal myoblasts, skeletal myotubes, and all types of nonneural cells in cultures of central and peripheral nervous system. This antibody did not stain chick skin epithelial cells, or chick hepatocytes. It is worth stressing that the anti-58K does not stain the native 100 Å filaments or the CB-Col induced cables of 100 Å filaments in mouse or human cells. Curiously, the anti-58K does stain the stress fibers in PtK2 cells and in human WI-38 fibroblasts (see below).

Anti-55K stained abundant filaments and filament bundles in skeletal myotubes, smooth muscle and cardiac cells, and also in peripheral nervous system Schwann cells. Fibroblasts, replicating presumptive myoblasts, chondroblasts, pigment cells, hepatocytes, neurons or central nervous system glia were not stained. An example of the specificity of anti-55K is shown in Fig. 8. The postmitotic myoblasts and multinucleated myotubes are brightly stained, while the mononucleated cells, including fibroblasts and replicating presumptive myoblasts are negative (Fellini *et al.*, 1978a).

Staining of skeletal myogenic cultures with anti-55K and anti-58K demonstrated a particularly striking and novel sequence of events. The absence of 55K subunits from replicating presumptive myoblasts and the rapid accumulation of these subunits in postmitotic myoblasts and immature myotubes (Fig. 8)

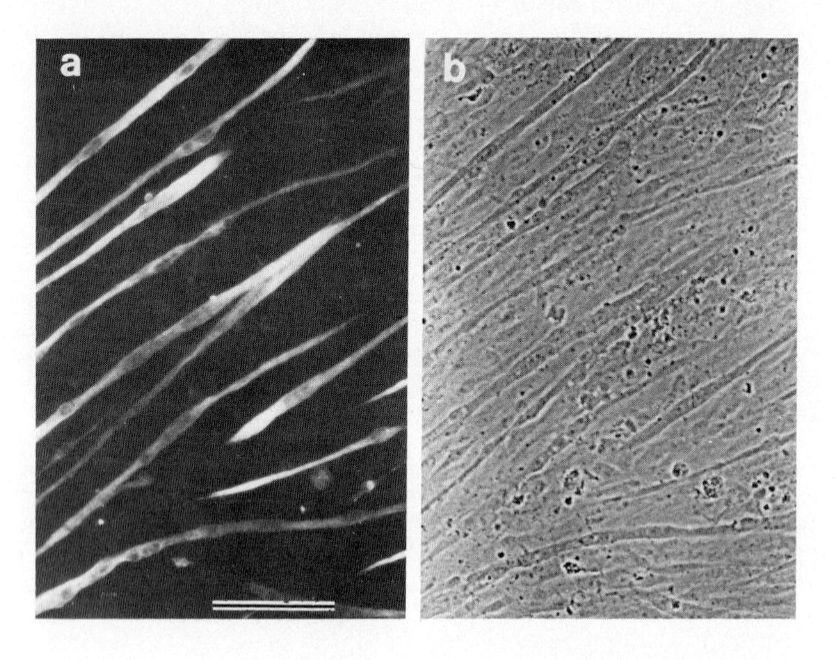

FIGURE 8. Fluorescence (a) and phase (b) views of a breast muscle culture, 3 days in vitro, *stained with anti-55K. Only the multinucleated myotubes are positive. The replicating mononucleated cells, which fill the area between myotubes, are negative. Bar = 100* μM.

represents a switch in the synthetic program analogous to that found for myosin heavy and light chains and other proteins characteristic of terminally differentiated skeletal muscle (Holtzer, 1978). At this stage, the myotubes as well as all mononucleated cells are also stained by anti-58K. Upon subsequent maturation, however, 58K subunits eventually disappear completely, and the anti-55K staining changes dramatically from bundles of longitudinally oriented filaments to a cross-striated pattern. The finding that anti-58K no longer stains longitudinally oriented 100 Å filaments in older myotubes, parallels observations made with the electron microscope, which reveal their disappearance as discrete morphological entities. The cross-striated pattern produced by anti-55K in mature myotubes raises many questions about the role of these transitory structures. Lack of resolution of the fluorescence microscope and the predelection for nonspecific binding of IgG to the I-Z band regions requires further work with anti-55K at the electron microscope level, to determine its precise localization. In any event, it will be interesting to relate our results using the anti-55K with the report of Lazarides and Hubbard (1976) for anti-desmin.

TONOFILAMENTS, AUTOANTIBODIES AND "NON-SPECIFIC" BINDING OF CONTROL SERA

The use of labelled antibodies is guaranteed to yield intriguing micrographs. Regrettably such micrographs often are difficult to interpret. This stems from failure to demonstrate that what is fluorescent under the microscope involved the interaction of the antigen and antibody being investigated, rather than involving "nonspecific" binding between macromolecules, such as hydrophic or electrostatic interactions. These and related problems were discussed years ago in Marshall *et al.* (1959), Holtzer (1960), and more recently in Fujiwara and Pollard (1976). These problems have been further compounded by the notion of sera from unimmunized animals containing low titers of nonprecipitating autoantibodies (e.g., Karsenti *et al*, 1977).

Osborn *et al.* (1977) have reported that sera from many unimmunized rabbits stain a fiber system in PtK2 cells, that they have referred to as composed of tonofilament-like, intermediate-sized filaments. We have confirmed their findings (Fig. 9). However, from the point of view of polymorphism it is worth stressing that we have never observed this "looping-fiber system" in any of the cells used in our studies and as shown in Fig. 10, when the anti-58K IgG was used at 1mg/ml on PtK2 cells it did not stain the "looping-fiber system" regard-

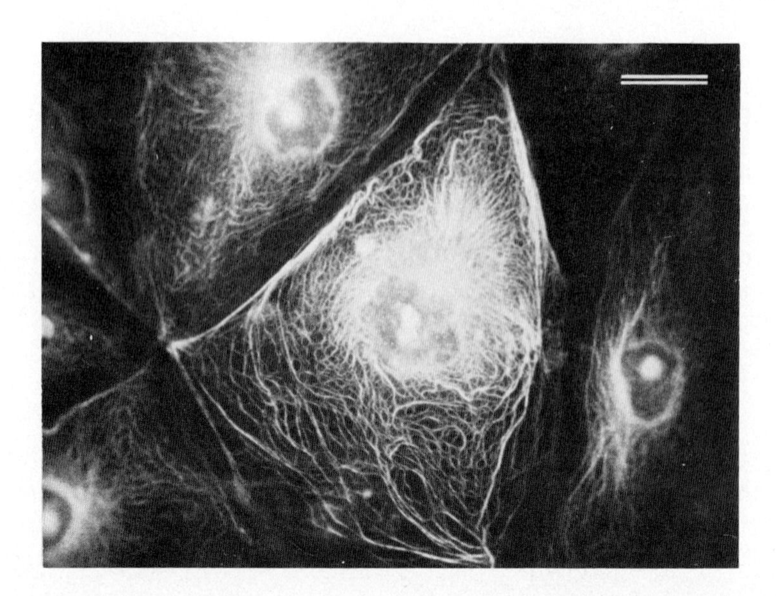

FIGURE 9. Fluorescence micrograph of a PtK2 cell fixed in -20°C methanol for 5 min followed by a brief rinse in acetone at 20°C and incubated with normal rabbit serum, 1/5 dilution) followed by fluorescein-conjugated goat anti-rabbit IgG (1 mg/ml) (Osborne et al.*, 1977). The "looped-fiber" system illustrated in this figure has not been observed in any of the numerous cell types used in these studies--fibroblasts, presumptive replicating myoblasts, postmitotic myoblasts and myotubes, chondroblasts, melanoblasts, brain cells, liver cells, or red blood cells. Following a wash at pH 9.6 the fluorescence of these stained fibers is totally abolished. The fluorescence of the 100 Å filaments in the cell types in which they occur is unaffected by such washes. Bar = 25 μM.*

less of fixation procedures used. Also noteworthy is the fact that the undiluted *whole* serum from which the anti-58K was prepared did stain the "looping-fiber system." Irrespective of what component in normal rabbit serum binds to the 100 Å filaments in PtK2 cells, our antibody clearly is directed against a different antigen. This conclusion is supported by the finding that, as judged under the phase and electron microscopes, the PtK2 cells were one of the few types of cells tested that could not be induced to form massive cables of 100 Å filaments with CB-Col. It is worth stressing that PtK2 cells, when fixed with formaldehyde and stained with anti-58K (1 mg/ml) exhibited fluorescent stress-fibers.

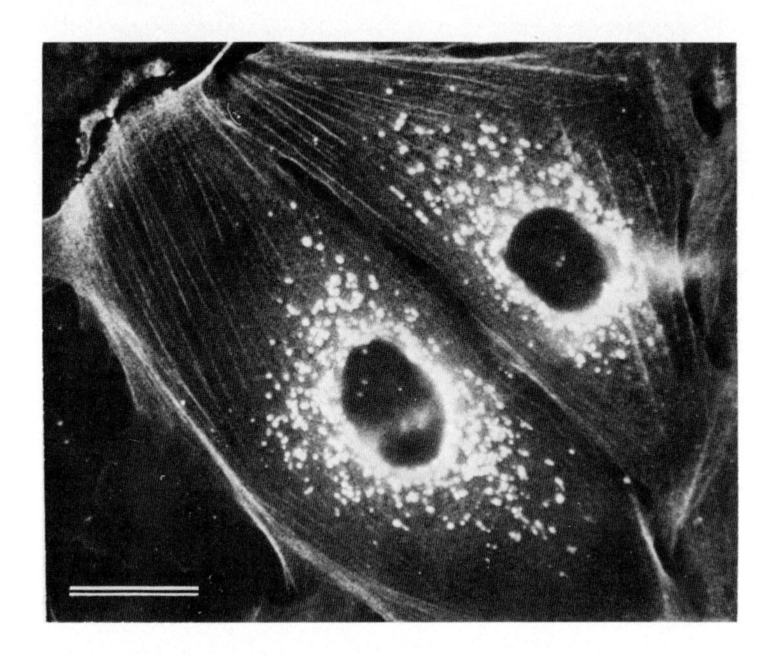

FIGURE 10. PtK2 cells fixed and stained with anti-58K IgG (1 mg/ml) as in Figs. 5-7. When the undiluted preimmune whole serum of the rabbit that yielded this active anti-58K antibody was used on PtK2 cells it stained the "looped-fiber" system. Note that the anti-58K IgG at 1 mg/ml stains the "stress-fibers" that have been described by many investigators using a variety of different antisera. It is to be stressed that the anti-58K antibody did not stain stress fibers in any chick cell type that was used in the experiments described here. Bar = 20 μM.

Most control IgG in our experiments did slightly "stain" the thickest CB-Col induced cables (Fig. 11). However, diluting the IgG concentration below 1 mg/ml abolished the fluorescence, whereas anti-58K stained the cables brilliantly at concentrations of 0.25 mg/ml. In addition, staining by control IgG, but not by the antibody, was virtually abolished by washing at pH 9.6 and diminished by inclusion of Triton X-100 in the neutral buffered wash. The staining of the fiber system in PtK2 cells by normal rabbit serum (Fig. 9) was also abolished by these treatments. These manipulations do suggest that 1) what we observed with the immune anti-58K was due to Ag-Ab reactions, and 2) if the staining by control IgG is due to autoantibodies they are of low titer and/or weakly bound and, in the case of PtK2 cells, probably directed against different antigenic sites.

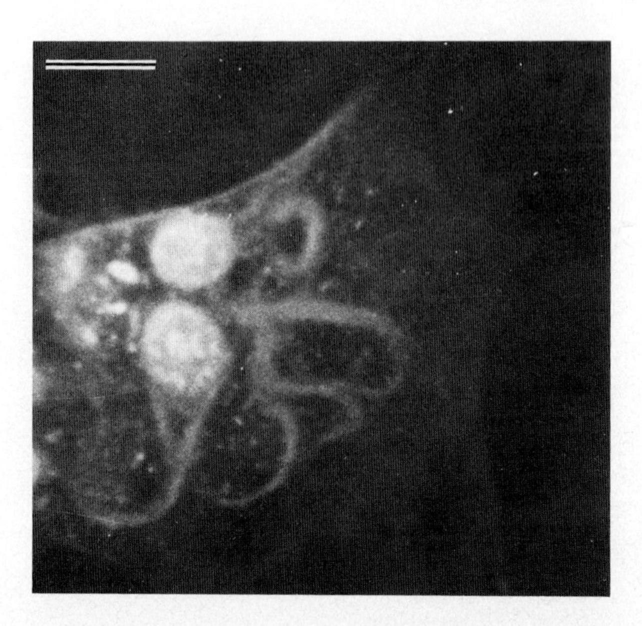

FIGURE 11. Fluorescence micrograph of a cable stained with the preimmune IgG of the rabbit that yielded an active anti-58K antibody. Note the weak positive staining of the thickest cables. How much of this weak fluorescence was due to "nonspecific" binding is difficult to determine. The photographic exposure of this cell was four times that used in the other figures. Even under these conditions, microspikes and fine extracellular processes were never observed. Bar = 20 μM.

Clearly much remains to be done on the immunological level before we have a clear understanding of the polymorphic nature of the different kinds of 100 Å filaments, and of how their subunits might relate to other structures in different types of cells.

BIOCHEMICAL CHARACTERIZATION OF THE 100 Å FILAMENTS

In the high-salt-insoluble material of every cell type thus far tested there is, in addition to a 43,000 MW protein, one or two proteins with molecular weights between 54,000 and 58,000 (Fig. 12). These proteins are present in surprisingly high concentrations. For example, roughly 15% of the total protein and 50-90% of the high-salt-insoluble residue of fibroblasts consists of actin and an associated 58,000 MW protein. Cells tend to fall into two classes with respect to these actin-associated proteins. The protein from *in vivo* gizzard,

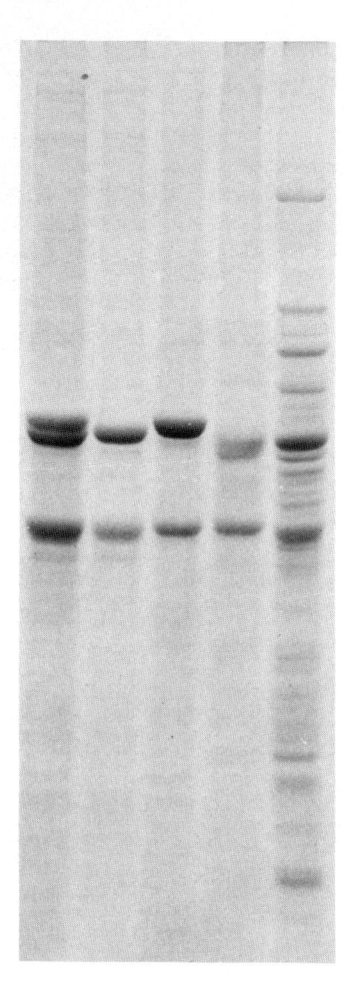

FIGURE 12. SDS-gel electrophoresis of high-salt insoluble residues. Tissues were extracted for 48 hr in 0.6 M KCl (two changes) and an additional 24 hr in 0.6 M Kl. The insoluble residues were washed in PBS, solubilized in an SDS-sample buffer, and electrophoresed in 10% polyacrylamide according to Laemmli (1970). In each case the major polypeptide components are actin, at 43K daltons, and a polypeptide of either 55K or 58K. (a) Fibroblasts plus gizzard. (b) Gizzard. (c) Fibroblasts. (d) Embryonic brain. (e) Embryonic liver.

muscle, brain, and liver is approximately 55,000 daltons, whereas the protein from cultured fibroblasts and *in vivo* red blood cells is approximately 58,000 daltons. *In vitro* and *in vivo* cardiac, skeletal and gizzard cells have both classes; the 58K protein in all these instances is almost certainly due to the fibroblast 100 Å filaments. The proteolytic digest patterns of these proteins reveals a high degree of polymorphism (Fig. 13). Comparison of these digest patterns with that of purified tubulin (not shown), suggests that the 55K protein in high-salt-insoluble residues of brain is probably tubulin, whereas the 55K and 58K components from other tissues is different from tubulin.

SUMMARY

The 100 Å intermediate-sized filaments present in a great variety of cells contain a 55,000 - 58,000 dalton subunit. Sequential treatment by cytochalasin-B and Colcemid induces, in many types of cells, the reversible aggregation of these filaments into massive, tortuous cables. Examination of many kinds of cells with antibodies to 1) the 58K subunit of fibroblasts, 2) the 55K subunit of smooth muscle, and 3) the subunit of brain filaments, reveals that the 100 Å filaments in different cell types are not all identical. Considerable structural polymorphism is also evident in peptide maps of the 55K and 58K proteins from a variety of sources. Smooth, skeletal and cardiac muscle cells contain a subunit different from that in nonmuscle cells. Neurofilaments have a unique antigenic determinant not found in any other cell type. The antigenic determinant(s) of fibroblast subunits exist in many different cell-types, and may suggest similarities in primary structure, in addition to differences. Tonofilaments of epithelial cells appear to be antigenically unrelated to the 100 Å filament subunits. Clearly, there are species differences among 100 Å filaments, suggesting that they are much less conserved evolutionarily than actin.

The observation that the longitudinally-oriented 100 Å filaments in developing myotubes are destined to disappear as discrete organelles, though the anti-55K binds to the I-Z region of mature myofibrils, raises intriguing questions regarding their function.

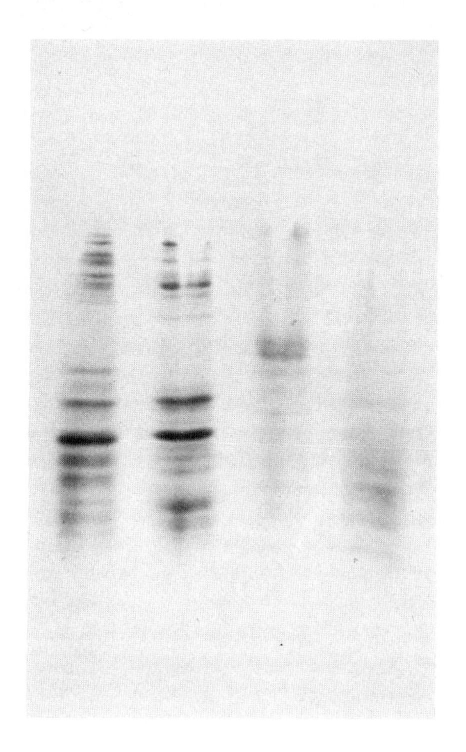

FIGURE 13. Proteolytic digest peptides displayed on a 15% SDS-polyacrylamide gel. High-salt insoluble residues were electrophoresed as in Fig. 11 and gel slices containing polypeptides of 55K or 58K daltons were excised and digested with 0.5 μg Pronase (Calbiochem) as described by Cleveland et al. *(1977). The displayed peptide patterns indicate that the 55-58K polypeptide of each tissue investigated is specific for that tissue. (a) 58K polypeptide from cultured fibroblasts. (b) 55K polypeptide from gizzard. (c) 55K polypeptide from embryonic brain. (d) 55K polypeptide from embryonic liver.*

ACKNOWLEDGMENTS

This work was supported by grants from NIH and the Muscular Dystrophy Association of America.

REFERENCES

Bennett, G. S., Fellini, S. A., Croop, J. M., Otto, J. J., Bryan, J., and Holtzer, H. (1978a). *Proc. Nat. Acad. Sci. U.S.A. 75,* 4364-4368.

Bennett, G. S., Fellini, S. A., and Holtzer, H. (1978b). *Differentiation 12,* 71-83.

Blose, S. H., Shelanski, M. L., and Chacko, S. (1977). *Proc. Nat. Acad. Sci. U.S.A. 74,* 662-665.

Brown, S., Levinson, W., and Spudich, J. A. (1976). *J. Supramolec. Struc. 5,* 119-130.

Chi, J. C. H., Rubinstein, N., Strahs, K., and Holtzer, H. (1975a). *J. Cell Biol. 67,* 523-537.

Chi, J., Fellini, S., and Holtzer, H. (1975b). *Proc. Nat. Acad. Sci. U.S.A. 72,* 4999-5003.

Cleveland, D. W., Fischer, S. G., Kirschner, M. W., and Laemmli, U. K. (1977). *J. Biol. Chem. 252,* 1102-1106.

Cooke, P. H. (1976). *J. Cell Biol. 68,* 539-556.

Croop. J., and Holtzer, H. (1975). *J. Cell Biol. 65,* 271-285.

DePetris, S., Karlsbad, G., and Pernis, B. (1962). *J. Ultrastruc. Res. 7,* 39-55.

Fellini, S. A., Bennett, G. S., and Holtzer, H. (1978a). *Am. J. Anat. 153,* 451-458.

Fellini, S. A., Bennett, G. S., Toyama, Y., and Holtzer, H. (1978b). *Differentiation 12,* 59-70.

Fujiwara, K., and Pollard, T. D. (1976). *J. Cell Biol. 71,* 848-875.

Goldman, R., and Knipe, D. (1973). *Cold Spring Harbor Symp. Quant. Biol. 37,* 523-534.

Goldman, R., Pollard, T. D., and Rosenbaum, J., eds. (1976). "Cell Motility." Cold Spring Harbor Laboratory, New York.

Holtrup, M. E., Raisz, L. G., and Simmons, H. A. (1974). *J. Cell Biol. 60,* 346-355.

Holtzer, H., and Holtzer, S. (1960). *Comptes Rendus des Travaux du Laboratoire Carlsberg 31,* 373-408.

Holtzer, H., Chacko, S., Abbott, J., Holtzer, S., and Anderson, H. (1970). *In* "Chemistry and Molecular Biology of the Intercellular Matrix" (E. A. Balazs, ed.), Vol. III, pp. 1471-1484. Academic Press, New York.

Holtzer, H., Sanger, J., Ishikawa, H., and Strahs, K. (1973). *Cold Spring Harbor Symp. Quant. Biol. 37,* 549-560.

Holtzer, H., Croop, J., Dienstman, S., Ishikawa, H., and Somlyo, A. P. (1975). *Proc. Nat. Acad. Sci. U.S.A. 72,* 513-517.

Holtzer, H., Fellini, S., Rubinstein, N., Chi, J., and Strahs, K. (1976). *In* "Cell Motility" (Goldman, R., Pollard, T.D., and Rosenbaum, J., eds.), pp. 823-839. Cold Spring Harbor Laboratory, New York.

Holtzer, H. (1978). *In* "Stem Cells and Tissue Homeostasis" (B. Lord, C. Potten, and R. Cole eds.). Cambridge University Press. pp. 1-25.
Ishikawa, H., Bischoff, R., and Holtzer, H. (1968). *J. Cell Biol. 38*, 538-555.
Ishikawa, H., Bischoff, R., and Holtzer, H. (1969). *J. Cell Biol. 43*, 312-328.
Karsenti, E., Guilbert, B., Bornens, M., and Avrameas, S. (1977). *Proc. Nat. Acad. Sci. U.S.A. 74*, 3997-4001.
Laemmli, U. K. (1970). *Nature 227*, 680-685.
Lazarides, E., and Hubbard, B. D. (1976). *Proc. Nat. Acad. Sci. U.S.A. 73*, 4344-4348.
Marshall, J. M., Holtzer, H., Finck, H., and Pepe, F. (1959). *Exp. Cell Res. Suppl. 7*, 219-230.
Mautner, V., and Hynes, R. O. (1977). *J. Cell Biol. 75*, 743-768.
Moellmann, G., and McGuire, J. (1975). *Ann. N. Y. Acad. Sci. 253*, 711-722.
Osborn, M., Franke, W. W., and Weber, K. (1977). *Proc. Nat. Acad. Sci. U.S.A. 74*, 2490-2494.
Peters, A., and Vaughn, J. E. (1967). *J. Cell Biol. 32*, 113-119.
Sanger, J. W., Holtzer, S., and Holtzer, H. (1971). *Nature New Biol. 229*, 121-123.
Schlaepfer, W. W. (1977). *J. Cell Biol. 74*, 226-240.
Small, J. V., and Sobieszek, A. (1977). *J. Cell Sci. 23*, 243-268.
Stenman, S., Wartiovaara, J., and Vaheri, A. (1977). *J. Cell Biol. 74*, 453-467.
Uehara, Y., Campbell, G. R., and Burnstock, G. (1971). *J. Cell Ciol. 50*, 484-497.
Weber, K., Rathke, P. C., Osborn, M., and Franke, W. W. (1976). *Exp. Cell Res. 102*, 285-297.
Wisniewski, H. M., Shelanski, M., and Terry, R. (1968). *J. Cell Biol. 38*, 224-229.

Motility in Cell Function
Proceedings of the First John M. Marshall Symposium in Cell Biology

THE CONTRACTILE CYTOSKELETON OF AMOEBOID CELLS *IN VITRO* AND *IN VIVO*

D. Lansing Taylor
Yu-Li Wang, Jeanne Heiple
and
Susan Hellewell

Cell and Developmental Biology
The Biological Laboratories
Harvard University
Cambridge, Massachusetts

INTRODUCTION

A. Amoeboid Movement

The ultimate goal of our research is the elucidation of the molecular basis of amoeboid movement, which was one of the research interests of John Marshall and his colleagues. Amoeboid movement is a fundamental life process exhibited by a wide variety of single celled organisms, as well as individual cells in multicelled organisms during at least some stage of development. Amoeboid cells include the giant free living amoebae, *C. carolinensis* and *A. proteus*, the amoeboid stage of *D. discoideum*, macrophages, leucocytes, many mammalian cells in culture, and cells "sorting-out" during development.

B Philosophy of Research on Amoeboid Movement

A complete understanding of the molecular basis of amoeboid movement will require the integration of information from five major areas of cell biology (Table 1). We have elected to initiate our study of amoeboid cells from the "inside-out" by first characterizing the contractile cytoskeleton and the secondary messengers involved in regulating cytoplasmic structure and contractility. Establishing the molecular basis of

ISBN 0-12-551750-5

TABLE 1. Amoeboid movement results from the Integration of at Least Five Major Cell Biological Processes

(1)	*Environmental signal - primary messenger*
(2)	*Signal reception at cell surface*
(3)	*Signal processing by plasmalemma*
(4)	*Intracellular signal - second messenger*
(5)	*Activation of contractile cytoskeleton*

cytoplasmic structure and contractility will facilitate investigations on the role of the primary messenger and signal processing by the cell surface and membrane on the initiation and regulation of cellular movement.

We have been correlating information on the contractile cytoskeleton at three levels of organization and complexity (Table 2). Our philosophy has been to prepare model systems of amoeboid movement, to characterize the cytoskeletal and motile properties of the models, and then to purify the pertinent proteins. The ultimate goal is the reconstitution of a regulated contractile cytoskeleton from purified proteins and the localization of specific proteins and secondary messengers in living cells.

The giant free living amoebae, *C. carolinensis* and *A. proteus* and the amoeboid stage of the cellular slime mold *D. discoideum* have been chosen as the primary amoeboid cells for investigation. The giant amoebae are extremely large cells which permit simplified single cell investigations. The giant amoebae also exhibit rapid and dramatic changes in cytoplasmic structure that accompany cell movements. Furthermore, neither cytoplasmic microtubules nor 10 nm intermediate filaments have been demonstrated which simplifies the molecular interpreta-

TABLE 2. Our Research on the Secondary Messenger and the Contractile Cytoskeleton of Amoeboid Cells has been Divided into Three Levels of Organization

(1)	*Models of contractile/cytoskeletal processes* in vitro
(2)	*Isolation, characterization and reconstitution of contractile/cytoskeletal proteins*
(3)	*Localization and structural dynamics of contractile/cytoskeletal processes* in vivo *(molecular cytochemistry)*

tions. Unfortunately, these cells have not been grown in axenic culture and culturing large numbers of cells has been difficult. In contrast, *D. discoideum* can be used for biochemical investigations since it can be easily grown in large numbers. In addition, a primary signal for cell movement (c AMP) has been identified in *D. discoideum* so that future studies on "signal processing" during movement will be simplified.

C. *Historical Perspectives on Cytoplasmic Structure and Contractility*

The concept that cytoplasmic structure or consistency was related to cell movement dates back many years. Dujardin (1835) identified the "living substance" in many protozoans and coined the term "Sarcode" to define this "living contractile jelly." Subsequently, von Mohl (1846) introduced the term "protoplasm" to characterize the semifluid constituents of plant cells that surrounded the nucleus. Finally, Schultze (1861) equated Dujardin's Sarcode with protoplasm after investigating a large number of different cell types. Thus, an apparently universal contractile substance with variable structure was identified before the end of the 19th century. However, the relationship between cytoplasmic structure and contractility took many years to establish and the evolution of this concept was spiced with debates between proponents of the fluid and semisolid nature of protoplasm (cytoplasm). For a more detailed description of the historical development of the concepts of cell structure and contractility the reader is directed to Allen (1961) and Taylor and Condeelis (1979).

The terminology and methodology used to define and to characterize the properties of cytoplasm have varied over the years depending on the prevailing concept of the cytoplasmic consistency (fluid or semisolid). When cytoplasm was thought to be essentially fluid (Berthold, 1886; Butschi, 1892; see, also, Allen, 1961; and Taylor and Condeelis, 1979, for review), the physical parameter of *viscosity* was utilized to quantitate cytoplasmic structure. However, only simple fluids obey Poiseuille's law and are defined as Newtonian fluids. Some fluids do not obey Poiseuille's law and possess more than one value of viscosity and are called non-Newtonian fluids. Cytoplasm from several types of cells has been shown to be a non-Newtonian fluid so that simple measurements of viscosity are of little value (Allen, 1961; Taylor and Condeelis, 1979).

Physical terms and methods previously applied to three dimensional cross-linked structures were applied to cytoplasm after the simple fluid nature of cytoplasm was abandoned. Subsequently, the elastic and viscoelastic properties of cytoplasm were identified (Chambers and Chambers, 1961; Allen, 1961; Taylor and Condeelis, 1979). However, cytoplasm was suspected to exhibit a range of structures from the complex fluid to solid states so that the colloidal terms "sol" and "gel" were ultimately used to define local cytoplasmic structure (Seifriz, 1936; Freundlich, 1937; Marsland and Brown, 1936; Condeelis and Taylor, 1979). Hyman (1917) used the colloidal chemical terms "sol" and "gel" to describe the variation in cytoplasmic consistency detected in amoebae. The cell cortex or ectoplasm was described as a "gel" and the streaming cytoplasm or endoplasm as a "sol" of variable viscosity. The changes in cytoplasmic structure were described as an integral part of the contractile events responsible for movement (Hyman, 1917; Pantin, 1923; Mast, 1926).

Recent advances in the field of cell structure and motility were made possible by a large number of investigators whose work is not cited in this limited review of our own studies. Their work has been invaluable to our own research and has been reviewed recently (Taylor and Condeelis, 1979). The present discussion is aimed at correlating the relationship between cytoplasmic structure and contractility at the three levels of organization presented in Table 2.

THE CONTRACTILE CYTOSKELETON STUDIED *IN VITRO*

A. *Single Cell Models*

A single cell model of cytoplasmic structure and contractility was prepared by Taylor *et al.* (1973) in Professor R. D. Allen's laboratory. Single specimens of *C. carolinensis* were physically demembranated in the presence of solutions designed to mimic rigor, relaxation, and contraction solutions for vertebrate striated muscle. In the presence of the low calcium ($<10^{-6}$ *M*) rigorlike solution the membraneless cytoplasm remained as nonmotile droplets. The nonmotile droplets contained fibrils consisting of parallel arrays of actin. A strain birefringence assay demonstrated that the cytoplasmic fibrils were highly viscoelastic suggesting the cross-linking of the actin filaments. Thick filaments identified morphologically (Taylor *et al.*, 1973) and biochemically (Condeelis, 1977) as myosin were observed in close association with the actin filaments (Taylor *et al.*, 1973; Moore *et al.*, 1973). The addition of a relaxation solution (Mg-ATP, *ca.*$<10^{-6}$ *M* Ca^{++}) decreased the viscoelasticity, while the contraction solution (*ca.*$\geq 10^{-6}$ *M* Ca^{++}) induced contractions of the cytoplasmic fibrils which performed work by moving fragments of glass. These early experiments identified the role of low calcium ion concentrations (*ca.* $<10^{-6}$ *M*) in maintaining a highly structured but nonmotile state of the cytoplasm, and elevated calcium ion concentrations (*ca.* $\geq 10^{-6}$ *M*) in initiating contractions. At least part of the viscoelastic properties of the cytoplasm were explained as the static interactions between actin and myosin. However, other structural processes could not be ruled out.

The relationship between contractions and cytoplasmic streaming was demonstrated when the cytoplasm from single amoebae was isolated into a contraction solution containing a threshold free calcium ion concentration (Flare solution) (Taylor *et al.*, 1973). The pH and the free calcium ion concentrations were buffered so that only transient gradients of these parameters would be possible within the cytoplasmic droplets. The droplets of cytoplasm formed membraneless "pseudopods" that extended from the central mass out into the flare solution and then returned forming "loops" of streaming cytoplasm that actually caused the locomotion of the droplets. The streaming events observed were almost identical to the pattern in living cells where the endoplasm flows to the tips of advancing pseudopods and everts to form the ectoplasmic tube (Allen, 1961, 1973; Taylor *et al.*, 1973; Taylor, 1976). It was suggested that localized contractions of the cytoplasm caused the streaming activity in the cell models. Furthermore, in the absence of the plasmalemma, "pseudopodia" were activated in all directions by the direct effect of the calcium. It was hypothesized that the plasmalemma would play a role in anchoring the contractile machinery and in localizing the contractile events in intact cells.

Glycerinated models of the cytoplasmic fibrils and intact cells were prepared in order to characterize the basic contractile machinery. Unfortunately, the glycerination process caused a dramatic decrease in cytoplasmic viscoelasticity and calcium regulated contractility (Taylor *et al.*, 1976a). Subsequently, the role of divalent cations in maintaining cytoplasmic structure and contractility was studied in both glycerinated and unglycerinated fibrils. It was demonstrated that cytoplasmic viscoelasticity, contractility (Taylor *et al.*, 1976a) and the stability of actin and myosin filaments (Condeelis *et al.*, 1976) was dependent on the presence of divalent cations. In particular, the free calcium ion concentration could not be lowered below *ca.* 1.0×10^{-7} *M* at pH 7.0 without causing the simultaneous loss of viscoelasticity, contractility, and actin and myosin filaments (Taylor *et al.*, 1976a; Condeelis *et al.*, 1976). These results demonstrated that the amoeba contractile/cytoskeletal proteins were structurally labile and probably explained the small number of filaments observed in glycerinated amoeboid cells except under conditions that induced contractions before or during glycerination (see Taylor and Condeelis, 1979, for review).

B. Bulk Cell Free Extracts

The single cell models were of limited value since a biochemical analysis was required to identify the necessary components for the calcium regulated structural and contractile properties of cytoplasm. Therefore, bulk cell free extracts were prepared from *A. proteus* using a modification of the method introduced by Thompson and Wolpert (1963) and extended by Pollard and Ito (1971). The bulk extracts from *A. proteus* were assayed for calcium regulated changes in structure and contractility, and the role of morphological changes in actin and myosin during the transition from the nonmotile state to the contracting-streaming state was investigated (Taylor *et al.*, 1976b).

The ionic regulation of the bulk extract cytoplasmic structure and contractility was identical to that demonstrated in single cell models. At free calcium ion concentrations of *ca.* 1.0×10^{-7} *M* at pH 7.0 the extracts formed a nonmotile mass termed a "gel" in keeping with the terminology reintroduced by Kane (1975). The subsequent elevation of calcium to the micromolar level induced contractions of the gelled mass monitored as large increases in both turbidity and birefringence. Ultrastructurally, the gelled material was loosely organized in a random configuration of "amorphous" material and short twiglike filaments that could not be identified morphologically as actin without labeling with HMM. The contracted extract contained large numbers of readily identifiable F-actin filaments. The term "transformation" was used to describe this change in the cytoplasmic structure during the transition from the sol-gel-contracted states. The polymerization of actin and/or changes in the association between actin and accessory proteins were suggested to be involved in the transformation (Taylor *et al.*, 1976b).

The contracted pellets were analyzed by SDS gel electrophoresis in order to determine the major proteins concentrated during the "contractile" event. The contracted pellets contained the amoeba myosin, actin, 80,000 daltons protein, and at least one low molecular weight protein *ca.* 32,000 daltons (Taylor *et al.*, 1976b).

A bulk extract model system capable of forming the nonmotile mass but not capable of contracting was prepared by centrifuging the extract to remove the myosin thick filaments. Contractility was restored only when the myosin containing fraction was added back to the extract.

The pH of the extracts was identified as a very important parameter in controlling both cytoplasmic structure and contractility since gelation and contractility were shown to be inhibited at pH's below 7.0. The effects of pH on the structure and contractility of the extracts was correlated with classical observations on the alteration in the intracellular amoeba pH in response to calcium. Reznikoff and Pollack (1928) demonstrated that the injection of calcium ions into cells pre-loaded with pH indicators caused a marked but transient decrease in the local pH. It was suggested by Taylor *et al.* (1976b) that a balance between local cytoplasmic pH and free calcium ion concentrations could be involved in controlling cytoplasmic structure and contractility. However, no decision could be made on the relative importance of pH and calcium (Taylor *et al.*, 1976b; Taylor 1976).

The small volume of extract that could be readily prepared from *A. proteus* encouraged us to seek another amoeboid cell for comparative studies at the biochemical level. The amoeboid stage of *D. discoideum* was chosen for the reasons discussed above. The structure and motility of extracts prepared from *D. discoideum* were almost identical to similar extracts prepared from *A. proteus* (Taylor *et al.*, 1977). Warming the extracts from either source at low free calcium ion concentrations at pH 7.0 caused nonmotile gels to form which contracted in response to either elevated pH (>7.0) or calcium (*ca.*>1.0×10^{-6} *M*). A gradient in the transformation from the sol to gel to contracting state was described as a necessary step to allow contractions to perform work in a particular direction. A mechanism of local gel breakdown with concomitant contractions was suggested to occur during movement.

The transformation from the relaxed to the contracted state was monitored by changes in turbidity and was shown to be reversible (Fig. 1). The ultrastructural studies initiated on extracts of *A. proteus* (Taylor *et al.*, 1976b) were extended to *D. discoideum* and the "transformation" of the actin containing structures in the extract was shown to be the same. In addition, a membrane-ectoplasm complex isolated from *C. carolinensis* (Taylor *et al.*, 1976b) and *D. Discoideum* (Taylor *et al.*, 1977) exhibited the same transformations of the actin containing structures as those observed in the extracts. It was concluded that several characteristics of amoeba cytoplasm were common to at least *A. proteus*, *C. carolinensis*, and *D. discoideum:* a) the major contractile proteins were similar, b) actin in association with accessory proteins was structurally dynamic exhibiting several supramolecular forms, and c) cytoplasmic structure and contractility were regulated in an unknown manner by the pH and calcium ion concentrations. The various forms of movement expressed by different amoeboid cells could be explained in part as a result of the site, rate, and extent of actin structural transformations and contractions (Taylor *et al.*, 1977).

Significant advances in our understanding of the phenomenology identified in earlier single cell models and the cruder cell extracts were made possible by preparing large volumes of a high speed supernatant of cell extracts from *D. discoideum*

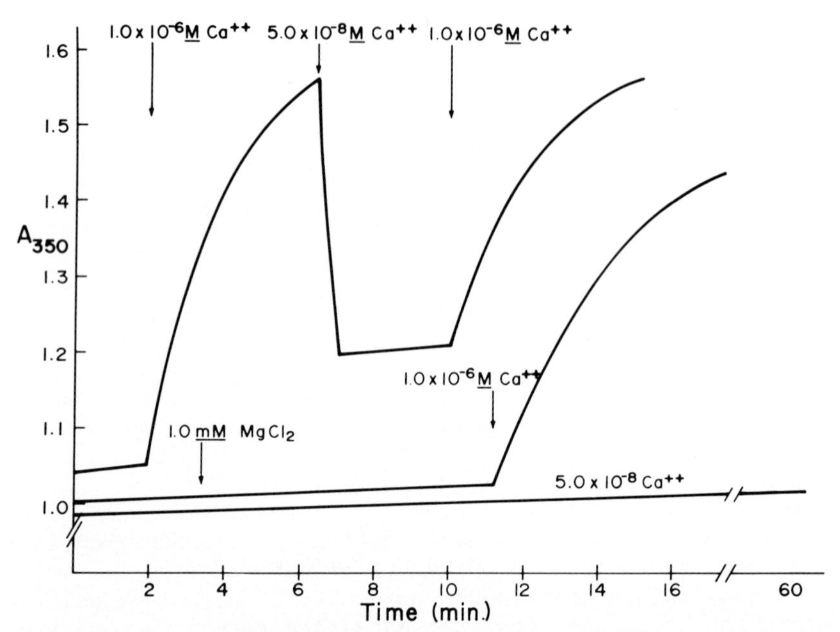

FIGURE 1. The cell free extract (S2) from D. discoideum *increased in turbidity when the free calcium ion concentration was raised to* ca. *1.0 x* 10^{-6}M, *pH 7.0 and decreased with the subsequent addition of EGTA which decreased the free calcium ion concentration to* ca. *5.0 x* 10^{-8} M, *pH 7.0. 1.0* mM *MgCl did not cause an increase in turbidity. (Reprinted with permission from Taylor* et al., *1977).*

(Condeelis and Taylor, 1977). The ionic regulation of gelation and contraction was characterized in detail and the effects of calcium and pH on both cytoplasmic structure (gelation) and contraction were quantitated in detail (Fig. 2). In particular, the first quantitative assay for gelation in bulk extracts was introduced and was identical to the method applied to single cell models (Taylor *et al.*, 1973) and living cells (Francis and Allen, 1971).

The separation of the phenomena of gelation and contraction was accomplished by precipitating the *D. discoideum* myosin out of the extract. The myosinless extract gelled under conditions of low free calcium ion concentrations and the gel broke down or "solated" when the calcium was raised to *ca.* 10^{-6} *M*. Calcium regulated contractions were restored when purified *D. discoideum* myosin was added back to the extract (Condeelis and Taylor, 1977). These experiments supported our working hypothesis that gelation and contraction are antagonistic processes that are regulated by the same secondary messengers (calcium and/or pH). Interestingly, agents that induced "solation" of

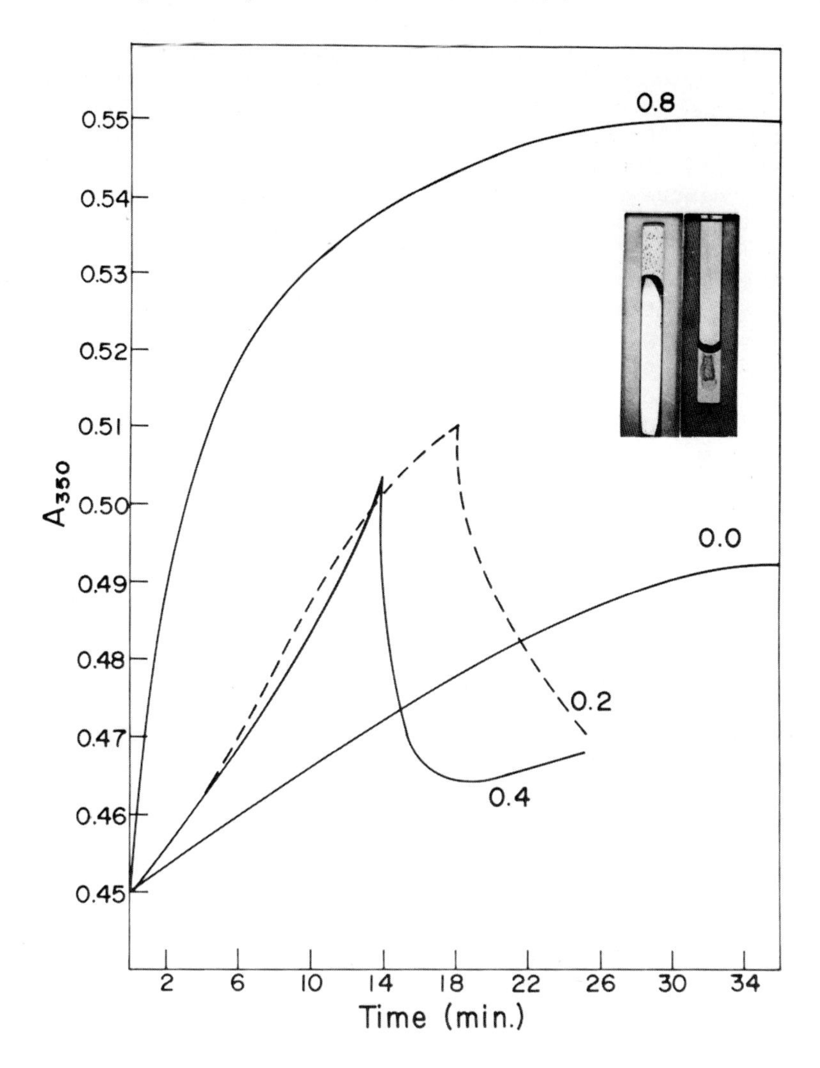

FIGURE 2. The cell free extract (S3) from D. discoideum *increased in turbidity and visibly contracted when the Ca/EGTA ratio was raised above ca. 0.2 and the rate of contraction increased directly with the Ca/EGTA. The extract was warmed from 4° to 25°C in the presence of the Ca/EGTA indicated. The inserts of cuvettes show (left) the extract with Ca/EGTA = 0.1 after 18 min and (right) with Ca/EGTA = 0.4 after 18 min. The drop in turbidity was due to the contracting extract "pulling out" of the light beam. (Reprinted with permission from Condeelis and Taylor, 1977).*

the myosin containing gels including low temperature, high pressure, cytochalasin B, as well as elevated calcium concentrations and pH resulted in contractions during or after the breakdown of the gel. These observations led us to speculate that the gelation events could serve both a cytoskeletal role and a regulatory role in controlling the interaction between myosin and actin. This concept was discussed in a subsequent review (Taylor, 1977a). An unresolved question also arose from the fact that myosin could be removed from extracts of *A. proteus* by centrifugation (Taylor *et al.*, 1976b) while *D. discoideum* myosin could not be pelleted out of the extract. The physical state of myosin in extracts and cells remains as an important unanswered question.

Recently, Hellewell and Taylor (1979) prepared a model system from *D. discoideum* that contained fewer proteins than the high speed supernatant from the bulk cell extract (Condeelis and Taylor, 1977) but maintained all of the phenomenological characteristics including calcium and pH regulated gelation and contraction. Contracted pellets from the cell extract were solubilized and the myosin was removed by precipitation. This relatively pure fraction (actin, 95,000, 75,000, 55,000, 35,000, 30,000, 28,000, 25,000, and 18,000 dalton components) contained the calcium pH regulated components of gelation, and contraction (after the addition of purified myosin (Fig. 3). In addition, a calcium sensitive gelation factor (CGFI) was purified and regulated gelation was reconstituted by mixing actin and the CGFI (95K daltons).

THE CONTRACTILE CYTOSKELETON STUDIED *IN VIVO*

A. *Viscoelasticity and Contractility*

As stated in the introduction, our research philosophy has been to correlate as much information obtained *in vitro* to living motile cells. This has been extremely important for the investigations on cytoplasmic structure since the phenomenon of gelation is certainly not without the possibility of being an interesting artifact. Therefore, different regions of the cytoplasm of *C. carolinensis* were assayed for the ability to exhibit strain birefringence (Taylor, 1977b). This assay used previously on single cells (Francis and Allen, 1971) and single cell models (Taylor *et al.*, 1973) and more recently on bulk cell free extracts (Condeelis and Taylor, 1977) identified a gradient of viscoelasticity increasing from tail endoplasm of *C. carolinensis* to the anterior endoplasm at the tips of the advancing pseudopods. This viscoelasticity extended into the ectoplasm. It was concluded that the amoeba

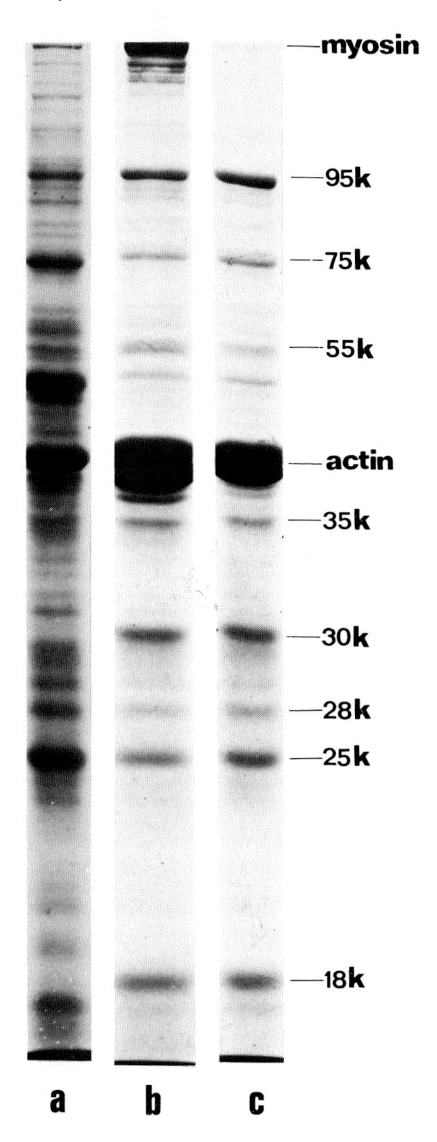

FIGURE 3. SDS-polyacrylamide slab gel electrophoresis of (a) high speed supernatant (S3) from the cell free extract of D. discoideum, *(b) contracted pellet formed by inducing the gelation and subsequent contraction of S3, and (c) model prepared from the contracted pellet by precipitating the myosin out of solution. The myosin-less model (c) exhibited calcium and pH regulated gelation and was capable of calcium pH regulated contractions after the addition of purified myosin.*

cytoplasm was a viscoelastic structure which varied quantitatively in time and position (Allen, 1961; Francis and Allen, 1971; Taylor, 1977b).

The ionic regulation was also tested *in vivo* by microinjecting contraction or relaxation solutions used in the previous studies *in vitro*. The calcium contraction solutions induced either localized or extensive intracellular contractions depending on the volume and calcium ion concentration. Although contractions could be induced anywhere in the endoplasm or ectoplasm, contractions were more readily elicited in the anterior endoplasm and ectoplasm than the posterior endoplasm. This gradient of contractility matched the gradient of cytoplasmic structure. Contractions were also induced by raising the intracellular pH above 7.0, while maintaining the free calcium concentration below *ca.* 1.0 μM. Furthermore, the injection of the relaxation solution (ATP, *ca.* $< 1.0 \times 10^{-6}$ Ca^{++}) caused the cells to lose the distinction between endoplasm and ectoplasm and motility ceased. The induced contractions and "relaxations" were reversible which further supported the role of calcium and/or pH in regulating cytoplasmic structure and contractility.

B. *Intracellular Free Calcium*

The calcium sensitive bioluminescent protein, aequorin, was microinjected into *C. carolinensis* in an effort to localize and quantitate the cellular free calcium ion concentration using an image intensifier (Taylor *et al.*, 1975). Although no spontaneous luminescence was detected, the cells locomoted and luminesced when placed in a weak electric field. The luminescence was continuous and maximal in the tail, while sporadic and weaker at the tips of advancing pseudopods. These complex results under aphysiological but nonlethal conditions were similar to the results obtained by Nuccitelli, *et al.* (1977) on freely moving cells using an extracellular probe of electrical currents. The aequorin and electrical measurements taken together suggest a two site rise in the intracellular free calcium ion concentration both in the tail and the tips of advancing pseudopods. However, no significant interpretation will be possible until direct measurements of the free calcium ion concentration are made on unperturbed cells.

C. *Intracellular pH*

The dramatic effect of changes in pH around pH 7.0 on the structure and contractility of both cytoplasmic extracts and living cells prompted us to assay the intracellular pH by updating the classical methods using pH sensitive dyes (see Chambers and Chambers, 1961).

Heiple and Taylor (1979) have measured the pH of both cytoplasmic extracts and single cells using a ratio photometer (Taylor, Zeh, and Heiple (1979). The dye phenol red exhibited two absorption peaks whose relative intensities varied with pH. The optical densities at the two peak wavelengths were measured simultaneously and the ratio of these O.D.'s were plotted versus pH (Fig. 4). Standard curves in buffers and cell extracts were then used to calibrate the intracellular pH. The initial measurements were designed to test the methodology and to detect any differences in pH between the front and rear halves of *C. carolinensis*. The spatial resolution was sacrificed in these early experiments to maximize the signal. A lower average pH has been detected in the tails of *C. carolin-*

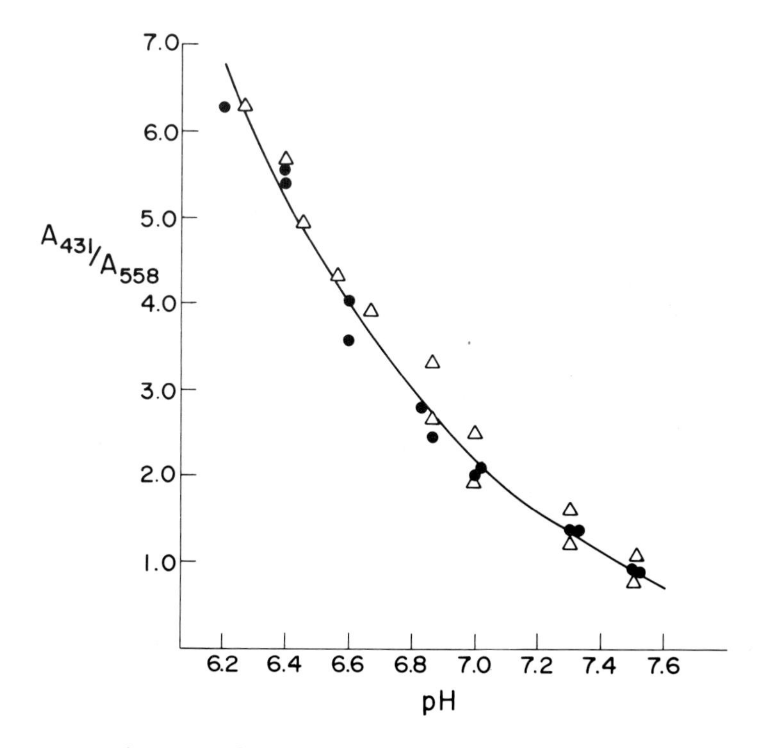

FIGURE 4. Standard curve of pH generated in the presence of a phosphate buffer (●) and the high speed supernatant (S3) from the cell free extract of D. discoideum *(Δ). The pH sensitive dye phenol red was used at a final concentration of 1.0 mM. The measurement of the ratio of optical densities at the two peak wavelengths was used to characterize the pH.*

ensis by *ca.* 0.1 to 0.2 units (*ca.* pH 6.8 to 6.9) in comparison with the anterior halves (*ca.* pH 7.0). These results are preliminary and more experiments with greater spatial resolution will be required to map intracellular pH conclusively. Two major experimental problems have been the confinement of the pH sensitive probes to the cytoplasm but not in organelles, and the maximization of the signal to noise ratio in single cells. These problems are currently under investigation.

D. *Localization of Specific Proteins* In Vivo

Actin labeled with 5-iodoacetamidofluorescein has been incorporated into living cells including *C. carolinensis* and the acellular slime mold *P. polycephalum* in order to determine the distribution, supramolecular form and functional state of actin *in vivo* (Taylor and Wang, 1978). Vertebrate striated muscle actin was labeled and then successfully compared to unlabeled actin with reference to its polymerizability and activation of muscle ATPase (Table 3). The labeled actin was judged a good reporter of actin activity based on several control experiments. 1) The labeled actin was incorporated into the high speed supernatant of *D. discoideum* extracts before gelation and contraction were induced in order to follow the distribution of actin during the sol-gel-contract cycle (Taylor and Wang, 1978). Densitometry on SDS gels of the contracted pellets from samples containing only endogenous unlabeled actin were compared with optical measurements on samples containing 10% w/w labeled actin. The labeled and unlabeled actin were distributed identically in these experiments demonstrating that the labeled actin was utilized during contraction. 2) Single cell streaming models (see Section II, A) were prepared from *C. carolinensis* following the microinjection of labeled actin into the living cells at a ratio of *ca.* 1:10 w/w labeled actin to endogenous amoeba actin. The fluorescence was maintained in the membraneless streaming loops only when

TABLE 3. A Summary of the Spectroscopic and Functional Properties of Actin Labeled with 5-iodoacetamidofluorescein

Dye/protein	*0.5 - 0.7*
Absorption peak - buffer A	*495*
Emission peak - Buffer A (corrected)	*521*
Viscosity - F-actin	*87% of control*
Actin activation - Myosin ATPase	*80-100% of control*

functional labeled actin was injected but not when denatured labeled actin or labeled bovine serum albumin was injected. 3) Single *C. carolinensis* were preloaded with both rhodamine isothiocyanate labeled bovine serum albumin and the fluorescein derivative labeled actin. An intracellular contracted knot was then induced by microinjecting the contraction solution at one tenth of the cell volume (see Section III,A). The cell was subsequently scanned with a photometer for rhodamine and fluorescein fluorescence and the results indicated that the labeled actin was enriched in the contracted region, while the bovine serum albumin was more randomly distributed (Taylor and Wang, 1978). These control experiments supported the notion that the labeled actin became incorporated into the cellular pool of actin.

The endoplasm and ectoplasm of *C. carolinensis* became relatively uniform in fluorescence when the functional labeled actin was microinjected. No discrete birefringent or fluorescent fibrils were observed except for a few microspikes in the tails where the membrane was highly folded. In contrast, the endoplasm of *P. polycephalum* exhibited uniform fluorescence, while distinct fluorescent bundles were detected in the ectoplasm after the incorporation of labeled actin (Fig. 5). The pattern of fluorescence observed in both *C. carolinensis* and *P. polycephalum* was consistent with the known patterns of birefringence and ultrastructure in these organisms.

Our initial results with incorporating functional, labeled contractile and cytoskeletal proteins into living cells suggest that this new technique of molecular cytochemistry *in vivo* could be applied to other proteins and other cells. The future development of this technology will focus on spectroscopic analysis of the labeled proteins in the purified state, in motile extracts and in living cells.

CONCLUSION

It has been determined that calcium and/or pH regulated changes in cytoplasmic structure (gelation) and contraction can be detected and assayed *in vitro* and *in vivo*. A working hypothesis has been developed to explain the dynamic events in cytoplasmic extracts. The cytoskeletal and contractile events follow separately but sequentially starting with the gelation of an initial low viscosity "sol" at pH 7.0 and a free calcium ion concentration below *ca.* 10^{-6} *M*. Upon either increasing the pH above 7.0 or the free calcium ion concentration to *ca.* 10^{-6} *M* the gel breaks down (decrease in actin-actin binding protein(s) cross-links), permitting myosin to actively move the actin filaments resulting in a contraction. Gel breakdown and

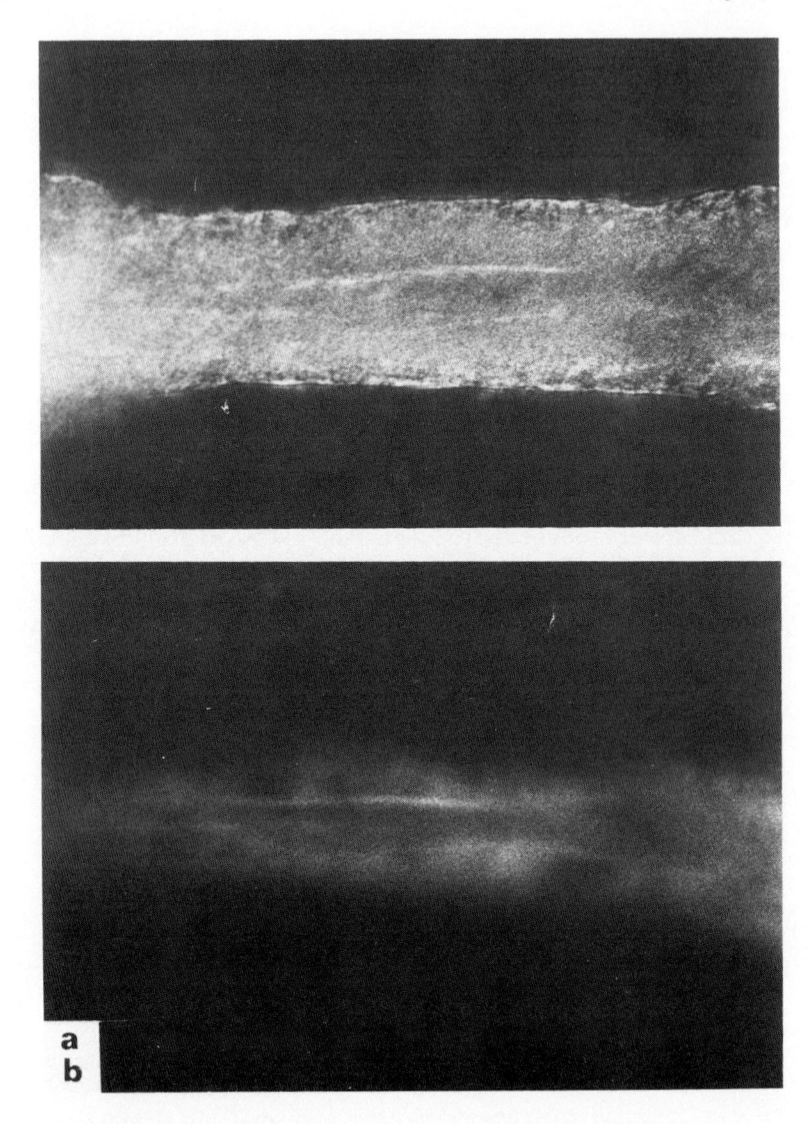

FIGURE 5. A region of a single living microplasmodium of P. polycephalum *injected with actin labeled with 5-iodo-acetamidofluorescein. The ectoplasm of the Physarum exhibited distinctly birefringent (a) and fluorescent bundles (x500). (Reprinted with permission from Taylor and Wang, 1978).*

contraction induced by raising the pH above 7.0 slowly drives the pH below 7.0, presumably by the hydrolysis of ATP, where further contractions and gelation are inhibited. Gel breakdown and contraction induced by raising the free calcium ion concentration to *ca.* $\geq 10^{-6}$ *M* also drives the pH below 7.0 which inhibits further contractions and gelation. In addition, the presence of calcium specifically blocks gelation so that the combined effect of lowered pH and the presence of calcium minimizes the cytoplasmic structure at the site of contraction. A self-limiting contraction could occur from the relationship between the local calcium ion concentration and pH. Contraction results in the formation of a highly structured actin-myosin pellet surrounded by a large volume of solated extract. The contraction supernatant has been labeled as sol′ since it contains F-actin filaments unlike the original cold extract (sol). At present the steps required to cycle this activity have not been completely defined (Fig. 6). A highly schematic diagram of the transition from the gel to the gel breakdown and contracting state is shown in Figs. 7 and 8. The physical state of actin and all the possible cross-linking proteins have not been identified but the suggested antagonistic relationship between gelation and contraction is illustrated.

The immediate goals of our work on model systems are to identify the proteins responsible for the calcium and/or pH regulated gelation and contraction and then to reconstitute the system.

At the cellular level we would like to map a cell like *C. carolinensis* for local pH, Ca^{++}, specific proteins, and the interaction of proteins (i.e., localization of the sol, gel, and contracting states in living cells). Thus, an integration of studies at the level of purified proteins, model systems, and living cells could ultimately lead to an understanding of the molecular basis of amoeboid movement.

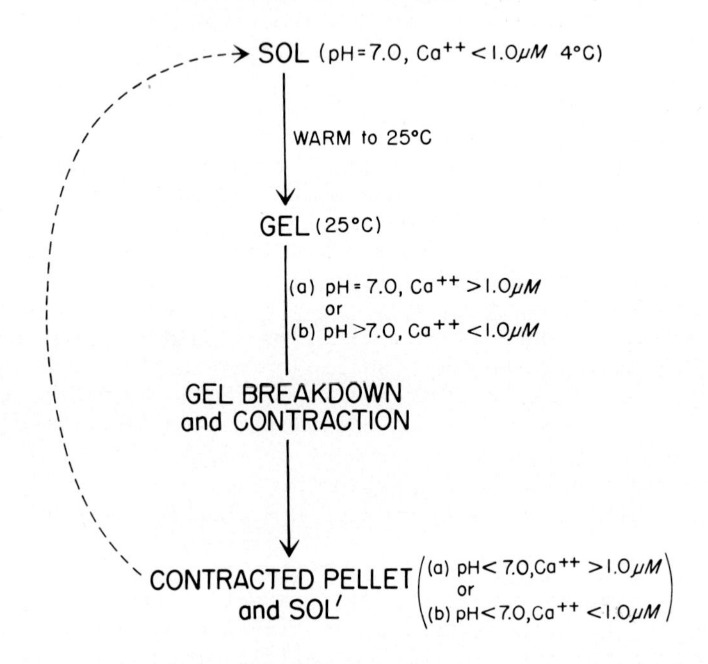

FIGURE 6. A working hypothesis for the sequence and control of the sol-gel-gel breakdown and contraction cycle in vitro. *See the text.*

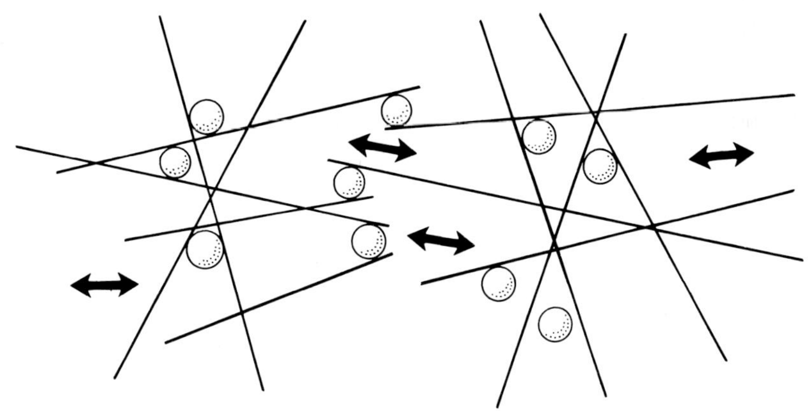

FIGURE 7. A highly schematic diagram of the "gel" containing actin cross-linked by as yet incompletely characterized actin binding protein(s). The physical state of the actin has not been adequately characterized.

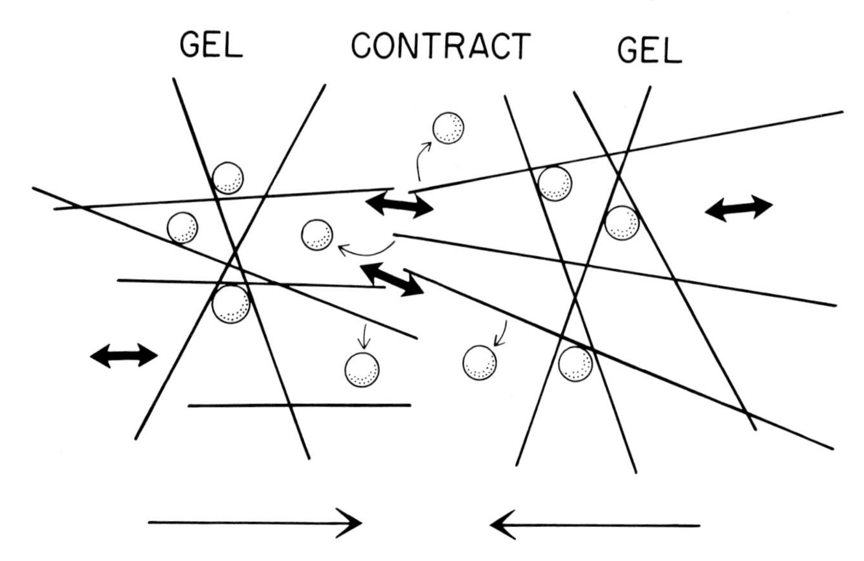

FIGURE 8. A highly schematic diagram of the gel breakdown and contraction induced by raising the pH or by raising the free calcium ion concentration to ca. 1.0 μM. The affinity of the actin binding protein(s) is diminished and the myosin actively pulls the free ends of F-actin filaments.

ACKNOWLEDGMENTS

The valuable collaboration with J. S. Condeelis, P. L. Moore, R. D. Allen, G. T. Reynolds, S. A. Hammond, J. A. Rhodes, and M. Rizzo on some of the work discussed is gratefully acknowledged.

REFERENCES

Allen, R. D. (1961). *In* "The Cell" (J. Brachet and A. E. Mirsky, eds.), Vol. II, pp. 135-216.
Allen, R. D. (1973). *In* "The Biology of Amoeba" (K. W. Jeon, ed.) p. 249. Academic Press, New York.
Berthold, G. (1866). "Studien über Protoplasma-mechanik." A. Felix, Leipzig.
Butschi, O. (1892). "Untersuchengen über mikroskopische Schäume und das Protoplasma." Englmann, Leipzig.
Chambers, R., and Chambers, E. L. (1961). "Explorations into the Nature of the Living Cell." Harvard University Press, Cambridge.
Condeelis, J. S. (1977). *J. Cell Sci. 25*, 387-402.
Condeelis, J. S., and Taylor, D. L. (1977). *J. Cell Biol. 74*, 901-927.
Condeelis, J. S., Taylor, D. L., Moore, P. L., and Allen, R. D. (1976). *Expt. Cell Res. 101*, 134-142.
Dujardin, F. (1835). *Ann. Sci. Nat. Zool. 4*, 343.
Francis, D., and Allen, R. D. (1971). *J. Mechanochem. Cell Motil. 1*, 1=6.
Freundlich, H. (1937). *J. Phys. Chem. 41*, 901.
Heiple, J., and Taylor, D. L. (1979). In preparation.
Hellewell, , and Taylor, D. L. (1979). *J. Cell Biol.* (In press).
Hyman, L. (1917). *J. Expt. Zool. 24*, 55-90.
Kane, R. (1975). *J. Cell Biol. 66*, 305-315.
Marsland, D. A., and Brown, D. E. S. (1936). *J. Cell Comp. Physiol. 8*, 167-178.
Mast, S. O. (1926). *J. Morph. Physiol. 41*, 347-425.
Moore, P. L., Condeelis, J. S., Taylor, D. L., and Allen, R. D. (1973). *Expt. Cell Res. 80*, 493.
Nuccitelli, R., Poo, M. M., and Jaffee, L. (1977). *J. Gen. Physiol. 69*, 743-763.
Pantin, C. F. A. (1923). *J. Mar. Biol. Assoc. United Kingdom 13*, 24.
Pollard, T. D., and Ito, S. (1970). *J. Cell Biol. 46*, 267-289.

Reznikoff, P., and Pollack, H. (1928). *Biol. Bull. (Woods Hole) 55,* 377-382.
Schultze, M. (1861). *Arch. Anat. 1,* 1-27.
Seifriz, W. (1936). "Protoplasm." McGraw-Hill, New York.
Taylor, D. L. (1976). *In* "Cell Motility" (R. Goldman, T. Pollard, and J. Rosenbaum, eds.). Vol. B, pp.
Taylor, D. L. (1977a). *In* "International Cell Biology 1976-1977" (B. R. Brinkley and K. R. Porter, eds.) pp. 797-821. The Rockefeller University Press, New York.
Taylor, D. L. (1977b). *Expt. Cell Res. 105,* 413-426.
Taylor, D. L., and Condeelis, J. S. (1979). *Int. Rev. Cytol., 56,* 57-144.
Taylor, D. L., and Wang, Y. L. (1978). *Proc. Nat. Acad. Sci. U.S.A.,* in press.
Taylor, D. L., Zeh, R., and Heiple, J. (1979). In preparation.
Taylor, D. L., Condeelis, J. S., Moore, P. L., and Allen, R. D. (1973). *J. Cell Biol. 59,* 378-394.
Taylor, D. L., Reynolds, G. T., and Allen, R. D. (1975). *Biol. Bull. 149,* 448.
Taylor, D. L., Moore, P. L., Condeelis, J. S., and Allen, R. D. (1976a). *Expt. Cell Res. 101,* 127-133.
Taylor, D. L., Rhodes, J. A., and Hammond, S. A. (1976b). *J. Cell Biol. 70,* 123-143.
Taylor, D. L., Condeelis, J. S., and Rhodes, J. A. (1977). *In* "Cell Shape and Surface Architecture" (J. P. Revel, U. Henning, and F. Fox, eds.). Vol. 17, pp. 581-603. Alan R. Liss, New York.
Thompson, C. M., and Wolpert, L. (1963). *Expt. Cell Res. 32,* 156-160.
Von Mohl, H. (1946). *Bot. Ztung. 4,* 73-78.

Motility in Cell Function
Proceedings of the First John M. Marshall Symposium in Cell Biology

THE MAINTENANCE OF PHOTORECEPTOR ORIENTATION

Alan M. Laties
and
Beth Burnside

Department of Opthalmology
Scheie Eye Institute
University of Pennsylvania School of Medicine
Philadelphia, Pennsylvania
and
Department of Physiology-Anatomy
University of California-Berkeley
Berkeley, California

Although this paper is primarily a scientific report, it is also in part a memoir since John Marshall played an initiating role in the work we present. In 1963, John Marshall had the only freeze-drying setup in the Anatomy Department when one of us (A.L.), then a postdoctoral fellow looking for a useful project to do, made the acquaintance in the Pharmacology Department of another scientifically underoccupied postdoctoral fellow. The two banded together to turn their excitement about the recently described histofluoremetric method for catecholamines to a practical result; but that result required freeze-drying. Hence the two sought out John Marshall. As was his custom, John Marshall was generous both with his time and his equipment. He let them share a corner of his laboratory to do freeze-drying. He also gave them keys to the laboratory so they could have access at night. The freeze-dryer of that time, much as a three month old, needed periodic refreshment: for the freeze-dryer it was dry ice. The freeze-drying and the rest of the histofluorometric method worked well and thanks to John Marshall, the two postdoctoral fellows produced a series of descriptive papers on autonomic innervation. Furthermore, there was an unexpected bonus to this work ---one that leads to the cytoskeletal properties of rods and cones.

ISBN 0-12-551750-5

I. GRADED DIFFERENTIAL ORIENTATION OF PHOTORECEPTORS

When tissue sections of the freeze-dried retina were observed in the fluorescence microscope, it became clear that the avoidance of liquid fixatives and lipid-extracting dehydrating agents yielded not only the sought after view of biogenic amines in the inner nuclear layer of the retina but also an unusual view of the photoreceptors. The curved outer segments seen in ordinary liquid-fixed tissue sections were replaced in the freeze-dried tissues by a remarkable order. Rods looked rodlike (Fig. 1): straight for their entire length, in perfect alignment one to another. More importantly,

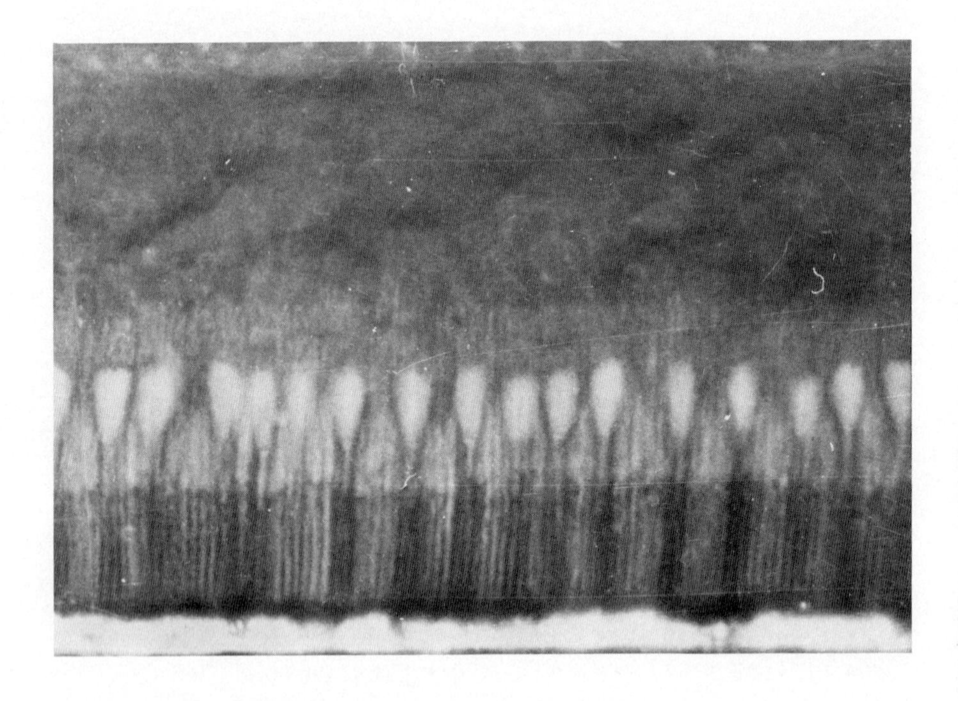

FIGURE 1. (a) In the parafoveal area of squirrel monkey retina rod and cone outer segments are perpendicular to the pigment epithelium and parallel one to another.

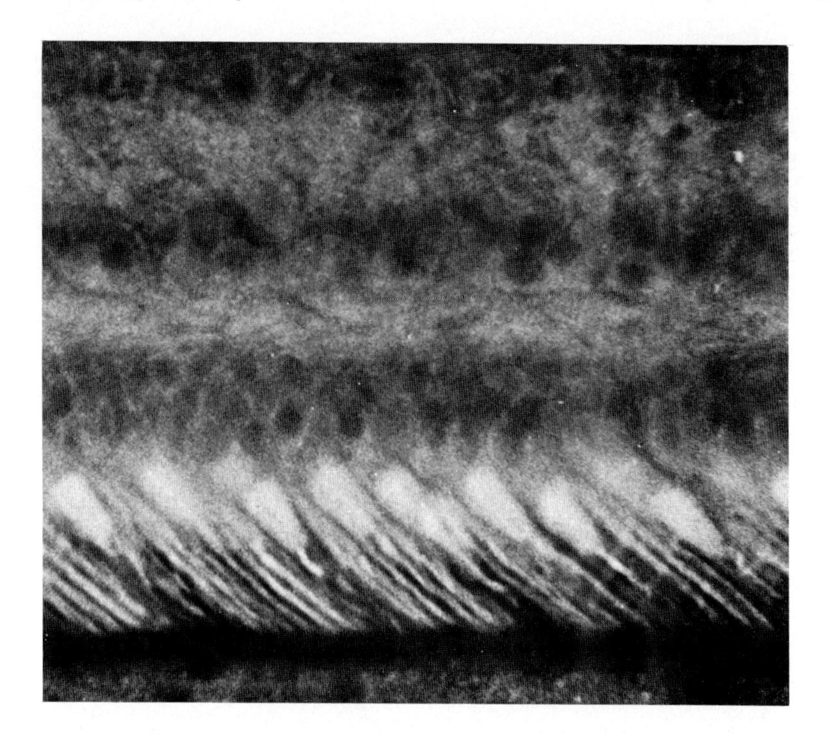

FIGURE 1. (b) Although still parallel one to another near the ora serrata, rod and cone inner and outer segments now tilt obliquely to face the entrance pupil of the eye. (10 μM tissue section. Fluorescence microphotographs x675).

in different regions of the eye it could be seen that photoreceptors had a precise orientation: at the back of the eye they were normal to a tangent; at the periphery they tilted 40-50° away from the normal (Fig. 1) (Laties, 1969). All the while they seemed perfectly parallel one to another in light microscopic sections. The photoreceptors exhibited a graded change in orientation such that their inner and outer segments were always coaxial to light rays entering the eye by way of the pupil (Fig. 2) (Laties *et al.*,1968; Laties and Enoch, 1971). Since rhodopsin is held in planar array on the individual leaflets which comprise the outer segments of rods (Schmidt, 1938) and since rhodopsin maximally absorbs light that strikes the leaflets at 90°, the photoreceptors were optimally oriented to absorb entering light rays. Thus, John Marshall's freeze-dryer had let us stumble onto a structural expression of molecular efficiency.

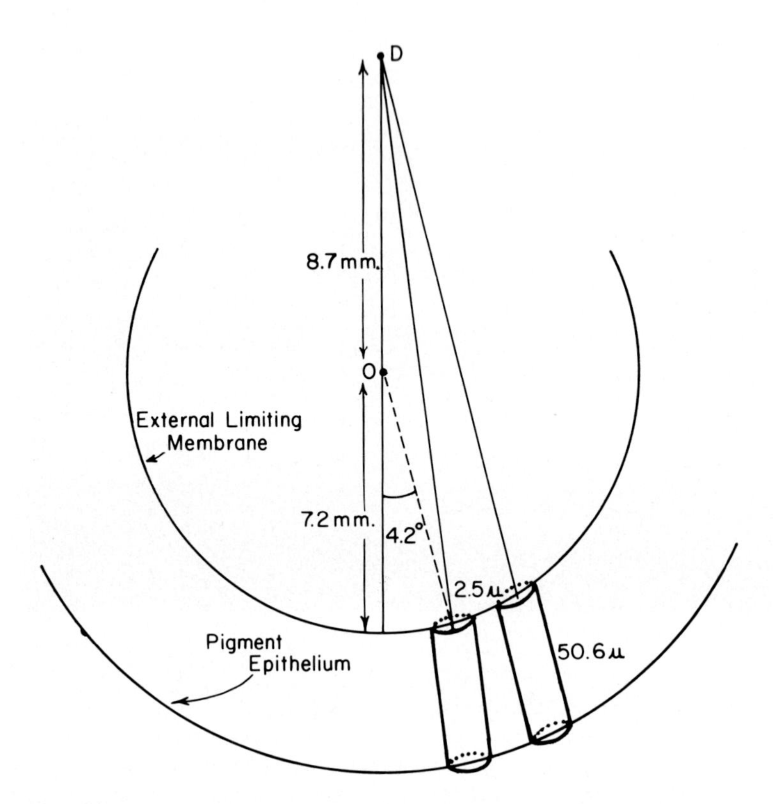

FIGURE 2. A schematic diagram illustrates the graded differential orientation of photoreceptors. Instead of pointing to the center of the eye (O), all photoreceptors are in line with an anterior point (D).

Despite the improved view of photoreceptors afforded by freeze-drying techniques, there still remained concern about the effects of quick-freezing and/or tissue sectioning on photoreceptor orientation. These concerns were largely allayed when Nicholas Webb, working in the biophysics unit at Kings College, recorded X-ray diffraction patterns from the intact frog eye. From the systematic skew of the diffraction patterns found as the beam moved across the retina, Webb (1972) was able to infer the same type and degree of graded differential orientation previously noted in the freeze-dried eye.

It should be added that psychophysical evidence of photoreceptor orientation already existed, for Stiles and Crawford (1933) had previously demonstrated that pencils of light entering the eye were perceived less well as they moved eccentrically toward the pupillary edge. Thus, one means for measuring visual efficiency is by the subjective perception of light, comparing the relative effectiveness of light rays presented at different orientations to the retina.

II. LIGHT, THE ORIENTING STIMULUS

Several observations using psychophysical measurements indicate that alignment of photoreceptors is in part a form of phototropism. When photoreceptors become disoriented in differing types of pathological conditions, there may thereafter be a progressive reordering such that Stiles-Crawford maxima are reestablished (Fankhauser and Enoch, 1962). For instance, as one example of this, the reestablishment of photoreceptor alignment has been documented in a patient after the surgical repair of a traumatic retinal detachment (Enoch *et al.*, 1973). Furthermore, there are early indications that the ability to regain a normal orientation is part of a more general phototropism: specifically, that as an adjustment to a changed direction of light source, photoreceptors can over time take on a new, orderly, and appropriate orientation to that light source. Such a reordering has been postulated by Dunnewold (1964) and, more recently, by Bond and MacLeod (1978). In each case, the Stiles-Crawford maxima in an eye with a displaced pupil was measured; in each case that maxima was displaced so that the photoreceptors were in line with the eccentric, incoming source of light.

III. CYTOSKELETAL FACTORS IN PHOTORECEPTOR ORIENTATION

Thus, what originally began as technical aid from a "muscle man" for a project in histochemistry led by fortunate chance to cytoskeletal and cytomotility questions. It also played a part in the collaboration of the two authors of the present review since one of us (B. B.) came to the Anatomy Department because of Penn's emphasis on muscle and cell motility and at Penn became attracted to the eye because of the range and scope of the cytoskeletal and cytomotility problems it presents.

Keeping in mind the foregoing geometrical and psychophysical observations, it is instructive to review the structure of rod and cone inner and outer segments to see how their form is maintained. In this vein, we may also consider possible mechanisms by which phototropism is achieved, that is, how do these rigid photoreceptors tilt to face an eccentric source of light.

The portion of the photoreceptor cell which intercepts light projects outward from the external limiting membrane. By convention, the part of the photoreceptor which contains the lamellar array of photoreceptor discs is called the outer segment. Suspended at some distance from the external limiting membrane by an inner segment, the outer segment contains the visual pigment rhodopsin. Acting as an optical funnel, the inner segment plays a physical role in photoreception. (Enoch, 1972). The two segments are joined by a slender connecting cilium. They function as a unit and that unit is fixed at one end upon the fulcrum of the external limiting membrane.

IV. CYTOSKELETAL ATTRIBUTES OF THE INNER SEGMENT

The dominant cytoskeletal structure of the inner segment is the microtubule. In transverse section microtubules are evenly distributed within the cytoplasm of rod and cone inner segments (Figs. 3, 4, and 8). Approximately 200 can be counted in a single cross section of a cone inner segment (Fig. 8). They extend the length of the inner segment as can be easily appreciated in longitudinal sections (Figs. 3, and 4). Although decidedly less common, some actin filaments can also be visualized within the inner segment as well (see Burnside, 1976). These are never as numerous or as easily visualized as the actin filaments present in the adjacent Mueller cell terminal projections (Fig. 7).

V. THE JUNCTION OF INNER AND OUTER SEGMENTS

In embryology the outer segment forms as a bud from the distal part of the connecting cilium. Thereafter, throughout life, it is the connecting cilium that joins the inner and outer segments (Fig. 9). Comprising nine doublets of tubules arranged in a ring, the connecting cilium is eccentric within the distal inner segment; tubules issuing from it end in one direction at variable distances along the outer segment (Steinberg and Wood, 1975) (Fig. 9) and in the other direction as a ciliary rootlet deep within the inner segment (see Fig. 8). Several strands of evidence have recently been presented to the effect that a polysaccharide-rich collar surrounds the connecting cilium (Hall and Nir, 1976; Bunt, 1978). Called

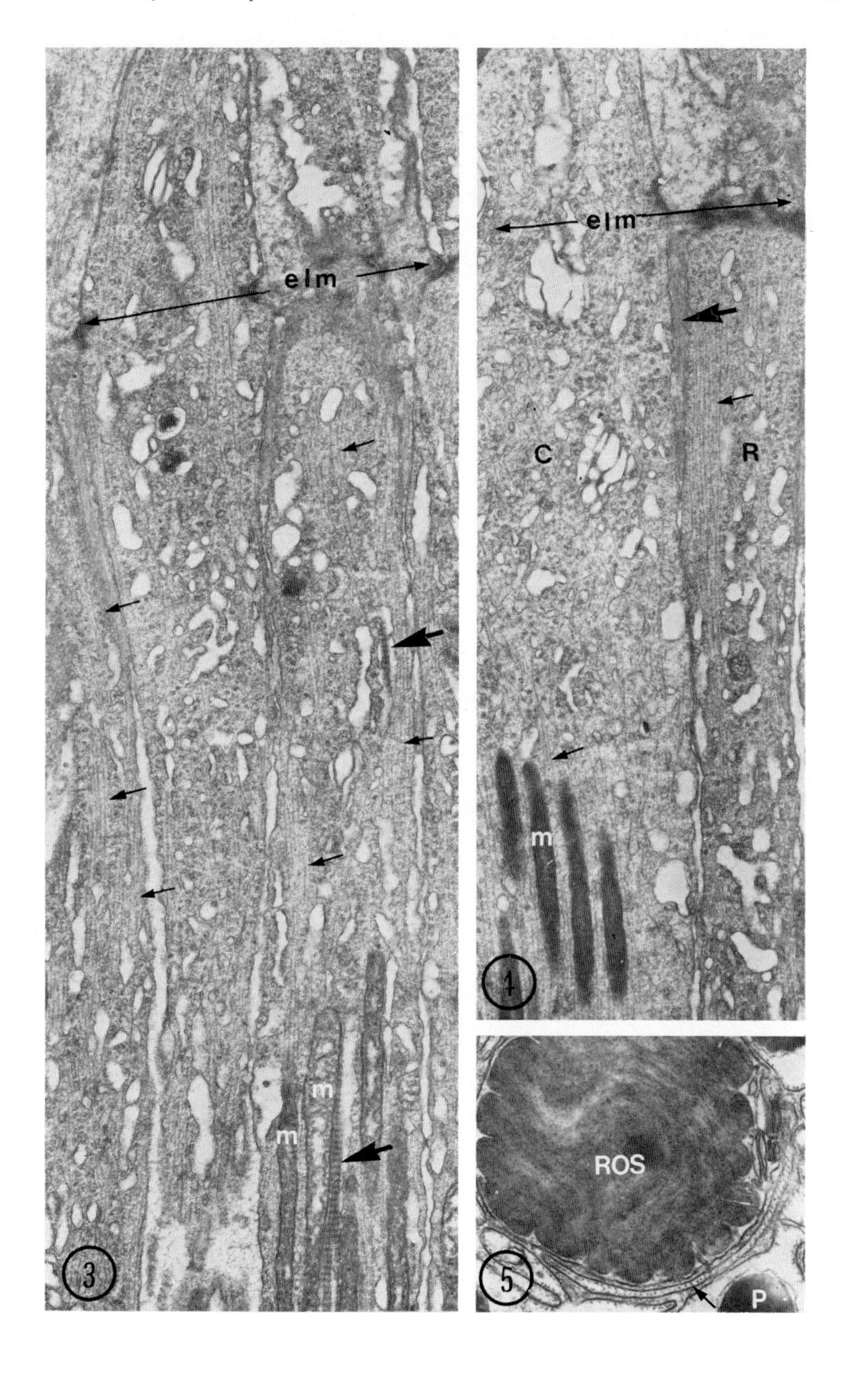
elm
m
m
3
elm
C
R
m
4
ROS
P
5

cilium-associated material, Bunt (1978) has pointed out the possibility that this material could act to stabilize the connecting cilium. In addition, extensions of the inner segment form a calyx about the proximal part of the outer segment (Fig. 9).

Further, in some species, there is a cuplike meeting of inner and outer segments. When looked at in the large, the junction area of the inner and outer segment, as Steinberg (1977) has observed, often forms a type of articulation.

VI. THE OUTER SEGMENTS

The outer segment, composed in the case of rods of a thousand or so evenly spaced membranous discs (Cohen, 1966), forms a structure analogous to a roll of coins, the plasma membrane in this instance playing the part of the paper wrapper. Just as in a roll of coins, the outside wrapping hampers lateral movement of individual discs and the rod outer segment holds its form. Structural stability is abetted by viscosity between adjacent discs. The combination of external wrapping and internal viscosity means that the rod outer segment could be expected to have considerable internal stability, requiring a marked shear force to dislocate rod discs.

FIGURE 3. Longitudinal section of inner segments of squirrel monkey rods near the region of the external limiting membrane (elm). The inner segments contain numerous microtubules (small arrows). In one rod two parts of the striated ciliary rootlet are visible (larger arrows). (Mitochondria = m) (x10,000).

FIGURE 4. Longitudinal section of inner segments of rod (R) and cone (C) near external limiting membrane (elm) in squirrel monkey retina. Numerous microtubules (small arrows) are present in both. Stiff microvilluslike processes of the Mueller cells project beyond the elm between the photoreceptors for several microns (large arrow) (x10,000).

FIGURE 5. Cross section of the tip of a rod outer segment (ROS) with embracing, leaflike processes of pigment epithelium (PE) (small arrow). Note the regular array of filaments cut in x-section in the PE process (small arrow). These filaments have been shown to contain actin (see Burnside, 1976). P-pigment granule. Squirrel monkey retina (x30,000).

VII. PIGMENT EPITHELIUM

Additionally, the pigment epithelial cells send apical processes to surround and to support the distal outer segment. Apical processes extend different distances in different species. For instance, in Rhesus monkeys they extend some 4 μm along the 28 μm long outer segment (Fig. 5). In all mammalian species so far examined, pigment epithelial apical processes appear to have structural stability themselves due to an orderly arrangement of membrane-bound actin filaments within them. (Fig. 5) (Murray and Dubin, 1975; Burnside and Laties, 1976).

The apical process of pigment epithelial cells have a special and unusual relationship to cone outer segments. As a general rule, cones have a lesser length than rods. Although there are substantial species and regional differences, it is common that cone outer segments do not reach the level of the pigment epithelium. Instead, when this occurs, the pigment epithelium sends out long apical processes to envelop the cone outer segment. They are already known to take part in the phagocytosis of shed cone material and may well, as Steinberg and Wood speculate, play a considerable role in the attachment of the retina (Steinberg and Wood, 1974; Steinberg *et al.*, 1977). To the degree that these long apical processes exert traction on the outer segment, they also constitute a mechanism for keeping the inner and outer segment of cones coaxial and for maintaining the orientation of the photoreceptive unit. The same attachment can as well be responsible for the small and reversible changes in photoreceptor orientation that follow the forward translation of the choroid and retina in marked accomodation (Blank *et al.*, 1975).

VIII. SUMMARY AND CONCLUSION

Thus, there is ample anatomical, X-ray diffraction, and psychophysical evidence to demonstrate that rods and cones within the retina are in a lawful, ordered array. If all are coaxial to the source of light, it follows that there is a graded differential orientation of the photoreceptive units within the retina. There is also evidence that under certain conditions, some physiological and some pathological, orientation of the photoreceptive unit can shift.

No single structure presently known accounts fully for the maintenance of the orientation of the photoreceptive unit. Instead, orientation derives from a series of special structural qualities working together; an attachment at the external limiting membrane, an array of microtubules in the inner

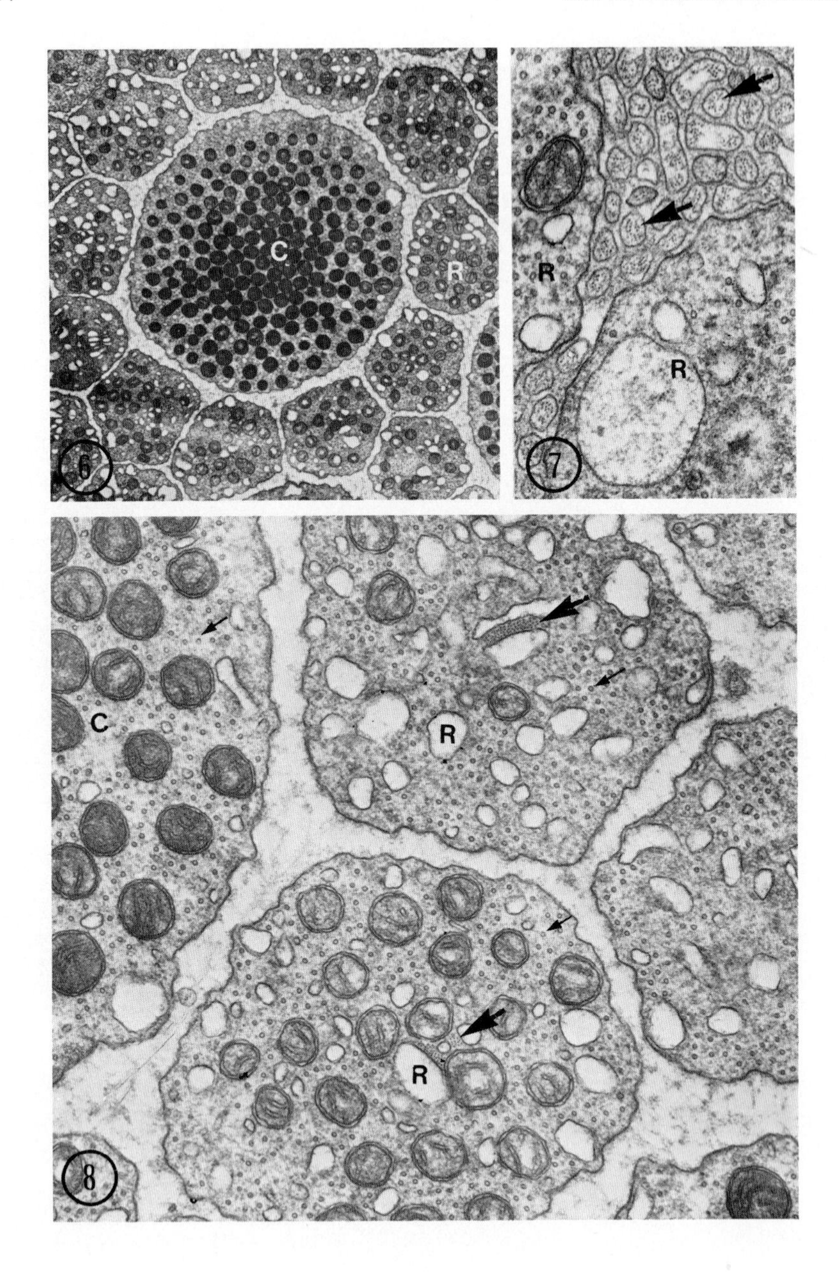

FIGURE 6. Low magnification micrograph of squirrel monkey rod (R) and cone (C) inner segments cut in cross section. Note that the cells are not attached to one another. They have an orderly arrangement. The dark round structures within them are filamentous mitochondria cut in cross section; these mitochondria are paraxially aligned in the photoreceptors. (x6000).

segment, a connecting cilium with microtubules extending in one direction to the outer segment and rootlets extending in the other direction deep into the inner segment, neatly stacked discs of the outer segment wrapped in plasma membrane, and a junction of pigment epithelium to photoreceptor outer segments.

Lastly, in order to explain the type of phototropism postulated by Dunnewold (1964) and Bond and MacLeod (1978), a tilt of 5° amplitude must be made by a photoreceptor. Reordering of a comparable degree from abnormal to normal orientation takes place after retinal reattachment (Enoch *et al.*, 1973). If, as supposed, the inner and outer segment pivot against the external limiting membrane, the single recognized agency available for such a movement is the microtubule of the inner segment. By elimination, each of the other identified cytoskeletal elements has to do with the form of a single photoreceptor (this includes packing of photoreceptors which is not discussed in the present report) rather than with a change in its attitude. Assuming for the moment that differential growth of microtubules alters photoreceptor orientation, the question arises: How is a signal for differential growth given? Since it is not clear from waveguide theory that displacement of a light source leads to an altered distribution of light either within the inner or the outer segment, a ready answer is not at hand. At present it simply is not known how such a signal is received or translated.

FIGURE 7. Cross section through the inner segments of two rods (R) just beyond the outer limiting membrane. The microvilluslike projections of the Mueller cells (large arrow) contain numerous actin filaments. Squirrel monkey retina (x45,000).

FIGURE 8. Cross sections of inner segments of rods (R) and cone (C) at a level distal to that in Fig. 7. Note the numerous and regularly spaced microtubules (small arrows) and the filaments of the centrally located ciliary rootlet (large arrows). Squirrel monkey retina (x29,000).

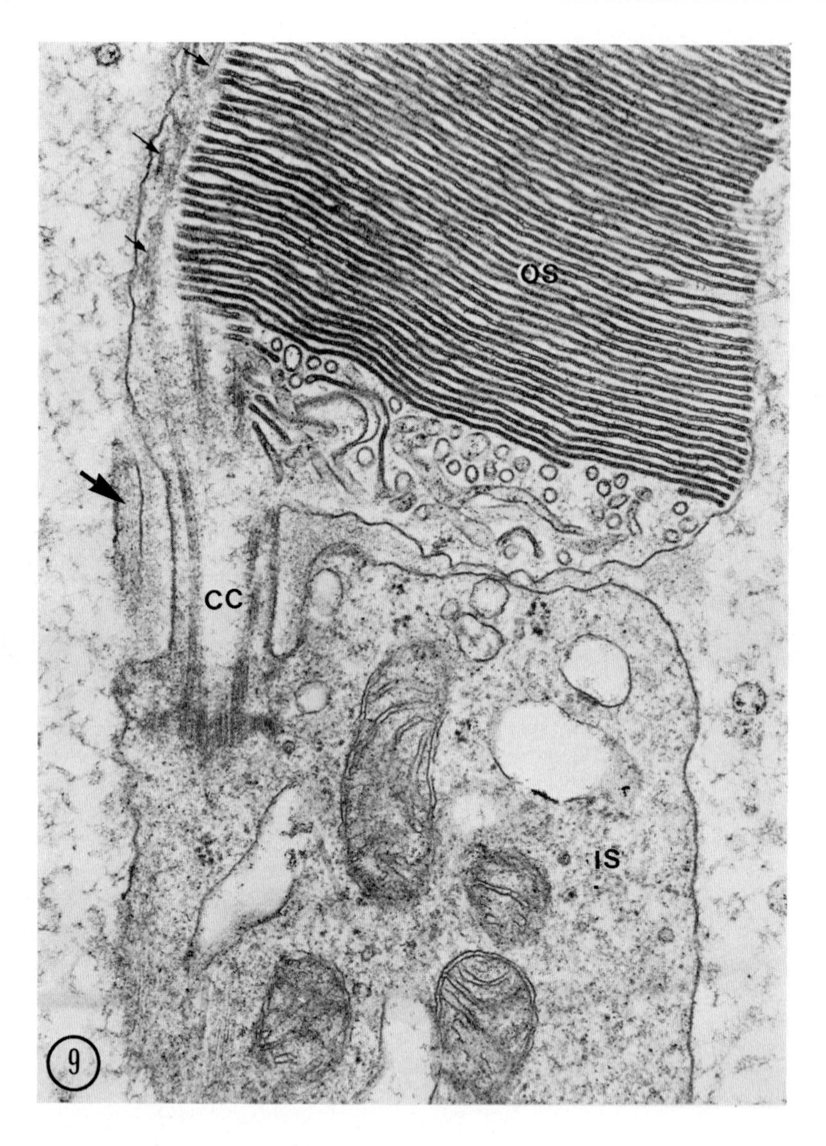

FIGURE 9. Longitudinal section through connecting cilium (CC) which joins the inner (IS) and outer segments (OS). Note that microtubules of the rudimentary cilium project alongside the discs in the outer segment (small arrows). Also visible is one of the microvilluslike projections which extend from the inner segment to cup the base of the outer segment (large arrow). Squirrel monkey rod (x60,000).

ACKNOWLEDGMENTS

This work was supported by National Eye Institute Grant 01194, an unrestricted grant from Research to Prevent Blindness, Inc., New York, and a grant from the Pennsylvania Lions Sight Conservation and Eye Research Foundation, Inc. (to A.L.) and NIH Grant GM-20109 (to B.B)

REFERENCES

Blank, K., Provine, R. R., and Enoch, J. M. (1975). *Vision Res. 15,* 499.

Bond, A. B., and MacLeod, D. I. A. (1978). (Submitted for publication.

Bunt, A. (1978). *Invest. Ophthalmol. 17,* 90-103.

Burnside, B. (1976). *J. Supramol. Struct. 5,* 257-275.

Burnside, B., and Laties, A. M. (1976). *Invest. Ophthalmol. 15,*570-575.

Cohen, A. I. (1966). *In* "The Retina: Morphology, Function and Clinical Characteristics" (B. R. Straatsma, *et al.* eds.). University of California Press, Berkeley.

Dunnewold, C. (1964). "On the Campbell and Stiles-Crawford Effects and Their Clinical Significance." Institute for Perception RVO-TNO, Soesterberg.

Enoch, J. M. (1972). *Am. J. Optom. and Arch. Am. Acad. Optom. 49,* 455-471.

Enoch, J. M., VanLoo, J., and Okun, E. (1973). *Invest Ophthalmol. 12,* 849.

Fankhauser, R., and Enoch, J. M. (1962). *Arch. Ophthalmol. 68,* 240-251.

Hall, M. O., and Nir, I. (1976). *Exp. Eye Res. 22,* 469.

Laties, A. M., Liebman, P. A., and Campbell, C. E. (1968). *Nature 218,* 172-173.

Laties, A. M. (1969). *Tissue and Cell 1,* 63-81.

Laties, A. M., and Enoch, J. M. (1971). *Invest. Ophthalmol. 10,* 69-77.

Murray, R. L., and Dubin, M. W. (1975). *J. Cell Biol. 64,* 705-710.

Schmidt, W. J. (1938). *Kolloidzeitschrift 85,* 137-148.

Steinberg, R. H., and Wood, I. (1974). *Proc. R. Soc. Lond. B 187,* 461-478.

Steinberg, R. H., and Wood, I. (1975). *J. Ulstrastruct. Res. 51*, 397-403.
Steinberg, R. H., Wood, I., and Hogna, M. J. (1977). *Phil. Trans. R. Soc. Lond. 277*, 459-474.
Steinberg, R. H. (1977). Personal communication.
Stiles, W. S., and Crawford, B. H. (1933). *Proc. R. Soc. Lond. Series B. 112*, 137048.
Webb, N. (1972). *Nature 235*, 44-46.

MITOSIS

Motility in Cell Function
Proceedings of the First John M. Marshall Symposium in Cell Biology

MOLECULAR MECHANISM OF MITOTIC CHROMOSOME MOVEMENT

Shinya Inoué
Daniel P. Kiehart
Issei Mabuchi
and
Gordon W. Ellis

Program in Biophysical Cytology
Department of Biology
University of Pennsylvania
Philadelphia, Pennsylvania
and
Marine Biological Laboratory
Woods Hole, Massachusetts

Dedicated with fond memories to John Marshall, great humanist, scientist, and teacher.

At this time, the three systems[1] which are considered active candidates for force generation in mitotic chromosome movement are a) acto-myosin, b) microtubule sliding, and c) microtubule assembly-disassembly.

In the following papers, Dr. Sanger will review those data that suggest the presence of actin in the spindle region and Dr. McIntosh will consider data implicating microtubule sliding as part of force production in chromosome movement. In

[1]*The Bajers propose that lateral interaction between microtubules plays a key role in chromosome movement ('73, "75). The interaction, especially between the kinetochoric and continuous microtubules must play an important role in force* transmission. *However, it is not clear to us how much the lateral interaction contributes to force* production *per se.*

ISBN 0-12-551750-5

this paper, we shall review microtubule assembly-disassembly and describe experiments which show that it is highly unlikely that an acto-myosin system is the force generator for chromosome movement.

I. SOME BASIC FEATURES OF MITOSIS

In contrast to the stable, force generating myofibrils of skeletal muscle, the mitotic spindle apparatus is a transient cytoplasmic structure. This apparatus assembles, functions, and disassembles in a matter of minutes as mitosis proceeds.

In the phase contrast or differential interference contrast microscopes, the spindle generally appears optically clear. In living cells, the spindle and astral rays thus appear as regions that exclude mitochondria, yolk granules, and other cellular inclusions. The spindle is generally not separated from the rest of the hyaloplasm[2] by a membrane despite some claims to the contrary (Wada, 1976). Rather, the spindle seems to be a region of the hyaloplasm, organized by transient arrays of microtubules.

To study spindle physiology, we have relied primarily on the polarizing microscope which reveals the anisotropic optical character of the spindle fibers and astral rays. These fibers manifest a weak but definite form birefringence that is positive. The birefringence can be accounted for by the array of oriented microtubules which structure the spindle fibers (Sato *et al.*, 1975). The fibers are resolvable with the light microscope.

As the cell organizes the mitotic apparatus and proceeds through anaphase and cytokinesis, birefringence allows us to 1) visualize and follow the transient fibrous organization of the mitotic spindle, and 2) quantitate the dynamic changes in the concentration of oriented microtubules as they are assembled or disassembled. The assembly-disassembly takes place in mitosis naturally, and can be modified by various alterations in the physical and chemical environment of the cell (Inoué, 1964; Inoué and Sato, 1967).

[2]*The inclusion-free ground substance of the cytoplasm, as used by Wilson (1928).*

Based on these studies we believe that the following six propositions are now reasonably well established. We propose that they be viewed as basic constraints in considering any mechanism for mitotic chromosome movement.

1) *In living cells, mitotic microtubules are in a labile equilibrium with a pool of tubulin dimers or oligomers.* Thus to a first approximation,

$$\text{TUBULIN} \underset{\substack{\text{Cold} \\ \text{Pressure} \\ \text{Colchicine} \\ \text{Ca}^{++}}}{\overset{\substack{\text{Glycols} \\ \text{D}_2\text{O}}}{\rightleftharpoons}} \text{MICROTUBULES.}$$

As reviewed elsewhere this equilibrium is not only seen in living cells but has also been observed during microtubule assembly in vitro (reviews Olmsted and Borisy, 1973; Inoué, 1976). Both *in vivo* and *in vitro,* the equilibrium appears to be dynamic (Bertalanffy, 1950) in nature. According to Weisenberg (personal communication), the equilibrium constants differ at opposite ends of the microtubule and produce a net flow of tubulin through the microtubule. Thus, tubulin appears to be incorporated at or near one end of the microtubule and released at the other end (Forer, 1965; Wilson and Margolis, 1978).

2) Despite the lability of spindle fibers, they exhibit mechanical integrity. From metaphase to anaphase, the mechanical integrity of chromosomal spindle fibers and their birefringence are coextensive in time and space. A birefringent fiber, therefore, mechanically links the individual chromosome, via its kinetochore, to the spindle pole. This has been clearly demonstrated in crane fly and grasshopper spermatocytes by Begg and Ellis through a combination of micromanipulation and polarization optical studies (Begg and Ellis, 1979; Nicklas, 1971; Nicklas and Staehly, 1967).

3) *In metaphase, chromosome-to-pole movement can be induced artificially by slow disassembly of microtubules if one spindle pole is anchored.* The velocity of this induced chromosome movement is proportional to the rate of microtubule disassembly, which is measured as the rate of birefringence decay. At higher rates of microtubule disassembly, chromosome velocity becomes proportionally greater until suddenly the chromosomes cease to move. We suppose that as microtubules disassemble too rapidly the structural integrity of the microtubule required for force transmission fails completely (reviewed in Inoué *et at.*, 1975; Salmon, 1975, 1976).

4) *In anaphase, chromosomal fiber microtubules disassemble,* most likely at or near the spindle poles (review Inoué, 1976). As in 3) above, *chromosome velocity is proportional to the rate of microtubule disassembly* (Fuseler, 1975).

5) *Chromosomes move in anaphase by two divisible events: anaphase-A and anaphase-B (Fig. 1).* In anaphase-A chromosomes

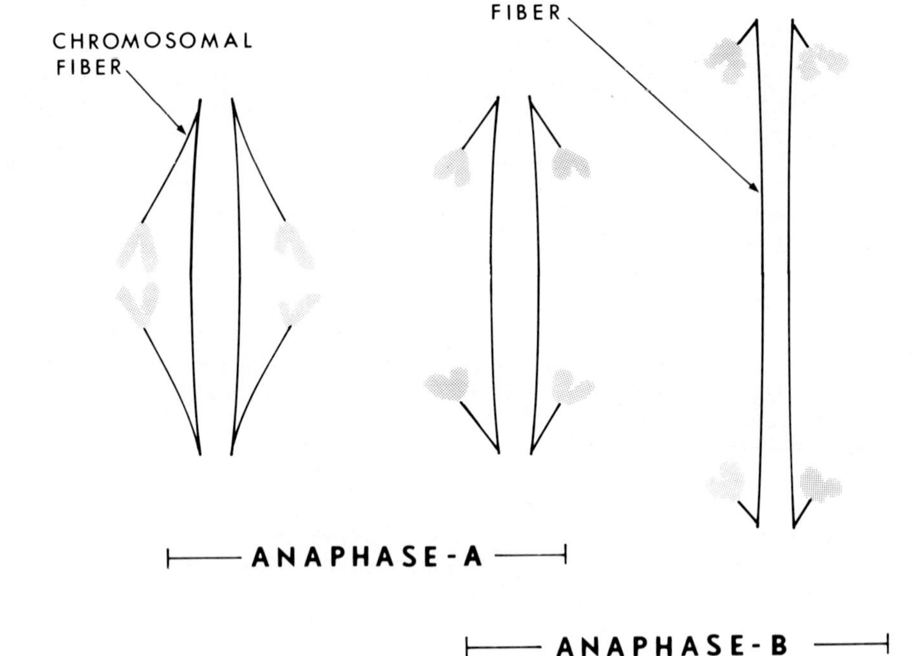

FIGURE 1. Schematic of anaphases-A and -B and location of spindle fibers. The fibers contain parallel arrays of long microtubules; at least some of the chromosomal fiber microtubules are attached to the kinetochores on the chromosomes.

move to the spindle poles as chromosomal fibers shorten. In anaphase-B the chromosomes are moved farther apart by the extension of the pole-to-pole distance. Generally these modes of anaphase chromosome movement overlap each other. In certain cases the two occur independently. For example in meiosis of oocytes, chromosomes separate by chromosome-to-pole motion without any spindle elongation (Fig. 3a,b). In certain protozoa, pole-to-pole elongation takes place only after chromosomes have reached the spindle poles (Inoué and Ritter, 1975, 1978).

6) *Microtubule disassembly is essential for chromosome-to-pole movement.* The disassembly rate-limits anaphase-A chromosome velocity (Fuseler, 1975, Salmon, 1975; Inoué *et al.*, 1975).

Given these and related observations, we have proposed (Inoué, 1959; Inoué and Sato, 1967; Inoué, 1976) that chromosomes are transported away from centrosomes, etc., by growth of microtubules (which accompanies their assembly from tubulin), and that chromosomes are transported towards the poles by shortening of microtubules (which accompanies their disassembly).

As chromosomes are drawn poleward, the poles must be held apart. In turn, as the chromosomal fiber microtubules disassemble, they must remain anchored both to the kinetochore and to the pole of the supportive central spindle. In this respect a labile lateral interaction between kinetochoric and continuous microtubules may play an important role in chromosome movement.[1]

II. A POSSIBLE SOURCE OF CYTOCHEMICAL ARTIFACT

As will be reported by the next speakers, much attention has recently been given to cytochemical use of fluorescent antibodies and other compounds which exhibit affinities to specific "contractile" proteins. However, the interpretation of these staining patterns for the localization of suspected participants in the mitotic process has a potential for error that arises from the structural aspects of dividing cells.

As has often been noted, the mitotic spindle during its formation excludes cytoplasmic particulates from its volume (e.g., Schrader, 1953, Fig. 1). Thus the relative concentration of hyaloplasm within the spindle is likely to be greater than that in the remaining cytoplasm. This results from the fact that the volume occupied by the microtubules in the spindle is typically only a little over 2% of the spindle volume (Sato *et al.*, 1975) while the volume occupied by nonhyaloplasmic constituents in the general cytoplasm may vary from a minimum of about 12% (osmotic dead space) to values over 80% (sedimentation studies) (Heilbrunn, 1952, pp. 130-131). The exclusion of most microscopic particulates from the spindle increases their relative concentration in the remaining cytoplasm while the particulates in turn displace hyaloplasm into the spindle region. Thus any molecular species distributed homogeneously in the hyaloplasm will be found in greater concentration in the spindle than in the peripheral cytoplasm.

A model to illustrate this phenomenon was constructed using plastic bags as the "cell membrane," glass beads as the "cytoplasmic particulates," fluorescein-containing glycerol as the "hyaloplasm," and a spindle model constructed of slender glass rods spaced to exclude the glass beads. Illuminated with near uv, the mitotic model shows a markedly increased fluorescence in the spindle region (Fig. 2).

An elevated fluorescence is also observed in the spindle region in a living cell injected with fluorescein labelled nonspecific IgG. This spindle fluorescence presumably originates from a similar exclusion of labelled IgG by cytoplasmic inclusions outside the spindle region (Kiehart, manuscript in preparation).

Consequently the appearance of a higher concentration of (flourescent) label in the spindle region does not of itself suggest a specific role for the target molecules in the mitotic mechanism.

III. DOES AN ACTOMYOSIN SYSTEM MOVE CHROMOSOMES?

As has been previously reported, we have been able to examine the role of myosin in chromosome movement by injection of an antibody into dividing starfish eggs (Kiehart *et al.*, 1977). The antibody, made against *Asterias* egg myosin, has been characterized by Mabuchi and Okuno (1977). It forms a single precipitin band against both purified egg myosin and crude egg homogenate. It also blocks the actin activated Mg-ATPase of myosin, but has no effect on the Ca^{++}- or Mg^{++}-ATPase activity of myosin in the absence of actin.

While cytokinesis is completely inhibited by injected antibody, mitosis is not inhibited (Fig. 3). Neither the poleward migration of the chromosomes nor the separation of half spindles is blocked (Kiehart *et al.*, manuscript in preparation). Even a fivefold increase of injected antibody over the dose required to prevent cleavage does not prevent chromosome movement.

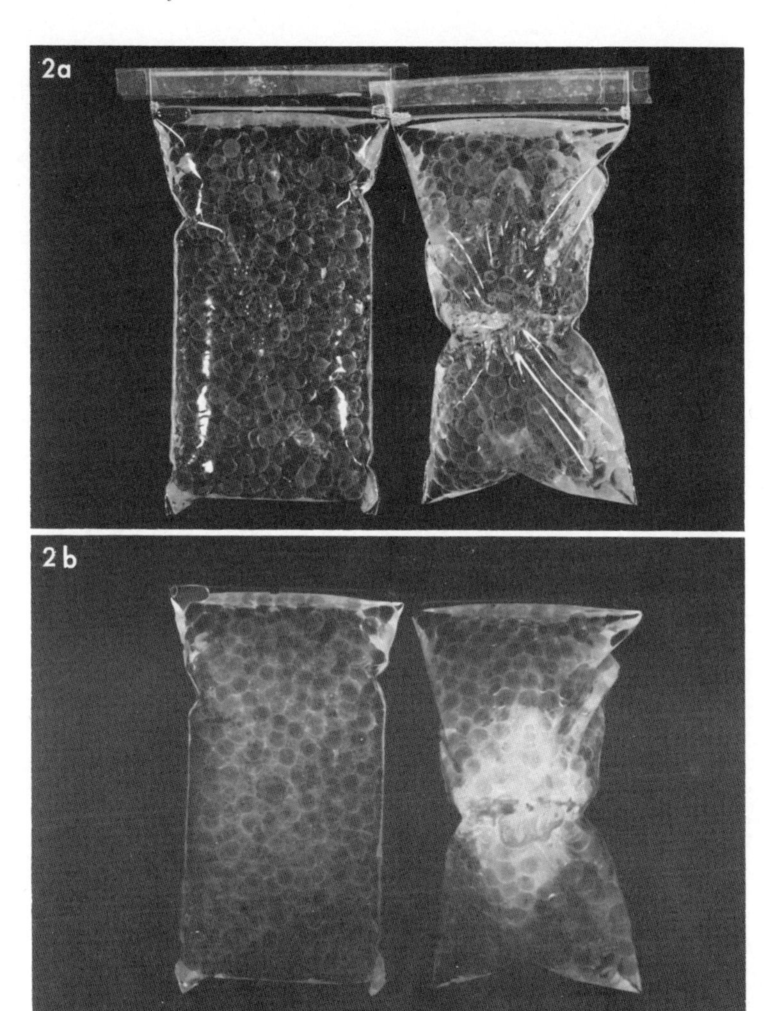

FIGURE 2. Model "cells" (plastic bags) showing on the left an interphase model and on the right a metaphase model. Each model contains 60 ml of glass beads (measured by displacement) and 100 ml of a 30 μM solution of fluorescein in glycerol. The metaphase model contains, in addition, a spindle model constructed of slender glass rods spaced to exclude the beads. (a) Models as seen in visible light. (b) Fluorescence of models under near ultraviolet illumination. The greater concentration of "hyaloplasm" in the granule-free spindle region of the metaphase model gives rise to an increased fluorescence, apparently of the spindle. Note the appearance of fluorescent "spindle fibers."

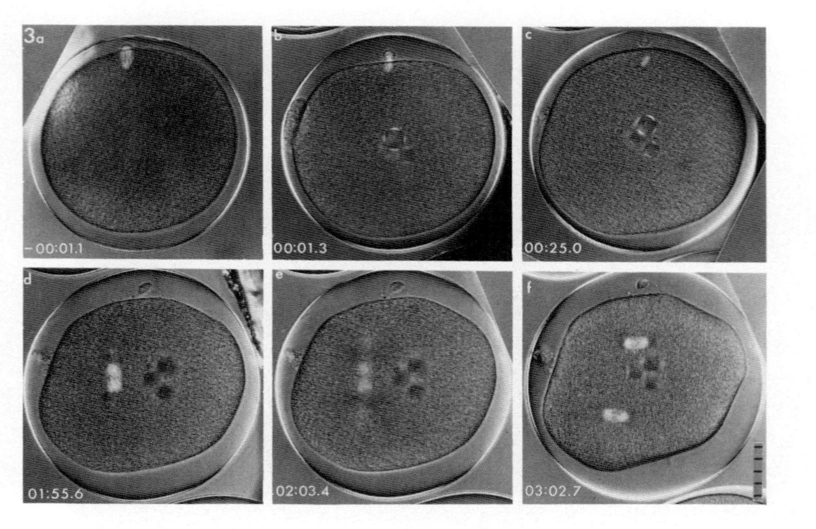

FIGURE 3. Microinjection of antibody against egg myosin blocks cytokinesis but not chromosome movement. A fertilized Asterias forbesi *egg in anaphase of first meiosis is injected with approximately 1.0 ng of rabbit IgG containing antibody against* Asterias amurensis *egg myosin. (a) Cell before injection. The birefringent, first meiotic spindle in midanaphase is clearly seen in polarized light. (b) Cell after injection. Oil droplets injected along with aqueous antibody solution mark the approximate site of injection. Meiosis progresses normally as chromosomes move to spindle poles (anaphase-A). Note characteristic lack of pole-to-pole elongation (anaphase-B) in this telophase cell. (c) 25 min after injection. The first polar body has already formed. Second meiotic spindles, one in the oocyte and one in the polar body, have appeared. Eventually meiosis proceeds but cytokinesis that would normally form the second polar body is blocked by anti-myosin. (d) Almost 2 hr after injection. The first metaphase* mitotic *spindle is seen. (e) In* mitotic *anaphase, normal pole-to-pole elongation (anaphase-B) occurs. Daughter nuclei are eventually formed at the poles, indicating that anaphase-A and -B proceed normally in the presence of anti-myosin. (f) 3 hr after injection. Two spindles appear in a common cytoplasm. First cleavage was blocked by anti-myosin, but mitosis proceeded normally. Time given in h:m. Scale interval 10 μm.*

For controls, cells were injected with myosin-absorbed antibody, or with buffer alone, or with non-immune IgG. These cells exhibited normal mitosis and cytokinesis. Antibody is not excluded from the spindle region, as in approximately 3 min after injection, fluorescein conjugated nonimmune IgG diffuses to homogeneity throughout the cytoplasm, with higher fluorescence found in the spindle. Thus, anti-myosin should be able to block an actin-myosin interaction anywhere in the cell.

By virtue of anti-myosin's inability to block chromosome movement, we conclude that an acto-myosin system is highly unlikely to contribute to force production in mitosis.

CONCLUDING REMARKS

In conclusion, while actin and myosin have been reported to be localized in the spindle, there is good reason to believe that all of the hyaloplasmic components would be found in increased concentration in the spindle region unless associated with excluded particulates or organelles. We have now performed functional tests which show that an actomyosin system is highly unlikely to be involved in either anaphase-A or anaphase-B. In contrast, the spacial distribution and the temporal behavior of microtubular bundles strongly implicate the microtubules as force transmitters and as regulators for chromosome movement. In addition, physiological experiments on spindle birefringence and chromosome behavior in living cells are most simply explained if chromosomes were transported by the very act of assembly and disassembly of microtubules.

ACKNOWLEDGMENTS

Supported in parts by grants from NSF-BMS 7500473, NIH-5 RO1 GM23475, NIH training Grant-5 TO1 HD00030 (D.P.K.), and NSF Grant-BMS 74-19934 (I.M. to Lewis G. Tilney).

REFERENCES

Bajer, A. S. (1973). *Cytobios 8,* 139-160.

Bajer, A., and Molè-Bajer, J. (1975). *In* "Molecules and Cell Movement" (S. Inoué and R. E. Stephens, eds.), pp. 77-96. Raven Press, New York.

Begg, D. A., and Ellis, G. W. (1974). *J. Cell Biol.* (in press).

Bertanlanffy, L. von. (1950). *Science 111,* 23-29.

Forer, A. (1965). *J. Cell Biol. (Mitosis suppl.) 25,* 95-117.

Fuseler, J. (1975). *J. Cell Biol. 67,* 789-800.

Heilbrunn, L. V. (1952). "An Outline of General Physiology," 3rd ed. W. B. Saunders Co., Philadelphia.

Inoué, S. (1959). *Rev. Mod. Phys. 31,* 402.

Inoué, S. (1964). *In* "Primitive Motile Systems in Cell Biology" (R. D. Allen and N. Kamiya eds.), pp. 549-598. Academic Press, New York.

Inoué, S. (1976). *In* "Cell Motility" (R. Goldman, T. Pollard, and J. Rosenbaum eds.), pp. 1317-1328. Cold Spring Harbor Laboratory, New York.

Inoué, S., and Ritter, H. Jr. (1975). *In* "Molecules and Cell Movement" (S. Inoué and R. E. Stephens, eds.), pp. 3-30. Raven Press, New York.

Inoué, S., and Ritter, J. Jr. (1978). *J. Cell Biol. 77,* 655-684.

Inoué, S., and Sato, H. (1967). *J. Gen. Physiol. 50,* 259-292.

Inoué, S., Fuseler, J., Salmon, E., and Ellis, G. W. (1975). *Biophys. J. 15,* 725-744.

Kiehart, D. P., Inoué, S., and Mabuchi, I. (1977). *J. Cell Biol. 75,* 258a.

Mabuchi, I., and Okuno, M. (1977). *J. Cell Biol. 74,* 251-263.

Nicklas, R. B. (1971). *In* "Advances in Cell Biology" (D. M. Prescott, L. Goldstein, and E. McConkey, eds.), Vol. 2, pp. 225-297. Appleton-Century-Crofts, New York.

Nicklas, R. B., and Staehly, C. A. (1967). *Chromosoma 21,* 1-16.

Olmsted, J. B., and Borisy, G. G. (1973). *Biochemistry 12,* 4282-4289.

Salmon, E. D. (1975). *Ann. N.Y. Acad. Sci. 253,* 383-406.

Salmon, E. D. (1976). *In* "Cell Motility" (R. Goldman, T. Pollard, and J. Rosenbaum, eds.). Cold Spring Harbor Laboratory, New York. pp. 1329-1341.

Sato, H., Ellis, G. W., and Inoué, S. (1975). *J. Cell Biol. 67,* 501-517.

Schrader, R. (1953). "Mitosis: The movements of chromosomes in cell division," 2nd ed. Columbia University Press, New York.

Wada, B. (1976). *Cytologia 41*, 153-175.

Wilson, L. and Margolis, R. L. (1978). *In* Cell Reproduction: in Honor of Daniel Mazia" (E. R. Dirksen, D. M. Prescott and C. F. Fox eds.), pp. 241-258. Academic Press, New York.

ACTIN AND THE MITOTIC SPINDLE

Joseph W. Sanger
Jean M. Sanger
and
Johanna Gwinn

Department of Anatomy
School of Medicine
University of Pennsylvania
Philadelphia, Pennsylvania

> "Both spindle and asters as seen in sections ordinarily show a beautiful fibrillar structure, consisting of delicate and closely crowded filaments which radiate from the spindle-poles, the astral rays spreading in all directions as they thread their way through the protoplasmic meshwork, and finally branching out in it to lose themselves insensibly."
>
> E. B. Wilson (1928)

The third edition of E. B. Wilson's classic book, "The Cell in Development and Heredity" was published 50 years ago and summarized what had been learned in the previous 50 years about the process of cell division. In these past 50 years, while many investigators have examined the problem of chromosome movement, its solution still remains elusive. The mechanism of chromosome movement, whether in meiosis or mitosis still remains a central unsolved problem in biology. The inclusion of this section on mitosis in a symposium devoted mainly to muscle is a reflection of John Marshall's interest in nonmuscle motility. That there is a relationship between muscle contraction and the pinching of an animal cell in two is widely accepted by a number of workers (Arnold, 1977).

ISBN 0-12-551750-5

That there is a relationship between muscle contraction and chromosomal movement is a possibility which has only recently been considered. Most of the impetus for this consideration has come from evidence that actin is present in some spindles. We will review some of the evidence which leads us to think that actin is a component of the spindle apparatus. However, even if we could be certain that actin is a normal component of spindles, we would still not know it if had a functional role in chromosome movement. Whether its function were contractile or cytoskeletal would still remain to be determined.

The spindle apparatus in contrast to a muscle myofibril is a "delicate" cell structure. Chromosome movement is clearly dependent on the presence of intact microtubules. If microtubules are destroyed by low temperature, high pressure, colchicine, or other drugs there is no movement of chromosomes. In fact the spindle apparatus falls apart. When these agents are removed, the spindle apparatus will reform and normal chromosomal movement can begin again.

Although microtubules are necessary for chromosome movement, do they supply the motive force? The reasons for first considering that they must was their abundance, alignment, chromosome attachment, and, most important, the observation that chromosome movement was dependent on their integrity. Despite all this suggestive evidence it is possible that another system supplies the force to move chromosomes while the microtubules supply the framework along which they move. The discovery that muscle proteins are in nonmuscle cells (see reviews by Pollard and Weihing, 1974; and by Korn, 1978) and that two of them may be localized in spindles has made this possibility the topic of active investigation.

Actin filaments decoratable with heavy meromyosin have been identified in electron microscopic analysis of thin sections of the spindles of glycerinated cells. Behnke *et al.* (1971) and Forer and Behnke (1972) using glycerinated cells treated with heavy meromyosin reported the presence of actin in meiotic spindles (crane fly testes), and Gawadi (1971, 1974) reported actin in mitotic spindles of locust spermatogonia. Hinkley and Telser (1974) also reported thin filaments decorated with arrowheads in mitotic spindles of neuroblastoma cells. More recently Forer and Jackson (1976) have found similar decorated filaments in the HMM-treated mitotic spindle in endosperm of *Haemanthus katherinae,* the African blood lily. Schloss *et al.* (1977) have also observed similar decorated thin filaments in PtK_1 calls. In recent electron microscopic work in our laboratory we have also found decorated actin filaments in PtK_2 spindles (We used FITC-HMM to decorate the actin).

Until 1974, only two cases were reported in which any filaments similar in diameter to actin had been identified in electron micrographs of routinely fixed spindles: Bajer and Molè-Bajer (1969) in *Haemanthus:* Müller (1972) in *Pales*. In neither case were the filaments clearly documented, possibly a reflection of their rarity and lability. More recently, McIntosh *et al*. (1975) using rat kangaroo (PtK_1) and Sanger (1977) using another strain of rat kangaroo cells (PtK_2) have also observed thin filaments in the mitotic spindle of routinely fixed cells. Perhaps more cases will be documented, once, more workers are looking for the filaments in the spindle. It should be remembered that Maupin-Szamier and Pollard (1978) have shown that actin filaments not stabilized with tropomyosin or heavy meromyosin are destroyed by osmium even after glutaraldehyde fixation. The work of Whalen *et al*. (1976) and Garrels and Gibson (1976) shows that there are several different types of actin α, β, and γ. If all are located within one cell. one type may be more sensitive to osmium than the others. Different fixation techniques, therefore, may preserve more actin than we can now see in cells.

Recently developed light microscopic methods for localizing actin have made it possible to view on a whole cell level the intracellular distribution of actin. The methods employ actin antibodies (Lazarides and Weber, 1974; Lazarides 1976) or fluorescent myosin fragments (Sanger 1975a; Sanger and Sanger 1976; Sanger 1977) (see Fig. 1). Using directly labelled fluorescent antibodies against actin, Lazarides and Weber showed that actin was localized in bundles in interphase cells and in the ruffling edges of motile cells. Controls showed the unlabelled actin antibody from rabbit to react specifically with actin on Ouchterlony plates. The labelled preimmune serum from the rabbit produced only a little diffuse staining in the cells while the commercially labelled anti-goat antibody did not stain the cells. Thus the staining achieved with the actin antibody represents the disposition of actin in the cell. A totally independent method of actin localization has been based on the binding of fluorescent labelled myosin fragments such as heavy meromyosin (HMM) and the subfragment-1 of myosin (S-1). The binding specificity of Fl-HMM and Fl-S-1 was confirmed on a variety of myofibrils: rabbit, chick, lobster, and scallop as well as on acromsomal caps of sperm, intestinal brush borders, ruffled membranes, nerve and chondrogenic cells, smooth and cardiac muscle cells---all known to contain actin (Sanger and Sanger, 1976). Motile interphase cells were found to possess not only fluorescent bundles of actin but also to have diffuse fluorescent staining in the pseudopodial regions.

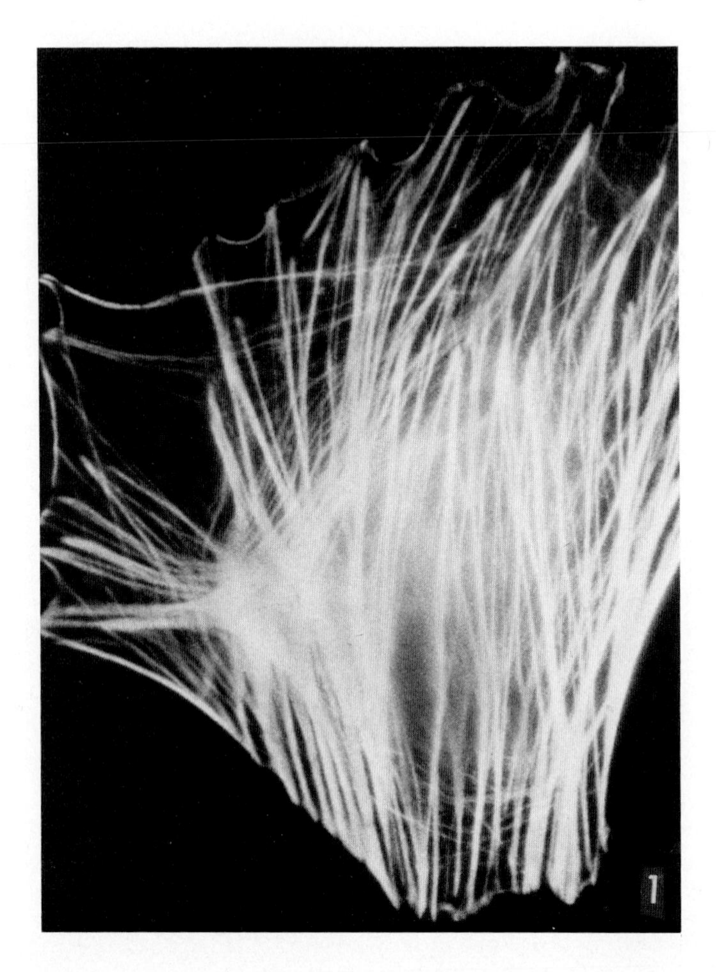

FIGURE 1. Control embryonic chick fibroblast stained with heavy meromyosin coupled to fluorescein isothiocyanate (Fl-HMM).

There is also evidence from fluorescent staining methods that actin is localized in mitotic spindles. Rat kangaroo cells do not round up as completely as most mitotic cells. Because they remain fairly flat during mitosis it is possible to identify the chromosomes and thus the progress and stages of mitosis. When these cells are glycerinated (Fig. 2) or fixed in formaldehyde and treated with detergent (Fig. 3) and then stained with Fl-HMM or Fl-S1, fluorescent staining is found in the spindle.

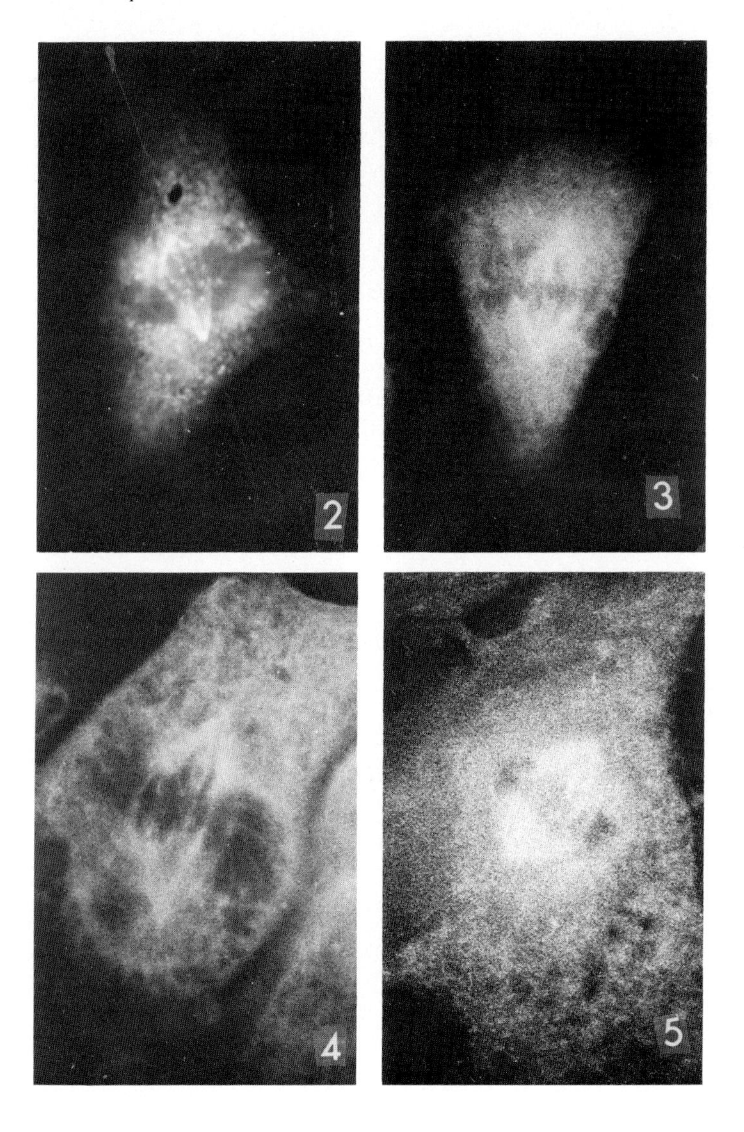

FIGURE 2. PtK_2 mitotic spindle. Cell was opened with glycerol and stained with Fl-HMM. This old procedure lasted two days.

FIGURE 3. PtK_2 mitotic spindle. Cell was fixed in formaldehyde, opened in nonidet P-40, and then stained with Fl-HMM. This new procedure lasted 1.5 hr.

FIGURE 4. Mitotic spindle of Salamander cell. Cell was fixed in cold acetone (-20°C) and then stained with tetra methylrhodamine isothiocyanate - HMM. This procedure lasted 40 min.

FIGURE 5. Mitotic spindle of embryonic chick cardiac fibroblast. Cell was fixed in formaldehyde and opened in cold acetone. The cell was then stained with Fl-HMM. This procedure lasted 1.5 hr.

Several other variations of staining the spindle were tried as well as using different cells. For example, salamander lung cells, primary culture, were fixed in cold acetone and then stained with Fl-HMM (Fig. 4). Chick cardiac fibroblasts were first fixed in formaldehyde and then placed in cold acetone. The cells were rehydrated and subsequently stained with Fl-HMM (Fig. 5). No matter how the cell was prepared (or what cell type) the stain in the spindle was confined to the fibers that connected the chromosomes with the centriolar regions. It was not present in astral fibers nor in the continuous spindle fibers that connect the poles and can be seen in the interzone between chromosomes during anaphase (Sanger and Sanger, 1976).

In our early staining of spindles (Sanger, 1975b) we used the procedure derived from muscle studies of glycerinating overnight, washing, and incubation of Fl-HMM for 10 to 24 hr. This procedure lasted for almost two days. The chance for postmortem extraction and movement of actin increases over such a long period of fixation and staining and, thus we have subsequently adopted many variations in our staining procedures. Our current procedure of using formaldehyde and nonidet or acetone takes no longer than 2 hr from fixing the cell to viewing it in the microscope. A comparison of Figs. 2 and 3 illustrates that the overall staining is about the same in both long-term glycerol treatment and short-term formaldehyde-nonidet. However, the kinetochores stand out in the glycerol procedure because of diminished staining in the fibers. In the formaldehyde procedure the staining along the fibers is uniform and the kinetochores no longer stand out. Other staining procedures geared to short periods of fixation and staining of the cells yield similar results as Fig. 3 (see Figs. 4 and 5). Recent and ongoing experiments on the types of proteins removed during the staining procedures indicates that we are removing part of the cellular actin (Fig. 6). This is a difficult problem to overcome. In Fig. 6 one can see that apparently no actin is extracted in the supernatant of cells fixed in cold ethanol (Fig. 1 was fixed in cold ethanol and subsequently stained with Fl-HMM). Yet if these ethanol treated cells are exposed to a low salt solution for 1 hr, as would be done for Fl-HMM staining, actin appears in the wash solution. Where is the actin coming from? Bray and Thomas (1976) have demonstrated that about 50% of the total actin in fibroblasts can be released after homogenization of the cells. This actin is nonsedimentable and appears to be in a monomer form (Bray and Thomas, 1976). We expect that monomer actin could be extracted by our procedures. It should be pointed out, though, that our procedures for staining cells do not involve homogenizing them so that presumably we are keeping much more actin in the cell.

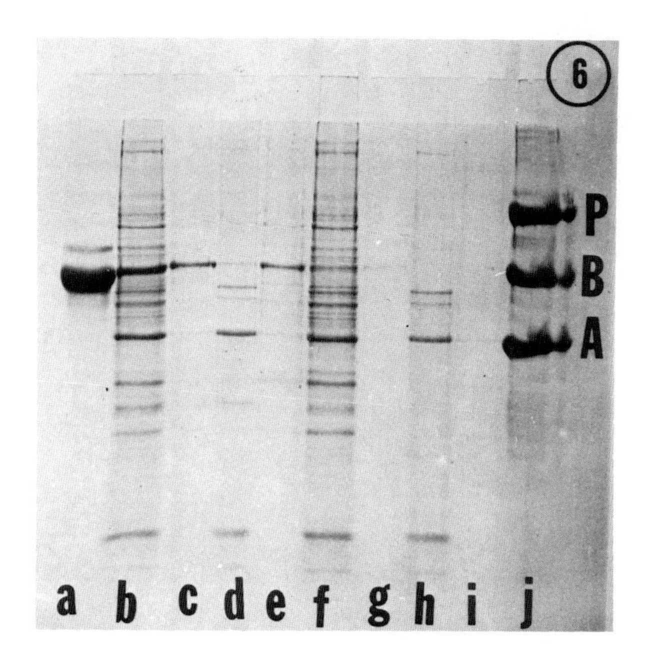

FIGURE 6. Slab gel of various fixation and extraction procedures on dishes of cells. These experiments were performed to demonstrate the effect of fixation procedure and myosin extraction on the cell's actin. (a) Sample of fetal calf serum: the major band is bovine serum albumin (BSA). All our cells are grown in 10% fetal calf serum in Eagle's medium. (b) Dish of cells extracted with glycerol and standard salt mix for a period of 1.5 hr. This is the procedure we now use for glycerol staining. (c) Supernatant from glycerol extracted dish. Note major band at BSA level, small band at actin level, and little else was extracted. (d) Dish of cells extracted with Nonidet P-40 (0.02% in 0.1 M KCl, 0.01 M PO_4, 0.001 M $MgCl_2$, pH 7.0). (e) Supernatant from Nonidet dish. Again notice major extracted band is at the BSA level and a small band at the actin level. More protein, however, is extracted with this procedure that with short term glycerination (see (c) above). (f) Dish of cells extracted with cold ethanol. (g) Supernatant of ethanol extracted dish. Note major band at BSA level. Note absence of actin. This procedure extracts the least protein. (h) Dish of cells that was first glycerol-extracted as in (b). The dish was then rinsed three times with a low salt solution and exposed to a Hasselbach-Schneider solution (myosin extracting solution) for 1.5 hr. Note continued presence of actin in myosin-extracted cells. (i) Supernatant from dish (h). Note small band at actin level. (j) Standards: Actin (A), bovine serum albumin (B), and paramyosin (P).

What is the possibility that HMM or S-1 induces actin filaments to form in the spindle? HMM can polymerize a solution of monomer actin even when the monomer concentration is very low (Yagi *et al.*, 1965; Kikuchi *et al.*, 1969). It is possible, then, that unpolymerized cytoplasmic actin may move into the spindle when we stain and be polymerized by the fluorescent HMM. If this is so, it is curious that the actin staining is confined to the chromosomal spindle fiber region. Nevertheless, the possibility that cytoplasmic contractile proteins move into the spindle during staining procedures is a problem which must be recognized.

Cande *et al.* (1977) have used indirect fluorescent antibody techniques to compare the localization of actin and tubulin in the mitotic spindle of the rat kangaroo cell (PtK_1). They demonstrated that actin is only found in the chromosomal spindle fibers. It was found to run continuously from the pole to the attachment on the chromosome at the kinetochore. Using conditions identical to those used for the actin antibody staining they added antibody against tubulin to sister cultures. The tubulin antibody stained not only the chromosomal spindle fibers, but also the astral rays and interzonal fibers between separated chromosomes. This clearly demonstrated that the two antibody staining reactions were specific and counters the criticism that fluorescent staining of cells is due to the sticking of the fluorescent compounds to dense fibrous networks in the cell. Furthermore in PtK_2 cells there is an extensive keratin filamentous system in the cells which is not stained by these actin and tubulin antibodies or the fluorescent myosin fragments (Sanger, 1977). These results further support the conclusion that the staining reactions are specific and suggest that actin is indeed a component of the spindle apparatus.

Of even greater interest is the agreement of the indirect actin antibody staining with the fluorescent myosin fragments staining results. That both stains should be confined to the chromosomal spindle fibers to the exclusion of the interzonal fibers and astral rays greatly strengthens the argument that actin is localized *in vivo* in this region. Furthermore, the tubulin indirect antibody staining shows that microtubules have not been removed from any particular region of the spindle. Thus, if actin moved into the spindle as a result of the staining technique, it seems unlikely that it would only become aligned along the chromosomal fibers. Electron microscopic evidence indicates that the interzonal region of the spindle does contain some actin fibers (Forer and Behnke, 1972; Gawadi, 1974). Whether their numbers are as great as those found in the chromosomal fibers cannot be judged until improved fixation methods become available that will preserve actin filaments to allow quantitation of results. Certainly, the light microscopic evidence indicates that there is much more actin in the chromosomal spindle fiber area (Sanger, 1975b; Cande *et al.*, 1977).

In addition to careful localization studies on both light and electron microscopic levels, we now need to test cell models of mitosis. A preliminary start has been made in this direction by Mabuchi and Okuno (1977) who have demonstrated that the injection of myosin antibody into a starfish blastomere at interphase blocked the formation of a cleavage furrow. However, if the antibody was injected near the beginning of cleavage the inhibition was reduced and cleavage usually took place. The antibody also affected the construction of the spindle. In 5 of the 17 blastomeres, spindle formation was stopped completely. In the other 12 blastomeres a smaller spindle was formed but chromosomal separation took place. However, instead of the normal 60 μm separation of daughter nuclei, the nuclei of myosin-injected cells separated by only 20 μm. Since the injection of preimmune serum had no effect on the spindle function, chromosome separation or nuclei separation, these experiments could be interpreted to indicate that myosin plays a role in the formation of a spindle and in the separation of poles (the opposite of what fluorescent staining of spindles would indicate). Mabuchi and Okuno used these observations to suggest that myosin was not involved in chromosome movement. Kiehart *et al.* (1977) report that microinjection of myosin antibody from one species when injected into oocytes of another species had no effect on chromosomal movement but did inhibit cytokineses. In these experiments, there was even a normal separation of daughter chromosome in contrast to the experiments of Mabuchi and Okuno (1977). This is ascribed to better microinjection technique. Before concluding that myosin (and actin?) has no role in chromosome movement, we must know how many types of myosin are present in the dividing cell. Myosin antibodies are very specific (Fugwara and Pollard, 1976; Rubinstein *et al.*, 1977). Could the antibody used have reacted with cleavage furrow myosin and not spindle myosin, if indeed two different types of myosin exist in these regions? Did the antibody crossreact with all the myosin in the cell? This is very important to know. If only one dimer of myosin per chromosome were left unreacted with antibody, this would provide enough force to permit chromosome movement (10^{-8} dynes). In the cleavage furrow, 100,000 times more force (10^{-3} dynes) is needed (Rappaport, 1971) and thus, many more myosin units would have to be unreacted with antibody before cleavage could occur. There is also an internal control for these microinjection experiments which should be examined. Saltatory movement is thought to be caused by an actin-myosin system (Rebhun, 1972). There was no discussion of whether or not the microinjection of myosin antibody stopped saltatory movement. If it does this would be strong evidence that myosin is not involved in chromosome movement. It would suggest that actin and myosin are responsible for the saltatory movement of particles in the spindle and have nothing to do with chromosome movement.

CONCLUSION

As workers at an earlier time in muscle research, we who are interested in mitosis are still searching our way through the crowded fibrillar elements of a beautiful structure. Unlike the stable structure of a myofibril, the spindle apparatus is very delicate reflecting its impermanence in the life of a cell. While it is clear that microtubules are a major component of this labile structure, other components have not been clearly identified. We feel that actin (and myosin) is one of these latter constituents but recognize that further studies are needed to substantiate this claim. Identification of the components of the spindle apparatus, however, is only the first step in determining what agents are responsible for chromosome movement.

REFERENCES

Arnold, J. M. (1976). *In* "The Cell Surface In Animal Embryogenesis" (E. Poste and G. Nicolson, eds.), pp. 55-80. Elsevier North-Holland, Amsterdam.

Bajer, A. S., and Molè-Bajer, J. (1969). *Chromosoma 27*, 448-484.

Behnke, O., Forer, A., and Emmersen, J. (1971). *Nature 234*, 408-440.

Bray, D., and Thomas, C. (1976). *Cold Spring Harbor Conf. Cell Prolif. 3*, 461-473.

Cande, W. Z., Lazarides, E., and McIntosh, J. R. (1977). *J. Cell Biol. 72*, 552-567.

Forer, A., and Behnke, O. (1972). *Chromosoma 39*, 145-173.

Forer, A., and Jackson, W. T. (1976). *Cytobiologie. 12*, 199-214.

Fugiwara, K., and Pollard, T. D. (1976). *J. Cell Biol. 71*, 848-875.

Garrels, J. I., and Gibson, W. (1976). *Cell 1*, 793-805.

Gawadi, N. (1971). *Nature 234*, 410.

Gawadi, N. (1974). *Cytobios. 10*, 17-35.

Hinkley, R., and Telser, A. (1974) *Espl. Cell Res. 86*, 161-164.

Kiehart, D. P., Inoué, S., and Mabuchi, I. (1977). *J. Cell Biol. 75,* 258a.

Kikuchi, M., Noda, H., and Maruyanna, K. (1969). *J. Biochem. (Tokyo) 65,* 945-952.

Korn, E. D. (1978). *Proc. Nat. Acad. Sci. U.S.A. 75,* 588-599.

Lazarides, E. (1976). *J. Cell Biol. 68,* 202-219.

Lazarides, E., and Weber, K. (1974). *Proc. Natl. Acad. Sci. U.S.A. 71,* 2268-2272.

Mabuchi, I., and Okuno, M. (1977). *J. Cell Biol. 74,* 251-263.

Maupin-Szamier, P., and Pollard, T. D. (1978). *J. Cell Biol. 77,* 837-852.

McIntosh, J. R., Cande, W. Z., and Snyder, J. A. (1975). *In* "Molecules and Cell Movement" (Inoué, S. and Stephens, R. E. eds.), pp. 31-75. Raven Press, New York.

Müller, W. (1972). *Chromosoma 38,* 139-172.

Pollard, T. D., and Weihing, R. (1974). *CRC Crit. Rev. Biochem. 2,* 1-65.

Rappaport, R. (1971). *Int. Rev. Cytol. 31,* 169-213.

Rebhun, L. (1972). *Int. Rev. Cytol. 32,* 93-137.

Rubinstein, N. A., Pepe, F. A., and Holtzer, H. *Proc. Nat. Acad. Sci. U.S.A. 74,* 4524-4527.

Sanger, J. W. (1975a). *Proc. Nat. Acad. Sci. U.S.A. 72,* 1913-1916.

Sanger, J. W. (1975b). *Proc. Nat. Acad. Sci. U.S.A. 72,* 2451-2455.

Sanger, J. W. (1977). *In* "Mitosis, Facts and Questions" (M. Little *et al.*, eds.), pp. 98-113. Springer-Verlag, Heidelberg.

Sanger, J. W., and Sanger, J. M. (1976). *Cold Spring Harb. Conf. on Cell Proliferation 3,* 1295-1316.

Schloss, J., Milsted, A., and Goldman, R. (1977). *J. Cell Biol. 74,* 794-815.

Whalen, R. G., Bulter-Browne, G. S., and Gros, F. (1976). *Proc. Nat. Acad. Sci. U.S.A. 73,* 2018-2022.

Wilson, E. B. (1928). "The Cell in Development and Heredity," 3rd ed. Macmillan, New York.

Yagi, K., Mase, R., Sakakibara, I., and Asai, H. (1965). *J. Biol. Chem. 240,* 2448-2454.

Motility in Cell Function
Proceedings of the First John M. Marshall Symposium in Cell Biology

STUDIES ON THE MECHANISMS OF CHROMOSOME MOVEMENT

J. Richard McIntosh

Department of Molecular, Cellular,
and Developmental Biology
University of Colorado
Boulder, Colorado

INTRODUCTION

During the formation of the mitotic spindle, the motions of individual chromosomes are complex, but the net result of the motile events is to arrange the already-doubled chromosomes at the midplane of the metaphase spindle. These organizational events involve first the orientation of the chromosomes so that sister chromatids point toward opposite poles and then the alignment of all chromosomes at the spindle equator. In most cells chromosome organization is mingled with the formation of the spindle itself; motility is in these cases coupled to controlled macromolecular assembly. During anaphase, the symmetry of the spindle is conserved as the chromatids separate and move toward opposite poles. Anaphase is more obviously a motile event than the earlier organizational movements of the chromosomes, but here too, there is a change in spindle organization: while the repositioning of the chromosomes is taking place, much of the spindle disappears, reflecting a macromolecular disassembly as the spindle completes its major purpose. The mitotic spindle is thus a particularly intriguing sort of motile machine. It forms for the occasion of cell division and disappears at the completion of its motile task.

From the early stages of its formation onward, the spindle exerts a force on each chromatid at the kinetochore (the chromosomal specialization to which a spindle fiber attaches). The direction of the force is defined by the direction of this spindle fiber (Nicklas and Staehly, 1967), and may generally be thought of as a pole-directed force, because the chromosomes orient so that each "kinetochore fiber" points toward one spin-

ISBN 0-12-551750-5

dle pole or the other. This pole-directed force is almost certainly responsible for the orientation of the chromosomes with one chromatid facing each pole (Östergren, 1959). It may also be responsible for the alignment of the chromosomes at the metaphase plate, because there are suggestions that the magnitude of the force is proportional to the distance between a metaphase chromosome and the pole it faces (Östergren, 1945; Bauer *et al.*, 1961; reviewed in Nicklas, 1977). During anaphase, the same force probably accounts for the shortening of the distance between chromosomes and poles and hence of the kinetochore fibers themselves.

Another aspect of spindle-associated motility is seen in the motion of the spindle poles. At the end of interphase, the poles are usually close together; they move apart either before or during spindle formation. During anaphase they generally move even farther apart in association with the spindle elongation that increases the distance of travel of the two sets of chromosomes. Forces that push the poles apart are a persistent feature of spindle action and must be considered in any analysis of the mechanisms of chromesome movement.

The fully formed spindle at metaphase displays several motile properties in addition to those obviously connected with chromosome segregation. The motions of chromosome fragments (Bajer, 1958) and of many unidentified objects within the spindle (Östergren *et al.*, 1960; Allen *et al.*, 1969; Nicklas and Koch, 1972) show that the spindle can exert pole-directed forces upon various bits of cytoplasm. Further, the motions of zones of reduced birefringence made on spindle fibers with an ultraviolet microbeam imply that the submicroscopic components of the spindle are flowing toward the poles, even when the chromosomes are essentially at rest (Forer, 1965).

This paper is an attempt to define some properties of spindles that are likely to be of mechanistic significance to generation of forces for chromosome motion and spindle growth.

SPINDLE COMPOSITION

The protein composition of the mitotic spindle could yield important insights into the mechanisms for chromosome movement, particularly in the light of the molecular conservativeness displayed by most motile machinery. Direct chemical analysis of spindles has thus far been of limited utility because there is no really successful method of isolation (reviewed in McIntosh, 1977). Isolated spindles clearly contain tubulin (Bibring and Baxandal, 1971; Chu and Sisken, 1976), and recent studies show a protein of electrophoretic mobility corresponding to one of the flagellar ATPases called dynein (Salmon and

Jenkins, 1977). A dyneinlike ATPase activity is found in several spindle preparations, but it is difficult to say whether it is a contaminant or a significant spindle component (Weisenberg and Taylor, 1968). Further evidence for dynein in spindles has recently been obtained with indirect immunofluorescence using antibodies prepared against flagellar dynein from sea urchin sperm tails (Mohri *et al.*, 1976). Immunofluorescence has also provided evidence that spindles contain the high molecular weight microtubule-associated protein from brain (Sherline and Schiavone, 1978) and the "Tau protein" from microtubules (Connolly *et al.*, 1977).

Spindles of glycerinated cells contain actin (Forer and Behnke, 1972; Gawadi, 1974; reviewed by Sanger, this volume). There is some evidence for myosin in spindles (Fujiwara and Pollard, 1976), but tropomyosin is apparently lacking (Sanger and Sanger, 1976). Immunofluorescence has also been used to identify the presence of a calcium-binding, regulatory protein (CDR) in spindles (Welsh *et al.*, 1978). The distribution of this troponin-C-like protein in spindles is intriguingly different from that of either tubulin or actin. Petzelt has identified a spindle-associated, Ca^{++}-activated ATPase, (Petzelt *et al.*, 1972), which is probably a membrane-bound enzyme, so there are several possible functions for the CDR. One could imagine that it is associated with the Ca^{++}-activated enzyme and is important for the regulation of microtubule assembly, or it could be part of a Ca^{++} regulatory pathway in the contraction of an hypothetical actomyosin system essential in chromosome movement.

The composition information may be summarized by stating that the spindle contains the proteins to allow chromosome movement by any of these mechanisms: by a regulated assembly and disassembly of microtubules, by a flagellalike microtubule sliding, or by a musclelike mechanochemistry. There is, therefore, no particular restriction placed on the acceptable models for chromosome movement by the current information about proteins in the spindle, and we are forced to look elsewhere to try to understand how chromosomes move.

PHYSIOLOGICAL EVIDENCE

The most useful sort of evidence for understanding chromosome movement would be that derived from experiments on spindle function. Colchicine and its analogues as well as other antimitotic drugs provide direct evidence for the importance of microtubules in chromosome movement (Inoué, 1952; Wilson and Bryan, 1974). There are no parallel experiments with drugs affecting actin to suggest any importance of this spindle component for chromosome movement. In fact, two experiments suggest actomyosin plays no role in chromosome movement. Cytochalasin B acts on some systems to disrupt actin-containing microfilaments and prevent an associated motile process, such as cell cleavage, from occurring (Schroeder, 1970). Cytochalasin B has no effect on chromosome movement in either lysed or unlysed cells. Both chromosome-to-pole and pole-from-pole motions occur normally in concentrations of the drug that completely block the action of the cleavage furrow (McIntosh, unpublished results). Kiehart *et al.* (1977) have shown that antimyosin injected into sea urchin cells can block cell cleavage but have no effect upon chromosome movement, further supporting the idea that actomyosin is unimportant for chromosome movement. These negative results are not rigorous demonstrations of the noninvolvement of actomyosin, because a given cell can contain more than one kind of actin (Whalen *et al.*, 1976), and more than one kind of myosin (Pollard, this volume). Different myosins tend not to cross react serologically (Fujiwara and Pollard, 1976), so spindle myosin could retain its function after antibody injection. Nonetheless, one must admit that there is at the moment no functional evidence for the action of actomyosin in chromosome movement.

There is ample evidence for the importance of microtubule assembly and disassembly in mitosis. Spindle assembly is obviously dependent upon microtubule growth and there are several examples of spindle elongation during anaphase where the final spindle length is about six times that at metaphase; here it is likely that individual spindle tubules elongate substantially (Roth and Daniels, 1962). It is also clear that microtubule disassembly is essential for chromosome-to-pole motion, because at least some of the tubules of the kinetochore fiber extend the distance from chromatid-to-pole, and the anaphase chromosomes cannot approach closer to the pole without a shortening of these tubules. There is no evidence for spindle tubule contraction (e.g., a thickening of the tubules during anaphase), and several observations suggest that kinetochore tubules disassemble at or near the poles during anaphase (Inoué, 1976). Salmon (1975) and Fuseler (1975) have observed that the rate of loss of spindle birefringence between chromosomes and poles is directly proportional to the rate of chromosome movement, suggesting strongly that the velocity of chromosome to pole movement is regulated by the rate of chromosome tubule disassembly.

Inoué and his co-workers argue that these observations support the concept that microtubule disassembly is causal for chromosome to pole movements (Inoué and Sato, 1967; Inoué, 1976). While it is clear that tubule disassembly is necessary for chromosome-to-pole movement, and it is probable from the evidence obtained by this group that tubule disassembly is permissive of chromosome movement and limits its rate, I think their evidence does not suggest that it is causal. Fuseler has confirmed the finding by Ris (1949) that within a fairly narrow temperature range, the velocity of chromosome movement is directly proportional to temperature. Increased temperature is known to shift the assembly-disassembly equilibrium of tubulin toward assembly both *in vivo* (Inoué, 1964), and *in vitro* (Weisenberg, 1972). I infer that during anaphase, the microtubule assembly equilibrium is itself being driven toward disassembly by some process whose temperature dependence is stronger than that of the disassembly process itself. One can readily imagine a motor that develops the pole-directed forces and pulls the chromosomes to the poles with a distance-dependent force. The rate of chromosome movement would then be regulated by a steady-state setup between the action of this hypothetical motor, and the power required to disassemble the microtubules of the kinetochore fiber at a given rate under a given set of physical conditions including temperature, pressure, and Ca^{++} concentration as variables. The viscous drag of the chromosome and its fiber are probably negligible in this steady state. If the motor includes microtubules as one essential component, this model fits all of the data from the Inoué group on rate of birefringence change and chromosome movement, it accounts for the apparent load independence of chromosome velocity (Nicklas, 1965), and it explains the anaphaselike movements seen by Inoué when a spindle shrinks. It leaves unexplained the nature of the hypothetical motor that is thought to pull the chromosomes to the poles. A goal of our research group is to try to identify that motor.

Several years ago we developed a system in which mammalian cells, strain PtK, were lysed while going through mitosis attached to a glass coverslip (Cande *et al.*, 1974). When the lysis conditions were chosen to favor polymerization of tubulin, and exogenous neurotubulin at equilibrium was included with the Triton X100 necessary for lysis, then anaphase chromosome movement would continue at approximately normal rates for 10-15 min after lysis (McIntosh *et al.*, 1975a). We have tried to use this system to find a "molecular monkey wrench" that would jam the works of an anaphase motor in an informative fashion, but our success has been limited. The only suggestive evidence we have obtained has been from the stopping of chromosomes by including an additional 0.1 *M* KCl in the lysis mixture, followed by a restart of both chromosome and pole motion with the addition of an extract from starfish sperm tails containing dynein. This has been a tricky result to repeat and we have never been sure of how much confidence to put in it (McIntosh *et al.*, 1975b).

More recently, Sakai and his co-workers have obtained slow and limited chromosome movement in truly isolated sea urchin spindles (Sakai *et al.*, 1976). This movement requires ATP; GTP will not serve. It is blocked by antiserum against sperm tail dynein from the same species, but is unaffected by antibodies against myosin. These results further the idea that dynein may be a component of the machinery that moves chromosomes, but one must still treat the idea with some skepticism. The velocity of chromosome motion in the isolated spindles is between 1/10 - 1/100 the normal value. The movement might not be a result of the physiological process, and thus the anti-dynein effect might be misleading.

In conclusion, we can say that there is some evidence that myosin is not involved in generation of forces for chromosome movement. There is fragmentary evidence that dynein is involved, but the evidence is open to criticism and is incomplete. There is good evidence that microtubule assembly-disassembly kinetics control the rate of chromosome to pole movement, but there is no direct evidence for the relation between tubule disassembly and force generation. There is, therefore, no clear answer at the present time as to the nature of the motors that develop mitotic forces.

STRUCTURAL STUDIES ON ANAPHASE MECHANISM

In the current state of ignorance about mitotic physiology, one useful undertaking is to characterize with all possible detail the structural rearrangements that occur within the spindle during anaphase and thus try to describe clearly the results of the mitotic motors.

We began our structural studies by counting the number of spindle microtubules in serial cross sections of mitotic cells at various stages in chromosome movement. Data from three types of mammalian lines WI-38 cells, CHO cells, PtK cells, have all been published (McIntosh and Landis, 1971; McIntosh *et al.*, 1975a; McIntosh *et al.*, 1975b). The essential features of these findings are diagrammed in Fig. 1. It must be noted, however, that these results are different from data obtained by Fuge in his similar analysis of spermatocytes of crane flies (Fuge, 1974). The results also differ in one point from those of Brinkley and Cartwright (1971) on PtK cells, but we believe this discrepancy to be due to technical differences only.

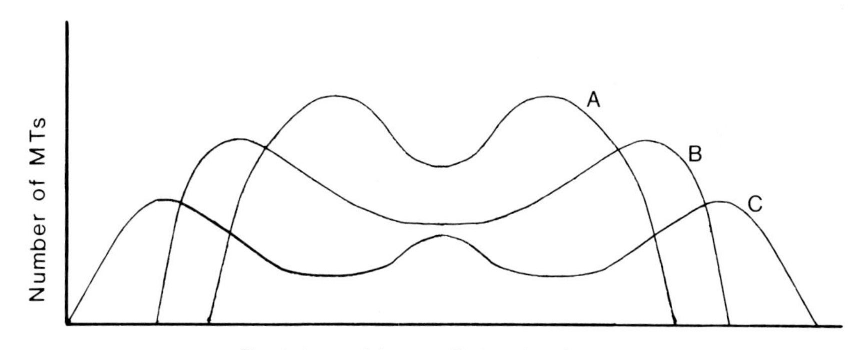

FIGURE 1. A summary of our findings on the relation between microtubule number and position along the spindle axis. Curve A represents metaphase; B and C show early and late anaphase. The spindle elongates; there is a general reduction in microtubule number; some of the tubules, both kinetochore and nonkinetochore, draw in to the poles; and a few tubules persist in the interzone between the separating chromosomes. A peak in the microtubule distribution develops at the midregion of the interzone during late anaphase - telophase.

Our model to account for the data is shown in Fig. 2. This is not a unique explanation of the data, but is is a simple one, and it conforms to all images of spindle structure we have seen during anaphase. The model cannot be a unique interpretation of the microtubule counts, because any given graph of tubule number can be explained by a large number of different microtubule arrangements, and more thorough analysis of spindle structure is required to distinguish unambiguously between the various possibilities.

The method we have chosen for this work is one of reconstructing the spindle in three-dimensions from serial cross sections. Preliminary analysis of portions of the mammalian midbody showed beyond question that this could be done (McIntosh *et al.*, 1975a), but that the problem was severe with mammalian cells which contain so many microtubules. For the early stages of the work, we have, therefore, addressed our attention to simpler, well-ordered spindles. Manton and her co-workers (Manton *et al.*, 1969) have shown that diatoms are elegant materials for the study of spindle structure. Pickett-Heaps and his group have analyzed the spindles of several diatoms and identified the general features of the spindles of these cells. We have used our cross section technology in collaboration with Pickett-Heaps' lab to analyze in detail the central spindle of the fresh water, penate diatom *Diatoma vulgare* (McDonald *et al.*, 1977). In this organism, as in other diatoms, the spindle is constructed as a central bar of clustered tubules that runs from one polar structure to another. This "central spindle" is seen from the reconstructions to be made from two tubule bunches, one from each pole, that interdigitate in the middle of the spindle. At metaphase the

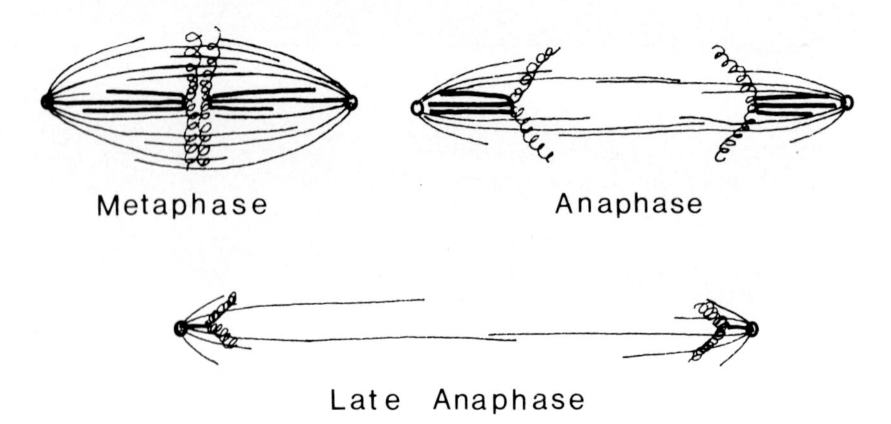

FIGURE 2. My understanding of the microtubule arrangements that give rise to the distribution graphs seen in Fig. 1.

length of this zone of overlap is about one-third the length of the pole-to-pole distance. The chromosomes are numerous and gather as a mass of chromatin at the equator. There are peripheral microtubules that fan out into this mass of chromatin, but do not appear to make any well defined connections to kinetochore structures (Fig. 3a). Anaphase may, as usual, be thought of in two parts: the migration of the chromosomes to the poles (Anaphase A) and an increase in the pole-to-pole distance (Anaphase B). During Anaphase A, the peripheral tubules shorten and disappear, but we have not been able to study this process clearly with the serial section reconstructions, because the peripheral tubules are oblique to the section plane and therefore hard to follow. Reconstruction of the central spindle at various stages of Anaphase A and B, however, shows that the tubules move as shown in Fig. 3b and 3c (McDonald *et al.*, 1977).

We have now completed development of an interactive graphics computer program in collaboration with B. M. Ross that allows us to extract considerably more information from the serial cross sections. This technology has been applied to the *Diatoma* pictures and a complete description of the results is now in preparation.

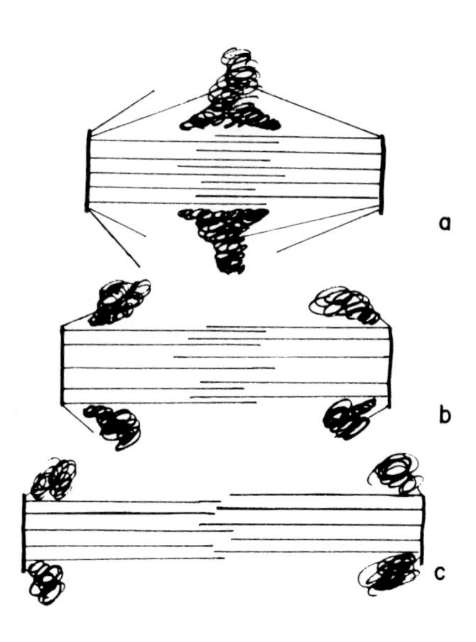

FIGURE 3. A diagrammatic representation of metaphase (a), midanaphase (b), and late anaphase (c) in Diatoma, *based upon the data of McDonald* et al. *(1977).*

Figure 4a and 4b are a stereo pair of an early anaphase central spindle. Each line represents a microtubule. The curves are second-order polynominals fit to the points that lie along each tubule as determined by tracking that tubule through serial sections. The spindle is shortened relative to its width and depth by a factor of two to improve the ease of viewing. The tilt from one stereo picture to the other is 8°. One can see from study of the stereo pair the way in which the tubules from each pole interdigitate in the midregion of the spindle or "zone of overlap." A similar view of a telophase spindle confirms the diagram seen in Fig. 3c (data not shown).

These structural studies demonstrate that the tubules of the central spindle move relative to one another during anaphase. One way they could move is through sliding. Another is through assembling subunits at one end and disassemblying

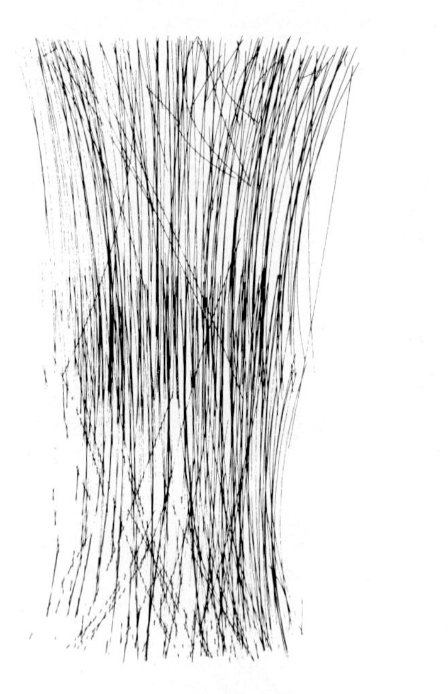

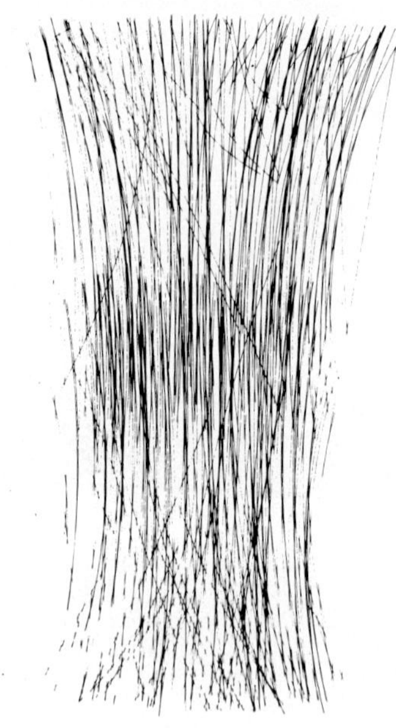

FIGURE 4. Stereo projections of a three-dimensional reconstruction of a midanaphase Diatoma *spindle. The spindle length has been shortened by a factor of two relative to its width and depth in order to spread out the tubules and ease viewing. The lines are second-order polynominals fit to the point lying along each tubule. The angular difference between the two projections is 8°.*

them from the other. While this latter possibility might seem at first sight to be farfetched, the formal possibility has been demonstrated by Wegner (1976) for actin, and Margolis and Wilson (1977) have clearly demonstrated the phenomenon for microtubules assembled *in vitro*.

Longitudinal images of the *Diatoma* central spindle show that there is an amorphous matrix material between the tubules in the zone of overlap. As the zone shortens during Anaphase B, the matrix remains uniformly concentrated in whatever remains of the zone of overlap. As the volume of the zone decreases with the shortening in the extent of overlap, the staining properties of this matrix darken. It is as if there were a given amount of the matrix and as the volume it occupied decreased, its concentration increased. A similar phenomenon has been described for mammalian cells (McIntosh and Landis, 1971). These facts favor a sliding model over a disassembly model, because the sliding could readily concentrate the matrix while disassembly requires a more elaborate model to account for the uniform increase in matrix staining.

I draw particular attention to this matrix because it is found at the midregion of the interzone of many spindles (mammal, diatom, nematode, and higher plant for example), and its nature is at present a complete mystery. It does not stain with antibodies to tubulin, microtubule associated proteins from brain, actin, myosin, α-actinin, CRP, or tropomyosin. It is, in a sense, a moving part of the spindle, and as such, deserves careful attention. Efforts are underway in our lab and in that of Ratner (Ratner *et al.*, 1977) to isolate and characterize this matrix material. It may well be a component of the sliding machinery for spindle elongation.

SUMMARY

There are not sufficient data to exclude any of the aforementioned models for the molecular mechanisms of chromosome motility. The current situation is confusing and rather unsatisfactory. It is easy to see that we need to make progress on the analysis of spindle structure and of spindle chemistry because even the most basic question, what is the motor that drives the chromosomes, is not yet answered. The scientific problem is difficult because a spindle must form, function, and disappear with every cell cycle, so genesis and motility are superimposed. The technical problems are formidable because the spindle is complex biochemically and large for electron microscopy. Nonetheless, the tools and approaches seem to be in hand to establish a good basis of fact for the understanding of chromosome movement, and I imagine that the next five years will witness a substantial improvement in the extent of our understanding.

REFERENCES

Allen, R. D., Bajer, A., and La Fountain, J. (1969). *J. Cell Biol. 43,* 4a.

Bajer, A. (1958). *Chromosoma 9,* 319.

Bauer, H., Dietz, R., and Robbelen, C. (1961). *Chromosoma 12,* 116.

Bibring, T., and Baxandall, J. (1971). *J. Cell Biol. 48,* 324.

Brinkley, B. R., and Cartwright, Jr., J. (1971). *J. Cell Biol. 50,* 416.

Cande, W. Z., Lazarides, E., and McIntosh, J. R. (1977). *J. Cell Biol. 72,* 552.

Cande, W. Z., Snyder, J. A., Smith, D., Summers, K., and McIntosh, J. R. (1974). *Proc. Nat. Acad. Sci. U.S.A. 71,* 1559.

Chu, L. K., and Sisken, J. R. (1977). *Exptl. Cell Res. 107,* 71.

Connolly, J. A., Kalnins, V. I., Cleveland, D. W., and Kirschner, M. W. (1977). *Proc. Nat. Acad. Sci. U.S.A. 74,* 2437.

Forer, A. (1965). *J. Cell Biol. 25,* 95.

Forer, A., and Behnke, O. (1972). *Chromosoma 39,* 145.

Fuge. J. (1974). *Protoplasma 82,* 289.

Fujiwara, K., and Pollard, T. D. (1976). *J. Cell Biol. 71,* 848.

Fuseler, J. W. (1975). *J. Cell Biol. 67,* 789.

Gawadi, N. (1974). *Cytobios 10,* 17.

Inoué, S. (1952). *Exptl. Cell Res. Suppl. 2,* 305.

Inoué, S. (1964). *In* "Primitive Motile Systems in Cell Biology" (R. D. Allen and N. Kamiya, eds.), p. 549. Academic Press, New York.

Inoué, S. (1976). *In* "Cell Motility" (R. D. Goldman, T. D. Pollard, and J. Rosenbaum, eds.), p. 1317. Cold Spring Harbor Press, New York.

Inoué, S., and Sato, H. (1967). *J. Gen. Physiol. 50,* Suppl., 259.

Kiehart, D. P., Inoué, S., and Mabuchi, I. (1977). *J. Cell Biol. 75,* 258a.

Manton, I., Kowallik, K., and Von Stosch, H. A. (1969). *J. Microsc. 89:3*, 295.
Margolis, R., and Wilson, L. (1977). *J. Cell Biol. 75*, 272a.
McDonald, K., Pickett-Heaps, J. D., McIntosh, J. R., and Tippit, D. H. (1977). *J. Cell Biol. 74*, 377.
McIntosh, J. R. (1977). *In* "Mitosis Facts and Questions" (M. Little *et al.*, eds.), p. 168. Springer-Verlag, Berlin.
McIntosh, J. R., Cande, W. Z., Snyder, J. A., and Vanderslice, K. (1975a). *Ann. N. Y. Acad. Sci. 253*, 407.
McIntosh, J. R., Cande, W. Z., and Snyder, J. A. (1975b). *In* "Molecules and Cell Movement" (S. Inoué and R. E. Stephens, eds.), p. 31. Raven Press, New York.
McIntosh, J. R., and Landis, S. C. (1971). *J. Cell Biol. 49*, 468.
Mohri, H., Mabuchi, I., Yazaki, I., Sakai, H., and Ogawa, K. (1976). *Devel., Growth, and Differ. 18*, 391.
Nicklas, R. B. (1965). *J. Cell Biol. 25* (No. 1, Pt. 2), 119.
Nicklas, R. B. (1977). *In* "Mitosis, Facts and Questions" (M. Little *et al.*, eds.), p. 150. Springer-Verlag, Berlin.
Nicklas, R. B., and Koch, C. A. (1972). *Chromosoma 39*, 1.
Nicklas, R. B., and Staehly, C. A. (1967). *Chromosoma 21*, 1.
Östergren, G. (1945). *Hereditas, 31*, 498.
Östergren, G. (1950). *Hereditas, 36*, 1.
Östergren, G., Mole-Bajer, J., and Bajer, A. (1960). *Ann. N.Y. Acad. Sci. 90*, 381.
Petzelt, C. (1972). *Exptl. Cell Res. 70*, 333.
Rattner, J. B., Krustal, G., and Hamkalo, B. A. (1977). *J. Cell Biol. 75*, 283a.
Ris, H. (1949). *Biol. Bull. 96*, 90.
Roth, L. E., and Daniels, E. W. (1962). *J. Cell Biol. 12*, 57.
Sakai, H., Mabuchi, I., Shimoda, R., Ogawa, K., and Mohri, H. (1976). *Devel., Growth, and Differ. 18*, 123.
Salmon, E. D. (1975). *J. Cell Biol. 65*, 603.
Salmon, E. D., and Jenkins, R. (1977). *J. Cell Biol. 75*, 295a.
Sanger, J. (1975). *Proc. Nat. Acad. Sci. U.S.A. 72*, 2451.
Sanger, J. W., and Sanger, J. M. (1976). *J. Cell Biol. 70*, 277a.
Schroeder, T. E. (1970). *Z. Zellforsch. 109*, 431.
Sherline, P., and Schiavone, K. (1978). *J. Cell Biol. 77*, R9.
Wegner, A. (1976). *J. Mol. Biol. 108*, 139.
Weisenberg, R. C. (1972). *Science 177*, 1104.

Weisenberg, R. C., and Taylor, E. W. (1968). *Exptl. Cell Res. 53*, 372.

Welsh, M. J., Dedman, J. R. Brinkley, B. R., and Means, A. R. (1977). *J. Cell Biol. 75*, 262.

Whalen, R. G., Butler-Browne, G. S., and Gros, F. (1976). *Proc. Nat. Acad. Sci. U.S.A. 73*, 2018.

Wilson, L., and Bryon, J. (1974). *Adv. Cell Molec. Biol. 3*, 21.

CONTRIBUTED PAPERS

Motility in Cell Function
Proceedings of the First John M. Marshall Symposium in Cell Biology

CHANGES IN MOLECULAR PACKING DURING SHORTENING OF *LIMULUS* THICK FILAMENTS

Rhea J. C. Levine
and
Maynard M. Dewey

The Medical College of Pennsylvania
Philadelphia, Pennsylvania
and
State University of New York at Stony Brook
Stony Brook, New York

We have previously determined that the paramyosin-containing (deVillafranca and Leitner, 1967; Levine *et al.*, 1972; Levine *et al.*, 1976) thick filaments of *Limulus* telson striated muscle decrease by 30-40% in length (Dewey *et al.*, 1973) and increase in diameter by a similar amount (Dewey *et al.*, 1977b), during sarcomere shortening below 7.0 μm (L_O) (Walcott and Dewey, 1978). We have isolated thick filaments preferentially long or short from this muscle, depending upon the conditions to which it was exposed prior to the isolation procedure (Dewey *et al.*, 1977c). Furthermore, we have caused isolated long filaments to shorten *in vitro* by the addition of Ca^{2+} to the EGTA-buffered Mg-ATP medium in which they were suspended (Dewey *et al.*, 1977a).

Optical diffraction patterns were obtained from electron micrograph plates of negatively stained *Limulus* thick filaments, either 1) isolated long or short or 2) induced to shorten *in vitro*. Diffraction patterns were made at both short and long specimen-to-film distances, in order to resolve both close- and long-spacing periodicities, respectively, along the filaments' lengths. At the short specimen-to-film distance, the meridional reflection at 14.5 nm and a layer line at 72.5 nm was visible on patterns from all of the filaments. Some filaments also showed a layer line at 55 nm (Fig. 1). Diffraction patterns could be separated into three classes, one class being those with layer lines indexing on orders of 145 nm, another with layer lines indexing on orders of 220 nm and a third show-

ISBN 0-12-551750-5

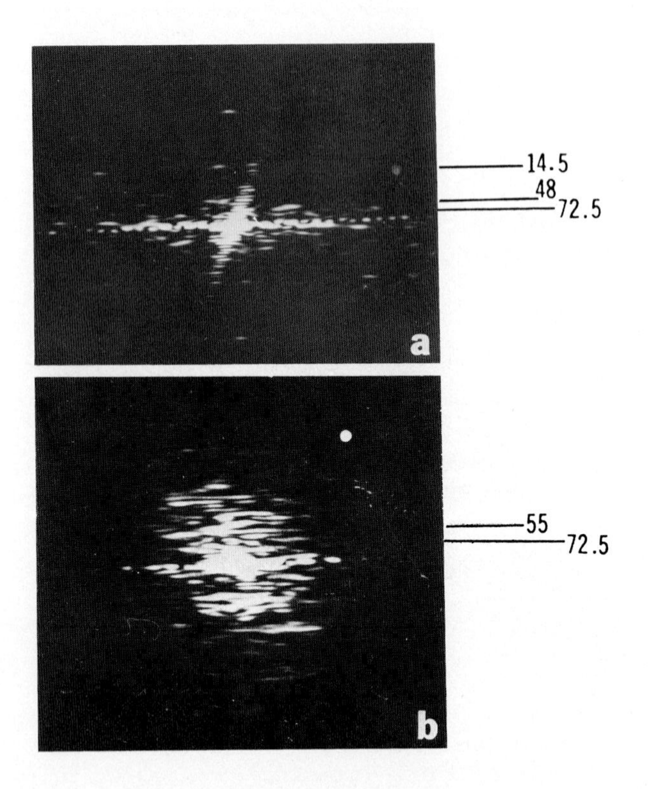

FIGURE 1. Optical diffraction patterns made from electron micrographs of two different isolated thick filaments at the short specimen to film distance. Both patterns (a) and (b) show meridional reflections at 14.5 nm^{-1} and layer lines at 72.5nm^{-1}. Pattern (b) also shows a layer line at 55 nm^{-1}.

ing both types of periodicities (Fig. 2).

Filament lengths were measured from the electron micrograph plates, within each class, and the mean filament lengths were compared among the three classes. The mean length of filaments showing reflections indexing on orders of 145 nm was 4.4 ± 0.537 μm (S.D.); that of filaments showing reflections indexing on orders of 220 nm was 3.2 ± 0.337 μm. Filaments which displayed both periodicities were intermediate in length, averaging 3.7 ± 0.538 μm (S.D.). The differences in mean length between short and intermediate and intermediate and long filament populations were statistically significant to $p < 0.02$, while the difference in mean length between the short and long filaments was significant to $p < 0.001$ (Fig. 3).

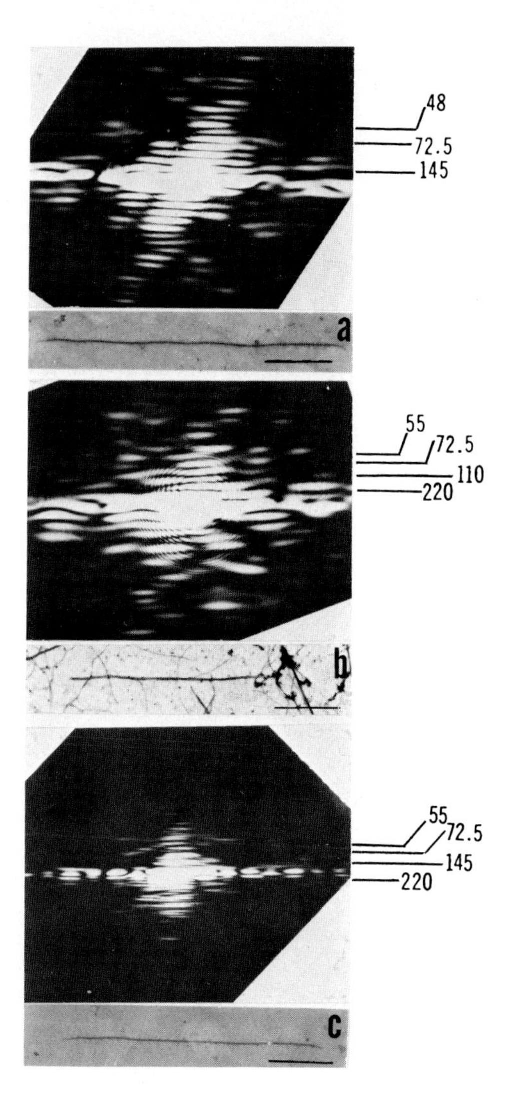

FIGURE 2. Optical diffraction patterns and the electron micrographs of the isolated filaments from which they were obtained, made at long specimen to film distances. Bars = 1 μm. Filament magnification = 10,000X. Numbers to right of diffraction patterns = nm^{-1}. (a) The optical diffraction pattern obtained from filament (a) shows reflections indexing on orders of 145 nm^{-1} only. (b) The optical diffraction pattern obtained from filament (b) shows reflections indexing on orders of 220 nm^{-1} only. (c) The optical diffraction pattern obtained from filament (c) shows reflections indexing on orders of both 145 nm^{-1} and 220 nm^{-1}.

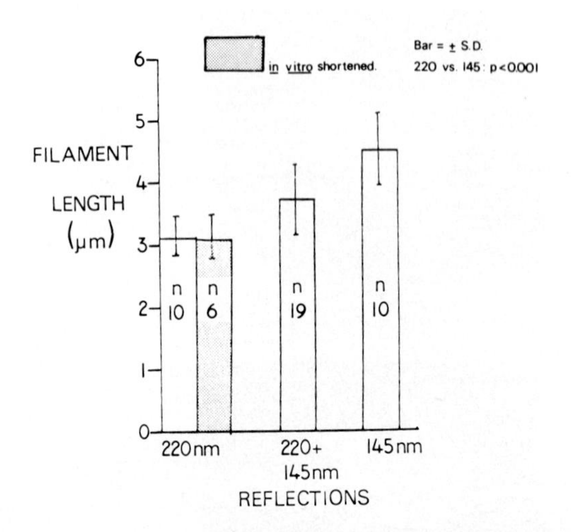

FIGURE 3. Bar graph relating length of isolated filaments to the long-spacing reflections obtained from optical diffraction patterns made at long specimen to film distances. n = *numbers of filaments both diffracted and measured in each class. No difference in either mean filament length or long-spacing reflections was seen in short filaments, whether they were isolated short, or shortened* in vitro.

These results agree with those from X-ray diffraction studies (Millman *et al.*, 1974; Wray *et al.*, 1974) in showing constancy of the 14.5 nm reflection, which we interpret as the axial interval between adjacent cross-bridge projections, at all filament lengths. If the long-spacing layer lines, which differ in long versus short filaments, reflect the repeat of the cross-bridge helices in these structures, then this repeat would occur at 10 x 14.5 nm in the long filaments (diameter = 23 nm) and 15 x 14.5 nm in the short (diameter = 33nm). A preliminary working model based on these data would predict the cross-bridge arrangement 1) as a 4-stranded helix having four projections in the same plane, occurring every 14.5 nm in long filaments, and 2) as a 6-stranded helix with six projections every 14.5 nm in short filaments. The same numbers for the strands were calculated, independently, from actin-myosin weight ratios (Pepe, personal communication). The shift in molecular packing occurring as the filaments shorten would thus result in a 33% increase in the number of projections along a segment of a short filament of increased diameter as compared with a segment of equal length of a long filament of 1/3 smaller diameter, thus providing a mechanism for the conservation of surface myosin molecules during filament shortening. The presence of both long and short periodicities in intermediate filaments, moreover, suggests that both types of molecular organization are present in different regions along each one-half thick filament as shortening proceeds.

ACKNOWLEDGMENTS

This work was supported by USPHS Grant HL15835 to the Pennsylvania Muscle Institute. R. C. L. holds RCDANS 70,476.

REFERENCES

deVillafranca, G. W. and Leitner, V. E. (1967). *J. Gen. Physiol. 50,* 2495.
Dewey, M. M., Levine, R. J. C., and Colflesh, D. E. (1973). *J. Cell Biol. 58,* 574.
Dewey, M. M., Baldwin, E., Colflesh, D., Walcott, B., and Brink, P. (1977a). *J. Cell Biol. 75,* 322a.
Dewey, M. M., Colflesh, D. E., Walcott, B., and Levine, R.J.C. (1977b). *Biophys. J. 17,* 160a.
Dewey, M. M., Walcott, B., Colflesh, D. E., Terry, H., and Levine, R.J.C. (1977c). *J. Cell Biol. 75,* 366.
Levine, R. J. C., Dewey, M. M., and deVillafranca, G. W. (1972). *J. Cell Biol. 55,* 221.
Levine, R. J. C., Elfvin, M. J., Dewey, M. M., and Walcott, B. (1976). *J. Cell Biol. 71,* 273.
Millman, B. M., Warden, W. J., Colflesh, D. E., and Dewey, M.M. (1974). *Fed. Proc. 33,* 1333.
Walcott, B., and Dewey, M. M. (1978). *Biophys. J. 21,* 55a.
Wray, J. S., Vibert, P. J., and Cohen, C. (1974). *J. Mol. Biol. 88,* 343.

Motility in Cell Function
Proceedings of the First John M. Marshall Symposium in Cell Biology

X-RAY DIFFRACTION STUDIES OF MYOSIN FILAMENT STRUCTURES IN CRUSTACEAN MUSCLES

John S. Wray

Biophysics Institute
Aarhus University
Denmark
and
Rosenstiel Basic Medical Sciences Research Center
Brandeis University
Waltham, Massachusetts*

Even so specialized a motile system as muscle exhibits great diversity of structure, which must be encompassed in a biologist's view of contractile mechanisms. Actin filaments appear to be highly conservative in structure, and diversity is more evident in size and cross-bridge arrangements of myosin filaments (Wray *et al.*, 1975). I summarize here X-ray observations on a group of crustacean muscles in which certain aspects of thick filament structure and its diversity are nicely illustrated.

In decapod crustaceans (lobsters, crayfishes, and their relatives), muscle fibers fall into two main classes: the fast muscles have short sarcomeres and a filament lattice resembling that in asynchronous insect flight muscles, while the slow fibers have thicker myosin filaments and a less regular filament lattice (Jahromi and Atwood, 1969). In the relaxed state, both give X-ray diffraction patterns in which myosin diffraction and actin diffraction can be distinguished (Wray *et al.*, 1978). Reflections due to cross-bridges weaken, and actin reflections are enhanced, when the bridges attach to actin in rigor. The cross-bridge layer lines establish the symmetry of the array of relaxed cross-bridges round the surface of the thick filament, as represented by the black spots in Fig. 1a. (These reflections are very few, and the cross-

*Present address.

ISBN 0-12-551750-5

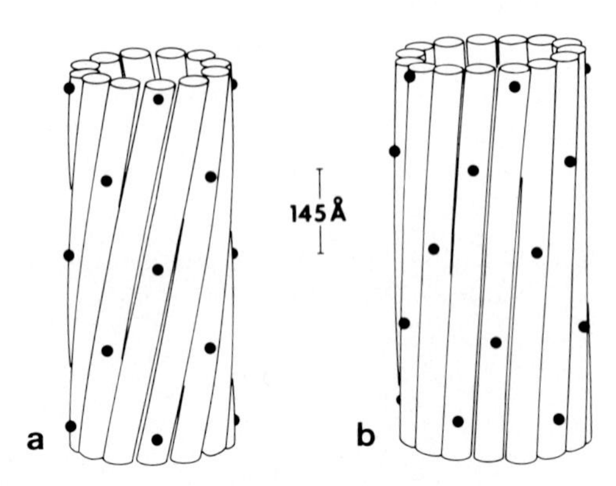

FIGURE 1.

bridges are evidently subject to large random fluctuations about their average positions.) But the myosin diffraction is not limited to the cross-bridge reflections; it also includes layer lines that do *not* change in rigor. These can be ascribed to the myosin backbone and can be interpreted as indicating that subfilaments of about 43 Å diameter connect up the surface lattice points in the particular way shown in Fig. 1a (except near the bare zone and ends of the filament). The rotational symmetries in Fig. 1 are consistent with present evidence, but are not yet established with certainty.

X-ray patterns from fast muscles were not all identical; the pitch of the most prominent myosin helix, deduced from the spacing of the cross-bridge reflections, was (for example) 300 Å in the fast body muscles of crayfishes, 310 Å in certain fast muscles of crayfish legs, and 318 Å in the fast muscles of the crayfish claw. The diffraction from the subfilaments varied correspondingly in spacing. The myosin filaments of many fast muscles thus form a family, all roughly resembling Fig. 1a but varying slightly in the amount of twist. This variation of myosin screw symmetry was the only indication in the X-ray pattern of the polymorphism of fast decapod muscles.

In X-ray patterns from slow muscles, the myosin diffraction was broadly similar in appearance, but indicated a rather different screw symmetry for the surface lattice (Fig. 1b). The myosin pitch was about 360 Å, and the pitch of the subfilaments correspondingly longer. (All slow muscles studied had the same screw symmetry: this surface lattice resembles the familiar Bear-Selby net of molluscan paramyosin, and slow muscles indeed contain larger amounts of paramyosin than fast muscles (Levine *et al.*, 1976).) The X-ray patterns did show features corresponding to the greater diameter of the myosin filaments, and the diffraction could be explained well by a higher rotational symmetry; the close similarity of intensities in patterns from fast and slow muscles suggests that the density variations defining the surface lattice are neverthless essentially conserved.

The tubular structures in Fig. 1 are a natural solution to the problem of assembling myosin molecules (Cohen, 1966). The 'head' ends can occupy identical environments at the filament surface if the molecules comprising each subfilament wind round each other. Necessary features of the model as derived here for crustacean muscles are that neighboring subfilaments are related by a 145 Å displacement, and that the repeat within each is 435 Å. Implications of the model are that molecules are staggered by 435 Å within the subfilaments, each of which would include three myosin rods in its 43 Å diameter.

These structures differ significantly from that favored by Squire (1973, 1975) as a general model for all myosin filaments. Essentially similar molecular interactions generate *subfilaments* here, but *layers* in Squire's model, as intermediate levels of filament organization. But amongst the diversity of crustacean muscles, those described here do illustrate a subsidiary point of Squire's argument that variations of screw symmetry and rotational symmetry are two ways in which different surface arrays of cross-bridges can result without large change in the local pattern of specific interactions in the filament backbone. A tubular structure is seen in thick filaments of many other muscles; appropriate transformations of the present subfilament model may describe a wide variety of filament backbones.

What functional significance has the polymorphism of screw symmetry in the fast muscles within a single animal? In the neuromuscular system of crustaceans, diversity of fiber types having different physiological properties compensates for the paucity of motor innervation (Atwood, 1976). One may expect a specialization of fibers (for example, in respect of speed, force, extent, and duration of shortening) to be accompanied by structural adaptations at many levels. Diversity of fiber types extends to the level of filament architecture, as established by the present observations. Possibly myosin-actin interaction depends on filament geometry sufficiently that slight change of screw symmetry can "tune" a muscle to optimal performance of its task.

ACKNOWLEDGMENTS

I thank EMBO for a Long-Term Fellowship. This work was also partly supported by Grant PCM76-10558 from the National Science Foundation.

REFERENCES

Atwood, H. L. (1976). *Progr. Neurobiol. 7,* 291-391.
Cohen, C. (1966). *In* Ciba Foundation Symposium, Principles of Biomolecular Organization" (G. E. W. Wolstenholme and M. O'Connor, eds.), pp. 101-129. Churchill, London.
Jahromi, S. S., and Atwood, H. L. (1969). *J. Exp. Zool. 171,* 25-38.
Levine, R. J. C., Elfvin, M., Dewey, M. M., and Walcott, B. (1976). *J. Cell Biol. 71,* 273-279.
Squire, J. M. (1973). *J. Mol. Biol. 77,* 291-323.
Squire, J. M. (1975). *Ann. Rev. Biophys. Bioeng. 4,* 137-163.
Wray, J. S., Vibert, P. J., and Cohen, C. (1975). *Nature 257,* 561-564.
Wray, J. S., Vibert, P. J., and Cohen, C. (1978). *J. Mol. Biol., 124,* 501-521.

Motility in Cell Function
Proceedings of the First John M. Marshall Symposium in Cell Biology

X-RAY DIFFRACTION FROM CHICKEN SKELETAL MUSCLE

Barry M. Millman

Department of Physics
University of Guelph
Guelph, Ontario, Canada

The experiments reported here have been done in collaboration with Dr. Frank Pepe, using muscle preparation methods similar to those he has used during preparation of specimens for electron microscopy (Pepe and Dowben, 1977). The muscles have been examined by low-angle X-ray diffraction in suitable saline solutions. The X-ray patterns have been used to assess the extent of tissue damage occurring during the fixation process, and to search for structural features additional to those seen in electron micrographs.

High resolution X-ray diffraction patterns were obtained using a 20 cm mirror-monochromator camera of design similar to Huxley and Brown (1967), from chicken breast muscle fixed overnight in 5% gluteraldehyde and examined in standard salt solution (100 m*M* KCl; 1 m*M* $MgCl_2$; 10 m*M* phosphate buffer: pH 7.0). These patterns showed reasonably good myosin layer-line reflections (Fig. 1). In the equatorial direction, the patterns showed 1,0 and 1,1 reflections with the 1,0 reflection considerably more intense than the 1,1 reflection (Fig. 2). The results indicate that these fixed muscles are in a relaxed state and have fairly well-ordered thick filament projections. Further X-ray diffraction studies are in progress on the effects of variations in fixation procedure on the muscle structure, and on the state of the muscle during later stages of preparation for electron microscopy.

New reflections have been observed from fresh and glycerol-extracted chicken breast muscle (and also from glycerol-extracted fish skeletal muscle). For these patterns, a moderate-angle toroidal camera (Elliott, 1965) was used. The patterns obtained show a broad equatorial reflection at about 4.5 nm (Fig. 3a) and diffuse off-axis reflections, particularly at equatorial spacings of 2.4 and 4.0 nm (Fig. 3b,c). These re-

ISBN 0-12-551750-5

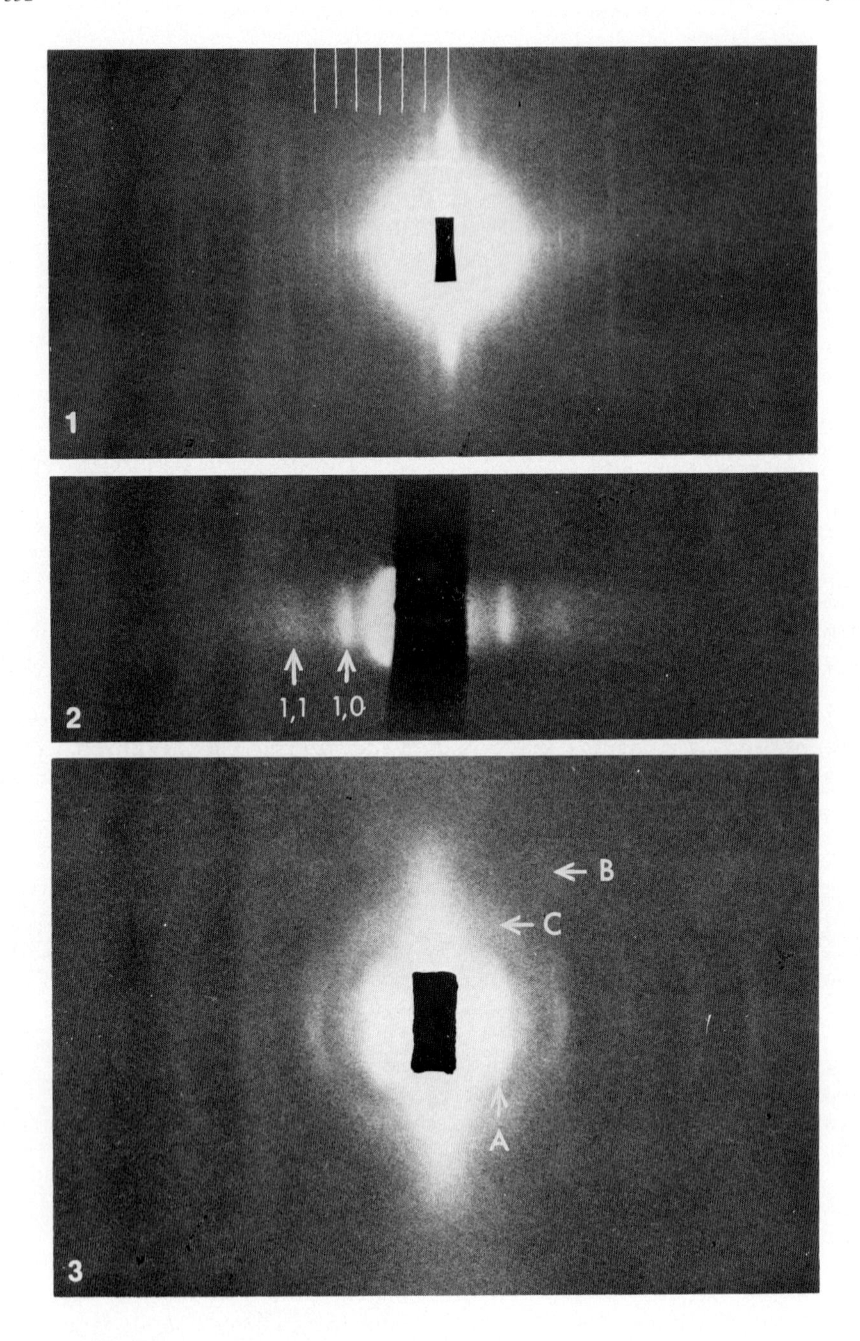

FIGURE 1. Meridional X-ray diffraction pattern from chicken pectoralis muscle fixed in 5% gluteraldehyde, taken with mirror-momochromator camera and a specimen-film distance of 22.3 cm. Indicated layer lines are orders of the 42.9 nm periodicity of the myosin filaments.

flections appear to come from the thick filament backbone: they remain unchanged after removal of the cell membranes with detergent (Triton X- 1% in standard salt solution) and are seen only when the myosin reflections at 7.2 and 14.4 nm are also seen on the meridian. The patterns indicate a basic unit cell of dimensions 8.0 nm (equatorial) by 10 nm (meridian) with substructural units staggered longitudinally. The equatorial reflection at 4.5 nm indicates that the filament is not constituted from a simple packing of myosin molecules (e.g., Pepe and Dowben, 1977). Studies are underway comparing the observed pattern to that predicted from model structures similar to the one proposed by Pepe and Dowben, or involving a staggered packing of α-helical protein chains (e.g., Longley, 1975).

ACKNOWLEDGMENTS

This research was carried out at the University of Pennsylvania, Philadelphia, with partial support from the Pennsylvania Muscle Institute as a visiting scientist. I would like to thank Dr. Kent Blasie for the loan of X-ray diffraction facilities.

FIGURE 2. Equatorial X-ray diffraction pattern from the same muscle strip as Fig. 1. Mirror-monochromator camera with a specimen-film distance of 26.3 cm. The 1,0 and 1,1 reflections from the hexagonal filament lattice are indicated.

FIGURE 3. X-ray diffraction pattern from glycerol-extracted chicken pectoralis muscle. Muscle fiber axis is horizontal. Toroid camera with a specimen-film distance of 6.8 cm. A, B, and C indicate diffuse reflections from the thick filament backbone.

REFERENCES

Elliott, A. (1965). *J. Sci. Instr. 42*, 312-316.
Huxley, H. E., and Brown, W. (1967). *J. Mol. Biol. 30*, 383-434.
Longley, W. (1975). *J. Mol. Biol. 93*, 111-115.
Pepe, F. A., and Dowben, P. (1977). *J. Mol. Biol. 113*, 199-218.
Squire, J. (1973). *J. Mol. Biol. 77*, 291-323.

Motility in Cell Function
Proceedings of the First John M. Marshall Symposium in Cell Biology

THE RANGE OF ATTACHMENT OF CROSS BRIDGES TO ACTIN IN RIGOR MUSCLES

John Haselgrove

Johnson Research Foundation
University of Pennsylvania
Philadelphia, Pennsylvania

During the active shortening of muscle, the actin and myosin filaments slide past each other while the myosin cross bridges attach and detach cyclically from the actin filaments. It is important to know how far the cross bridges are free to move and which actin monomers are potentially available for cross bridges to attach. Although there is little information on these parameters in living muscle, they are probably similar in rigor muscle, and will determine where the cross bridges attach in rigor. Reedy's (1968) electron micrographs of rigor insect flight muscle show the cross bridges attached to actin: in longitudinal sections the cross bridges form "double chevrons" in which two cross bridges attach every 38.5 nm.

I have used a computer to calculate the positions at which the cross bridges are expected to attach to actin in rigor muscle, and thus simulate Reedy's double chevrons. The actin filaments were described as the usual two stranded helix of monomers with a helix pitch of 2 × 38.5 nm (Hanson and Lowy, 1963), and the myosin filaments were described as six strand helices with 14.5 nm between cross bridges (Squire, 1972): the cross bridges were free to move round the myosin filaments by 30° to search for an adjacent actin filament. The cross bridges were permitted a total axial range of R nm, and to move round the actin filament by a total angle ϕ to search for an available actin monomer. Each cross bridge had just one head for attachment.

The general appearance of the computed attachment of cross bridges is governed predominantly by the parameters for the flexibility of the cross bridges (R and ϕ) and very little by the exact values describing the shape of the filaments. Fig. 1 shows some of the possible appearances of the actin filaments

ISBN 0-12-551750-5

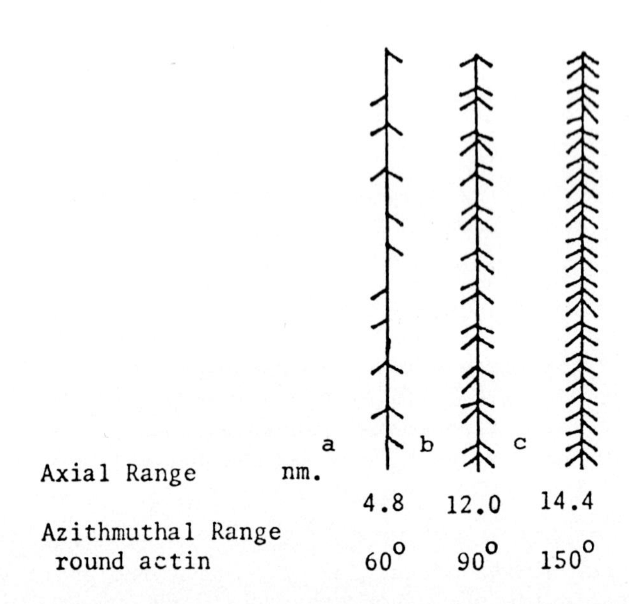

FIGURE 1. Computed appearance of actin filaments in rigor muscle with attached cross bridges, for different values of the axial and azimuthal range of the cross bridge. The cross bridges are angled at approximately 45°; the exact angle reflects the point of origin on the myosin filament.

with different values for the constraints placed on the cross bridges. If the cross bridges are constrained too much then very few can attach to actin (Fig. 1a), but if they are free to move then almost every one attaches (Fig. 1c). With appropriate choices of the parameters then the computed pattern does simulate the observed electron-microscope image very well (Fig. 1b). The values of R and ϕ shown in Fig. 1b which give an acceptable model are not unique: other values of R and ϕ also give double-chevron appearance and these are indicated by the area enclosed by the dotted line in Fig. 2.

Figure 2 indicates that the cross bridges probably have the ability to reach at least 7 nm and possibly as much as 14 nm axially and be able to reach 60 - 120° azimuthally round the actin filaments. Thus between three and five actin monomers are available at each turn of the actin helix for labelling by the cross bridges on the adjacent myosin filament. The actin monomers which are not available for labelling may be physically blocked by the troponin molecules, or may simply be out of range because the cross bridges are not flexible enough to reach right round to the side of the actin filament.

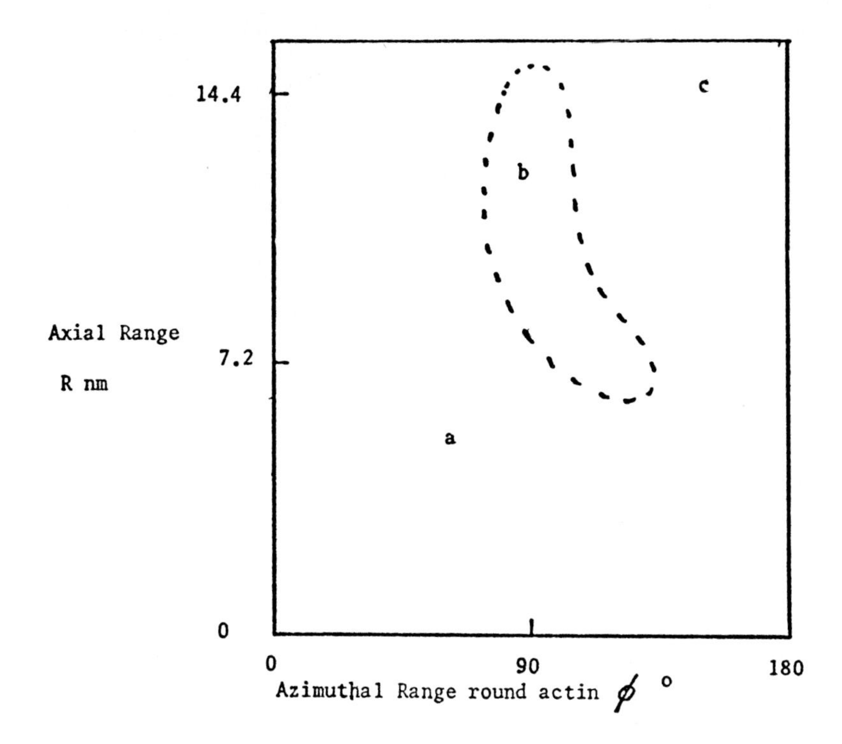

FIGURE 2. The dotted line indicates the area in which the values of R *and* ϕ *give computed models which simulate the observed cross bridge pattern. The letters indicate the pairs of values chosen for the models in Fig. 1.*

An extended account of this work has appeared in press (Haselgrove & Reedy)

ACKNOWLEDGMENTS

This work was supported by USPHS Grants HL 15835 to the Pennsylvania Muscle Institute and GM-12201.

Reedy, M. K. (1968). *J. Mol. Biol. 31*, 155.
Hanson, J., and Lowy, J. (1963). *J. Mol. Biol. 6*, 46.
Haselgrove, J. C., and Reedy, M. K. (1978). *Biophys. J. 24*, 713.
Squire, J. M. (1972). *J. Mol. Biol. 72*, 125.

Motility in Cell Function
Proceedings of the First John M. Marshall Symposium in Cell Biology

DICTYOSTELIUM MYOSIN: EFFECT OF RNA ON ITS AGGREGATION PROPERTIES

Peter R. Stewart[*]
and
James A. Spudich

Department of Structural Biology
Sherman Fairchild Center
Stanford University School of Medicine
Stanford, California

Myosin isolated from muscle cells is often associated with significant amounts of RNA. In the case of chicken muscle myosin, this RNA alters the chromatographic behavior of the protein (Baril *et al.*, 1966), and changes the properties of LMM paracrystals (Chowrashi and Pepe, 1977). We report here that RNA present in myosin prepared from *Dictyostelium discoideum* alters the aggregation state of filaments which this myosin forms at low ionic strength.

Myosin purified from *Dictyostelium* by the method of Clarke and Spudich (1974) contains 50-150 μg RNA/mg protein. This RNA contains a heterogeneous mixture of discrete size species, generally in the range 4S (approximately 2 x 10^4 daltons) to 20S (approximately 2 x 10^6 daltons); in different preparations of myosin the relative amounts of these species vary.

Precipitation of RNA-containing myosin by dilution into a low ionic strength buffer containing 2-10 m*M* Mg^{++} results in the formation of lateral and linear aggregates of short, bipolar thick filaments (Fig. 1a). In some instances, the aggregates appear continuous for lengths greater than several times that of individual bipolar thick filaments, indicating that something more than simple head-to-head aggregation may have occurred. These longer filaments show a transverse 14 nm repeat in certain regions.

[*] *Present address: SGS Biochemistry, Australian National University, Canberra, 1600, Australia.*

ISBN 0-12-551750-5

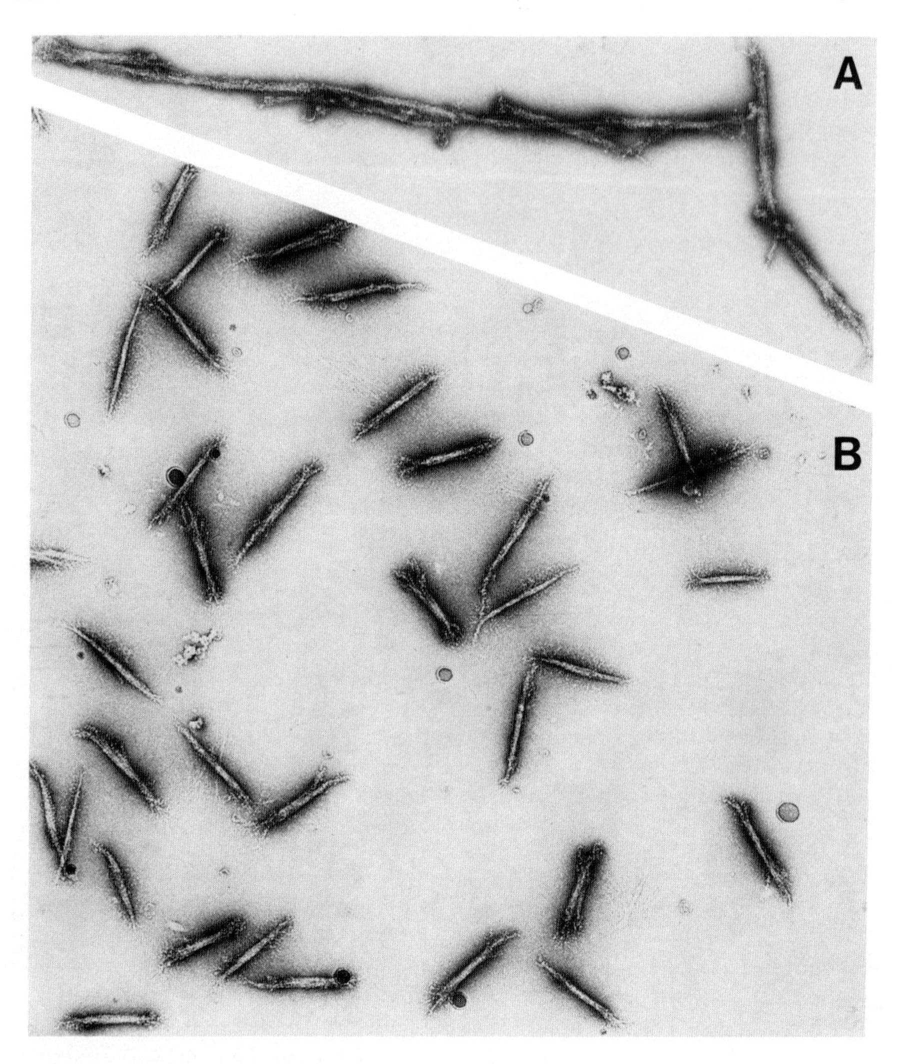

FIGURE 1. (a) Myosin was prepared by the method of Clarke and Spudich (1974) and stored in 0.5 M KCl, 10 mM Tris-Cl (pH 7.5), 1 mM EDTA, 1 mM DTT. It was then diluted (1 in 40, final protein concentration 0.1 mg/ml) into 10 mM Tris-maleate (pH 6.5), 10 mM $MgCl_2$. (b) Myosin further purified by chromatography on DEAE-cellulose (Mockrin and Spudich, 1976) was diluted as in (a). Grids were prepared by negative staining with 1% uranyl acetate, as described by Huxley (1963). Magnification (a) × 32,400; (b) × 27,000.

RNA can be removed from the *Dictyostelium* myosin preparations either by treatment with pancreatic RNase or by chromatography on DEAE-cellulose. The RNA content is then less than 5 μg/mg protein. Direct dilution of this RNA-free myosin into low salt plus Mg^{++} buffer yields individual bipolar thick filaments almost exclusively (Fig. 1b). These filaments have a mean length of 540 nm, are 35-40 nm wide, and often show a 14 nm transverse repeat along their length.

A range of RNA analogs and other polyanions or polycations have been added to RNA-free myosin to determine the specificity of the effect. Ribosomal RNA from *Dictyostelium*, high molecular weight DNA (salmon sperm) and heparin (a sulphated polysaccharide) also induce the effect at concentrations similar to those of RNA derived from myosin. Weaker effects are shown by tRNA from *Dictyostelium* or yeast, poly C, poly A, and poly aspartic acid. Aggregation is not significantly induced by poly AC, poly glutamic acid, poly lysine, ATP, pancreatic RNase or tubulin.

The question of whether a specific component of cellular RNA is associated with *Dictyostelium* myosin *in vivo* remains to be answered directly. It is clear, however, that the effect of RNA and other natural polyanions on the structure of *Dictyostelium* myosin are such that, unless specific steps are taken to remove naturally associated polyanions, caution should be exercised in the interpretation and comparison of structural (and perhaps catalytic) properties of this molecule. It is possible that the same may be true for myosin isolated from other cell types.

ACKNOWLEDGMENTS

This work was supported in part by United States Public Health Service Grant GM 25240-01 (JAS) and American Cancer Society Grant VC 121E (JAS).

REFERENCES

Baril, E. F., Love, D. S., and Herrmann, H. (1966). *J. Biol. Chem. 241*, 822-830.

Chowrashi, P. K., and Pepe, F. A. (1977). *J. Cell Biol. 74*, 136-152.

Clarke, M., and Spudich, J. A. (1974). *J. Mol. Biol. 86*, 209-222.

Huxley, H. (1963). *J. Mol. Biol. 7*, 281-308.

Mockrin, S. C., and Spudich, J. A. (1976). *Proc. Natl. Acad. Sci. U.S.A. 73*, 2321-2325.

Motility in Cell Function
Proceedings of the First John M. Marshall Symposium in Cell Biology

PARACRYSTALS FORMED FROM VERTEBRATE SKELETAL AND SMOOTH MUSCLE LMM: AXIAL REPEATS AND HYBRIDIZATION

Phyllis R. Wachsberger
and
Frank A. Pepe

Department of Anatomy
University of Pennsylvania
Philadelphia, Pennsylvania

Light meromyosin fragments were prepared by brief tryptic digestion of rabbit uterine myosin and rabbit and chicken skeletal myosins; studies of their aggregating properties, either individually or as mixtures were made. It is reasonable to assume that the LMM portion of the myosin molecule, which carries the solubility characteristics of myosin would have aggregation properties which are similar to the intact myosin molecule.

All three LMM's were freed of RNA contaminants either by column purification or by RNAase treatment. The LMM fragments were each dissolved in 0.6 *M* KCl, 0.01 *M* Imidazole, pH 7.0, and paracrystals were formed by dialysis into either a pH 7.0 buffer containing 0.15 *M* KCl, 0.01 *M* Imidazole or a pH 6.0 buffer containing 0.025 *M* KCl, 0.075 *M* Imidazole.

Figure 1 shows the following:

1) At pH 7.0, chicken skeletal LMM forms paracrystals with a 14 nm axial repeat, whereas rabbit skeletal LMM forms paracrystals with a 43 nm axial repeat. At pH 6.0, both chicken and rabbit skeletal LMM's form paracrystals with a 14 nm repeat. Therefore, there is a difference in aggregation properties between chicken and rabbit skeletal LMM's at pH 7.0 but not at pH 6.0.

2) Uterine LMM forms only small nonperiodic aggregates at pH 7.0. At pH 6.0, it forms paracrystals with an axial repeat of 14 nm. The 14 nm repeat pattern observed with uterine LMM, in negative as well as positive stain, typically consists of a

ISBN 0-12-551750-5

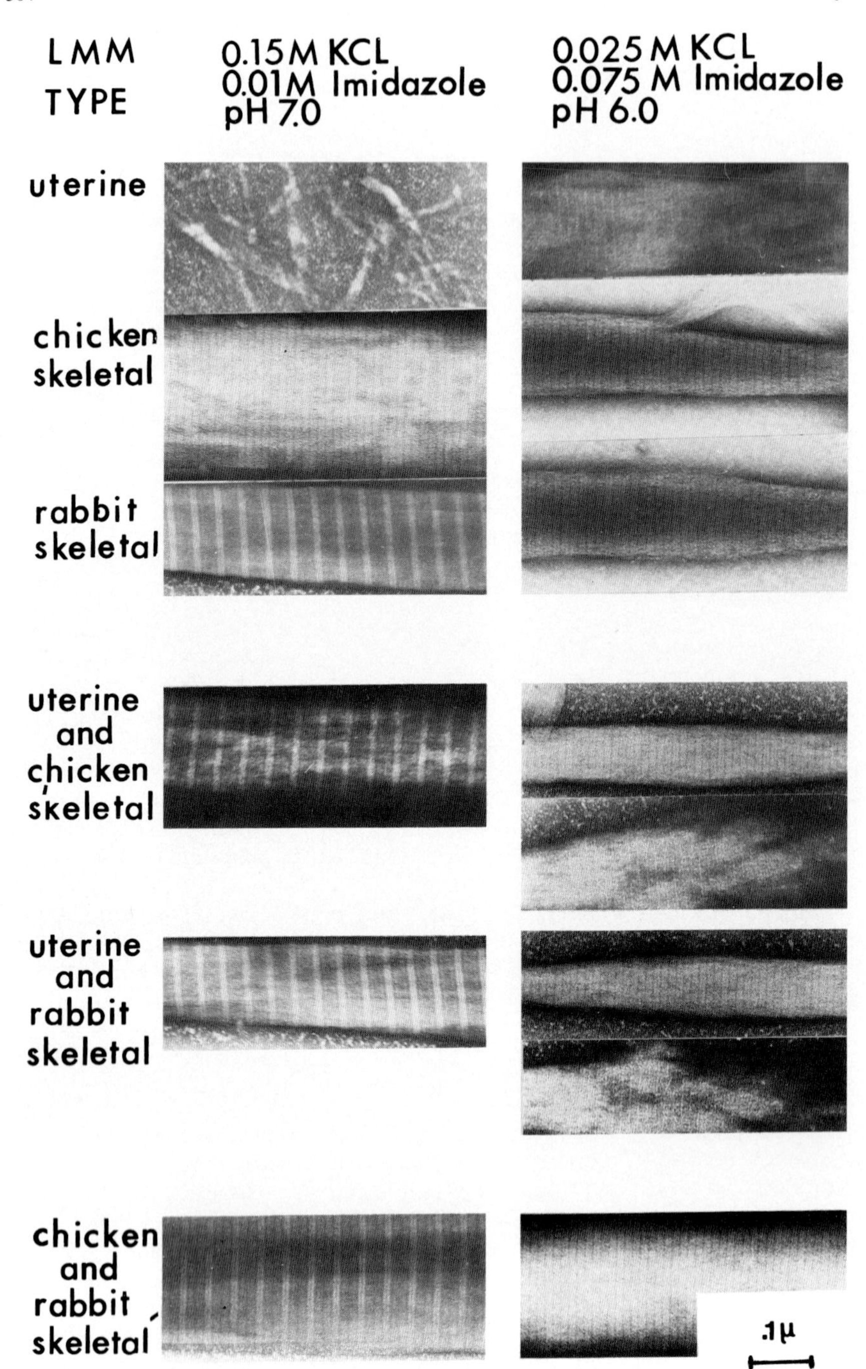

LMM TYPE
0.15 M KCL
0.01M Imidazole
pH 7.0
0.025 M KCL
0.075 M Imidazole
pH 6.0
uterine
chicken skeletal
rabbit skeletal
uterine and chicken skeletal
uterine and rabbit skeletal
chicken and rabbit skeletal
.1μ

FIGURE 1.

narrow light band every 14 nm. This pattern differs from the axial repeat of 14 nm seen with skeletal LMM at either pH 6.0 or pH 7.0; the skeletal 14 nm repeat always consists of a dark line every 14 nm. This difference in staining pattern may be indicative of differences in charge distribution of the rods of the skeletal and smooth muscle myosins.

3) At pH 7.0, although uterine LMM forms only small nonperiodic aggregates, and chicken skeletal LMM alone shows only a 14 nm repeat, mixtures of uterine and chicken skeletal LMM's form paracrystals with a 43 nm repeat only. Likewise at pH 7.0, mixtures of uterine and rabbit skeletal LMM form only paracrystals with a 43 nm repeat; the nonperiodic small aggregates of uterine LMM are not present. These observations are indicative of hybridization between uterine and skeletal LMM's. This is consistent with observations of the formation of hybrid filaments from vertebrate skeletal and smooth muscle myosin mixtures made by Pollard (1975) as well as Kaminer *et al.* (1976). We have similarly observed the formation of hybridized filaments from mixtures of uterine and skeletal myosins under these same conditions.

4) At pH 6.0, where uterine LMM gives paracrystals with a 14 nm repeat consisting of light bands and skeletal LMM gives a 14 nm repeat of dark bands, mixtures of uterine and either rabbit or chicken skeletal LMM's do not form hybridized paracrystals. Rather, a mixture of paracrystals characteristic of the uterine 14 nm repeat and the skeletal 14 nm repeat is observed.

5) Although, at pH 7.0, chicken skeletal LMM gives a 14 nm repeat and rabbit skeletal LMM gives a 43 nm repeat, mixtures of rabbit and chicken skeletal LMM's form paracrystals with a 43 nm repeat only. This is indicative of hybridization between the two types of skeletal LMM's. At pH 6.0, since both skeletal LMM's give 14 nm repeats consisting of narrow dark bands, it is not possible to say if hybridization is occurring.

ACKNOWLEDGMENT

This work was supported by USPHS Grant HL15835 to the Pennsylvania Muscle Institute.

REFERENCES

Kaminer, B., Szonyi, E., and Belcher, C. D. (1976). *J. Mol. Biol. 100,* 379.

Pollard, T. D. (1975). *J. Cell Biol. 67,* 93.

Motility in Cell Function
Proceedings of the First John M. Marshall Symposium in Cell Biology

FAST-TO-SLOW MYOSIN TRANSFORMATION IN PRE-EXISTING MUSCLE FIBERS DURING CHRONIC STIMULATION

Neal Rubinstein,[1]
Katshuhide Mabuchi,[2]
Frank Pepe,[1]
Stanley Salmons,[6]
John Gergely,[2,3,5]
Frank Sreter[2,3,4]

[1]Department of Anatomy
University of Pennsylvania School of Medicine
Philadelphia, Pennsylvania

[2]Department of Muscle Research
Boston Biomedical Research Institute
Boston, Massachusetts

[3]Department of Neurology
Massachusetts General Hospital
[4]Departments of Neurology and [5]Biological Chemistry
Harvard Medical School
Boston, Massachusetts

[6]Department of Anatomy
Medical School
University of Birmingham
Birmingham, United Kingdom

Continuous stimulation of a rabbit fast muscle at a frequency which imitates the firing rate of a motorneuron to a slow muscle changes the fast muscle's physiological and biochemical parameters to those of a slow muscle. This transformation includes the replacement of the fast type of myosin by the slow type of myosin, as demonstrated by changes in myosin ATPase activity and myosin light chains (Sreter *et al.*, 1973). Two hypotheses could explain the cellular basis of these changes. First, if the fibers were programmed to be fast or slow, but not both, a change from one muscle type to another would imply atrophy of one fiber type accompanied by *de novo* appearance of the other fiber type. Alternatively, pre-exist-

ISBN 0-12-551750-5

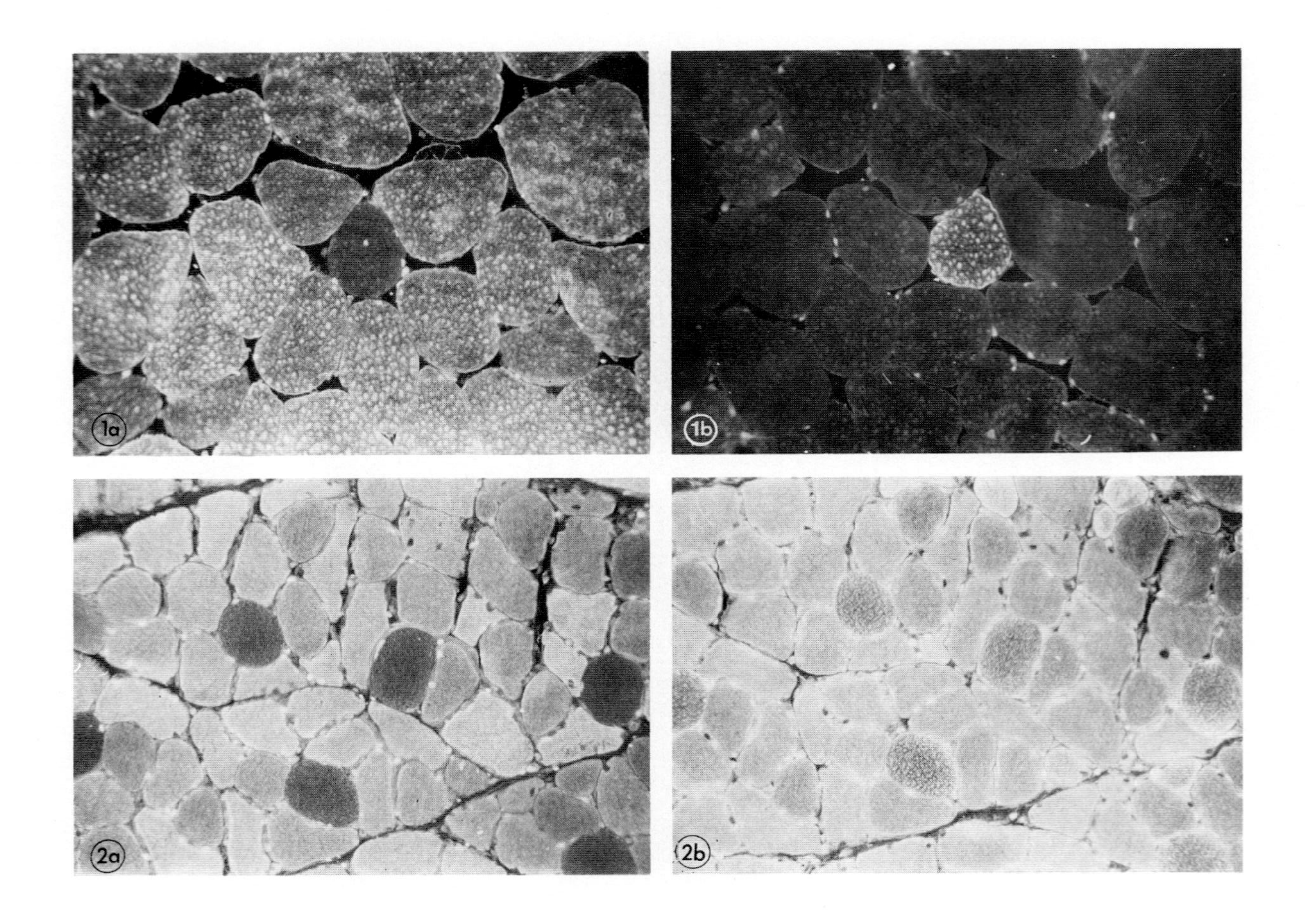

FIGURE 1.
FIGURE 2.

ing muscle fibers themselves could be changing from the expression of one set of genes to the expression of another.

We wished to distinguish between these possibilities. Since earlier work had suggested that each normal muscle fiber contained only one myosin type (Arndt and Pepe, 1975), we decided to look at the myosin content of individual fibers during transformation of a rabbit fast to a slow muscle by chronic stimulation. Fluorescein labelled antibodies against fast (AF) and slow (AS) muscle myosins of rabbits were prepared by procedures originally applied to chicken muscle (Arndt and Pepe, 1975). In the unstimulated peroneus longus muscle, most fibers stained only with AF (Fig. 1a); a small percentage stained only with AS (Fig. 1b); and serial sections demonstrated that no fiber stained with both antibodies. If the muscle transformation during chronic stimulation involved the appearance, disappearance, or change in size of individual, permanently-programmed fibers, each fiber should still contain only one myosin type. If, however, individual fibers were labile and could actually switch the type of myosin they were synthesizing, both myosins should be found in individual fibers during the transformation. In a muscle stimulated for three weeks, most fibers stained with both AF (Fig. 2a) and AS (Fig. 2b). With increasing time of stimulation, there was a progressive decrease in staining intensity with AF and an increase in staining intensity with AS within the same fibers. These results are only consistent with a theory that individual, pre-existing muscle fibers actually switch from the synthesis of fast myosin to the synthesis of slow myosin.

REFERENCES

Arndt, I., and Pepe, F. (1975). *J. Histochem. Cytochem. 23*, 159-168.

Sreter, F., Gergely, J., Salmons, S., and Romanul, F. (1973). *Nature 241*, 17-18.

Motility in Cell Function
Proceedings of the First John M. Marshall Symposium in Cell Biology

EVIDENCE FOR AN INTRINSIC DEVELOPMENTAL PROGRAM IN AVIAN AND MAMMALIAN SKELETAL MUSCLES

Raman K. Roy, Frank A. Sreter,
and
Satyapriya Sarkar

Department of Muscle Research
Boston Biomedical Research Institute
and
Department of Neurology
Harvard Medical School
Boston, Massachusetts

Many myofibrillar proteins such as myosin, tropomyosin (Tm), and troponin, etc., exist in distinct polymorphic forms which seem to be correlated with the functional type of striated muscle fibers (Perry, 1974). The polymorphism of contractile proteins in a skeletal muscle fiber does not appear to be invariant, as judged by observed changes during development (Amphlett *et al.*, 1976; Pelloni-Muller *et al.*, 1976; Roy *et al.*, 1976, 1979, 1978; Rubinstein *et al.*, 1977; Sreter *et al.*, 1975), and during adaptive response of the fiber following cross reinnervation and chronic electrical stimulation (for a review see Amphlett *et al.*, 1975; Salmons and Sreter, 1976). In this report we describe the developmental patterns of myosin light chains and Tm subunits in rabbit and chicken striated muscles. Our results suggest that an intrinsic developmental program coding for specific polymorphic forms of these proteins exists in all early embryonic skeletal muscle fibers regardless of whether the fibers mature into adult slow or fast types.

SDS-polyacrylamide gel electrophoretograms of Tm isolated from various striated muscles (Fig. 1) indicates: a) In adult skeletal fast muscles αTm is present either as the major component as in the case of rabbit fast muscles (gel 7; α/β = 80/20), or as the only subunit present in chicken breast muscles (gel 4). In contrast, the βTm is the predominant species found in these muscles during development, as shown in the case of 12-day old chick embryonic breast muscle (gel 2;

ISBN 0-12-551750-5

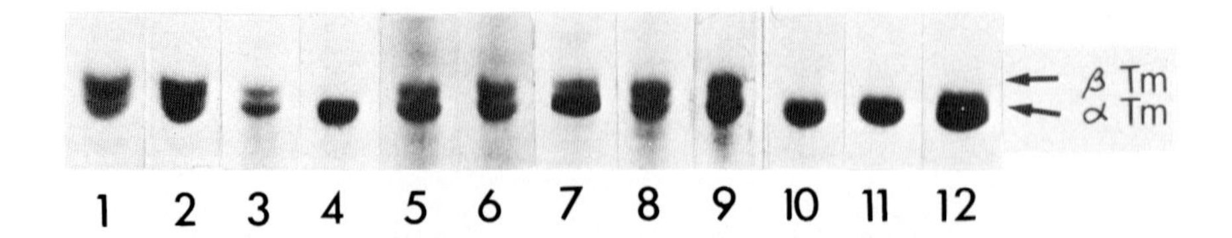

FIGURE 1. SDS-polyacrylamide gel electrophoresis of Tm isolated from skeletal and cardiac muscles. Proteins loaded per gel 6-15 μ*g.* Gels 1-6, *Tm of chicken skeletal muscles: embryonic leg (gel 1); embryonic breast (gel 2); adult leg (gel 3); adult breast (gel 4); ALD (gel 5); PLD 9gel 6).* Gels 7-9, *Tm of rabbit skeletal muscles: adult fast (gel 7); adult slow (gel 8); 1-day old skeletal (gel 9).* Gels 10-12, *Tm of cardiac muscles: adult and embryonic chicken cardiac Tm (gel 10); adult and embryonic rabbit cardiac Tm (gel 11); adult chicken cardiac Tm and rabbit cardiac Tm + rabbit skeletal αTm (gel 12).*

α/β = 20/80) and 1-day old rabbit skeletal muscles (gel 9; α/β = 50/50). b) The same developmental pattern is also observed in chicken leg muscle which consists of mixed fibers (gels 1 versus 3). c) The correlation that the speed of contraction of a striated muscle fiber is directly related to its αTm content (Perry, 1974) seems to be valid for adult rabbit skeletal fast and slow muscles (gels 7 and 8). However, two representative slow and fast avian muscles, chicken anterior latissimus dorsi (ALD) and posterior latissimus dorsi (PLD), respectively, give identical patterns (gels 5 and 6, α/β = 55/45), indicating that the above-mentioned correlation is not applicable for all muscles and species. d) The cardiac muscles of both chicken and rabbit contain only the αTm which was invariant during development (gels 10-12).

Evidence for multiple forms of Tm subunits, which differ in charge only, was obtained by electrophoresis in urea at pH 3.4 in the absence of SDS. As shown in Fig. 2, αTm of adult chicken breast muscles contains two distinct components (gel 1), neither of which comigrates with the single bands of rabbit skeletal α (gel 3) and β subunits (gel 4). Among the three bands obtained with adult chicken leg muscle Tm (gel 2), the middle one corresponds to βTm, as judged by its comigration with rabbit βTm (gel 5). The remaining two bands correspond to the two components of chicken skeletal αTm (gels 1 and 2). Both rabbit and chicken cardiac Tm give single bands which differ in mobilities (gels 6 and 8). While rabbit cardiac Tm

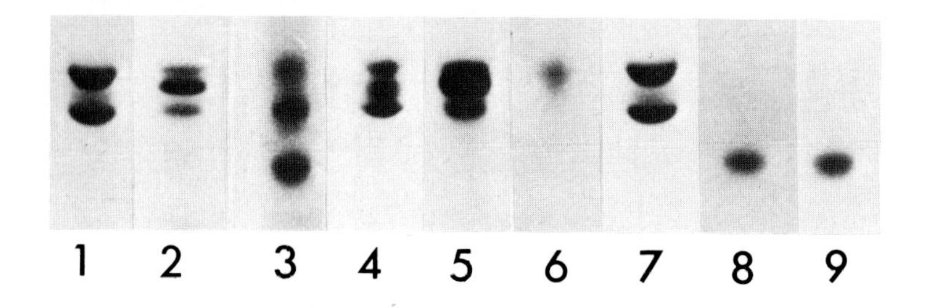

Figure 2. Urea-gel electrophoresis at pH 3.4 of skeletal and cardiac Tm. Gel 1, adult chicken breast Tm; gel 2, adult leg Tm; gel 3, chicken breast Tm and rabbit skeletal aTm; gel 4, chicken breast Tm and rabbit skeletal βTm; gel 5, chicken leg Tm and rabbit skeletal βTm; gel 6, chicken cardiac Tm; gel 7, chicken cardiac Tm and chicken breast Tm; gel 8, rabbit cardiac Tm; gel 9, rabbit cardiac Tm and skeletal αTm.

comigrates with rabbit skeletal α subunit (gel 9), only the more slowly moving component of chicken skeletal αTm (gel 1) comigrates with the chicken cardiac Tm (gel 7), indicating the complexity of Tm subunit patterns.

Myosins isolated from rabbit embryonic and adult fast muscles are very similar, as judged by identical mobilities of their three light chains (Fig. 3, scans A and C). The embryonic myosin, however, contains a decreased amount of LC_3 light chain (Sreter *et al.*, 1975). Rabbit slow myosin is clearly distinct from embryonic myosin, as shown by the absence of LC_3 light chain (scan B) and the nonidentity of their LC_1 and LC_2 light chains (scan D). Similar results were obtained for chicken embryonic and adult skeletal myosins.

These results indicate that the high βTm content and the fast myosin light chains LC_1 and LC_2 observed in all embryonic skeletal fibers are due to an intrinsic developmental program. Subsequent appearance of polymorphic forms of contractile proteins, e.g., slow myosin light chains; increased accumulation of LC_3 light chain; and changes in α/β ratio of Tm subunits, represent differential gene expression in different adult fibers in response to exogenous stimuli.

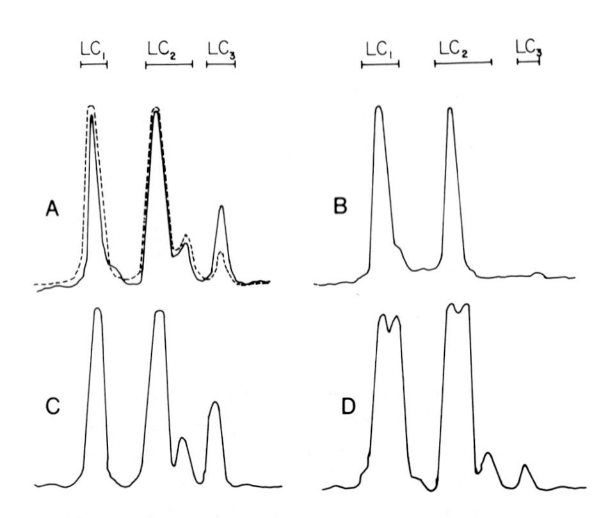

FIGURE 3. SDS-gel runs of adult and embryonic rabbit skeletal muscle myosins. Scan A: adult fast myosin ( ); embryonic skeletal myosin ( ). Scan B: adult slow myosin. Scan D: adult fast myosin and embryonic skeletal myosin. Scan D: adult slow myosin and embryonic skeletal myosin.

ACKNOWLEDGMENTS

This work was supported by grants from the National Institutes of Health (AM 13238 and AG 00262), and the Muscular Dystrophy Associations of America, Inc., and carried out during the tenure of a Research Fellowship from the Charles A. King Trust to R. K. Roy.

REFERENCES

Amphlett, G. W., Perry, S. V., Syska, H., Brown, M. D., and Vrbova, G. (1975). *Nature 257*, 602-604.

Amphlett, G. W., Syska, H., and Perry, S. V. (1976). *FEBS Lett. 63*, 22-25.

Pelloni-Muller, G., Ermini, M., and Jenny, E. (1976). *FEBS Lett. 67*, 68-74.

Perry, S. V. (1974). *In* "Exploratory Concepts in Muscle" (A. T. Milhorat, ed.), Vol. II, pp. 319-328. Excerpta Medica, Amsterdam.

Roy, R. K., Potter, J. D., and Sarkar, S. (1976). *Biochem. Biophys. Res. Comm. 70*, 28-36.

Roy, R. K., Sreter, F. A., and Sarkar, S. (1979). *Develop. Biol. 69,* 15-30.
Roy, R. K., Sreter, F. A., and Sarkar, S. (1978). *In* "Aging" (G. Kaldor and W. J. DiBattista, eds.), Vol. 1. pp. 23-48. Raven Press, New York.
Rubinstein, N. A., Pepe, F. A., and Holtzer, H. (1977). *Proc. Natl. Acad. Sci. U.S.A. 74,* 4524-4527.
Salmons, S., and Sreter, F. A. (1976). *Nature 263,* 30-34.
Sreter, F. A., Balint, M., and Gergely, J. (1975). *Develop. Biol. 46,* 317-325.

Motility in Cell Function
Proceedings of the First John M. Marshall Symposium in Cell Biology

ULTRASTRUCTURAL LOCATION OF ANTIBODIES TO TROPONIN COMPONENTS IN DYSTROPHIC SKELETAL MUSCLE

Frank J. Wilson

Department of Anatomy
College of Medicine and Dentistry of New Jersey
Rutgers Medical School
Piscataway, New Jersey

Tamio Hirabayashi

Institute of Biological Sciences
The University of Tsukuba
Ibaraki 300-31, Japan

Mary J. Irish

Department of Anatomy
College of Medicine and Dentistry of New Jersey
Rutgers Medical School
Piscataway, New Jersey

The effects of hereditary avian muscular dystrophy on the three major components of troponin were analyzed by antibody staining techniques. Dystrophic chickens of strain 308 and normal New Hampshire birds were used in this study. Segments of I bands were prepared from glycerinated pectoral muscles by methods described in a previous communication (Irish *et al.*, 1977). Antisera against each of the three troponin components were reacted with the I segment preparations. Each antiserum was specific for its homologous antigen and showed no immunochemical cross reactions with the other two troponin components. Antiserum against troponin-T was prepared against that protein

ISBN 0-12-551750-5

of 44,000 molecular weight isolated and purified from chicken breast muscle. The antibody treated I segments from normal and dystrophic muscles were then reacted with an anti-immunoglobulin and negatively stained with 1% aqueous uranyl acetate.

Normal I segments treated with anti-troponin-T showed the antibody to be distributed along the lengths of the thin filaments with a periodicity of 38.6 nm ± 0.8 nm (mean ± standard deviation of the mean), and I segments from dystrophic muscle displayed a periodic distribution of antibody at intervals of 38.3 nm ± 0.9 nm ($P>0.01$) (Fig. 1a,b). When normal I segments were reacted with anti-troponin-I, the periodicity of antibody distribution was 38.1 nm ± 0.8 nm (Fig. 1c). Dystrophic I segments treated with anti-troponin-I had a periodicity of 38.6 nm ± 1.9 nm ($P>0.01$) (Fig. 1d). Thus, there were no significant differences in the locations of anti-troponin-I and anti-troponin-T in normal and dystrophic pectoral muscles. Normal I segments reacted with anti-troponin-C showed a periodicity of 38.4 nm ± 0.7 nm (Fig. 1e). In contrast to these controls the I segments from dystrophic muscle did not possess a distinct periodicity when reacted with anti-troponin-C (Fig. 1f).

When we analyzed the thin filament proteins by polyacrylamide gel electrophoresis with sodium dodecyl sulfate (SDS), we observed that although troponin-C was present in the thin filaments of dystrophic muscle, the protein showed a slight change in its electrophoretic mobility when compared with normal controls (Fig. 2). Moreover, the gel of normal muscle proteins exhibited one tropomyosin band while that of dystrophic muscle had two bands in the tropomyosin region. Thus, it appears that there has been a shift in the genetic expression of tropomyosin from an α-type in normal muscle to an α/β type in dystrophic muscle.

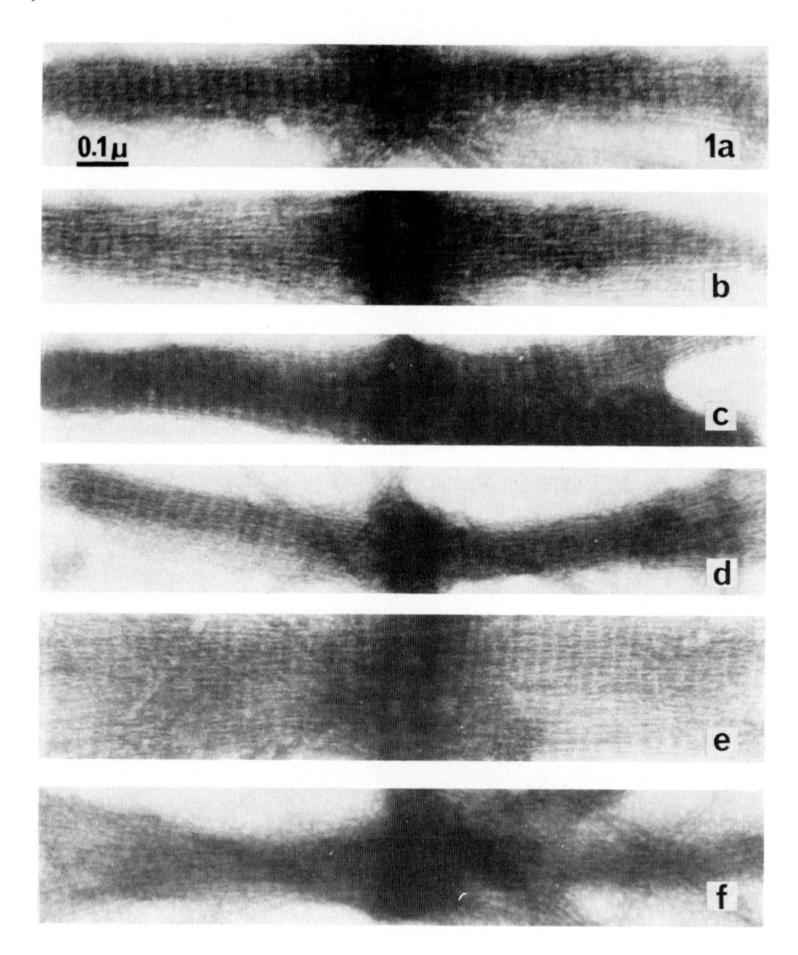

FIGURE 1. Segments of I bands reacted with antibody to troponin components (x 67,000). (a) Normal I segment and anti-troponin-T. (b) Dystrophic I segment and anti-troponin-T. (c) Normal I segment and anti-troponin-I. (d) Dystrophic I segment and anti-troponin-I. (e) Normal I segment and anti-troponin-C. (f) Dystrophic I segment and anti-troponin-C.

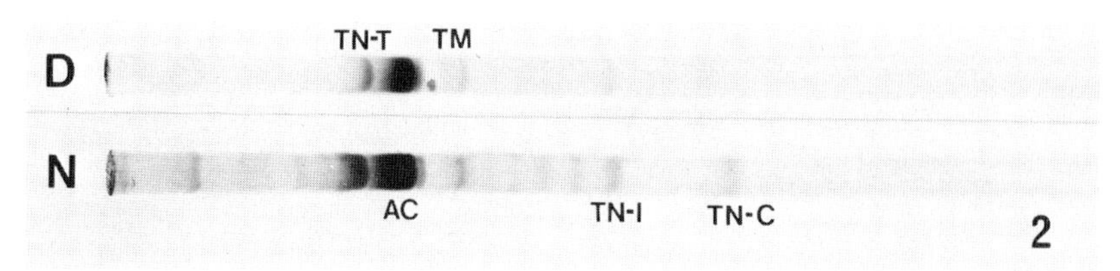

FIGURE 2. Gel electrophoresis of thin filament proteins from dystrophic (D) and normal (N) chicken pectoralis muscles. 10% acrylamide, 0.1% SDS, 100 mM sodium phosphate buffer, pH 7.0. AC is actin; TM, tropomyosin; TN, troponin.

With regard to troponin-C there are at least three explanations consonant with our results. The data may reflect 1) a quantitative reduction in troponin-C in the I segments of dystrophic muscle, or 2) a qualitative difference between troponin-C's of normal and dystrophic muscles, or 3) a blockage of the antibody binding sites of troponin-C by some unidentified material in the dystrophic muscle.

ACKNOWLEDGMENTS

This research was supported by grants from the Muscular Dystrophy Association, Inc., and the American Heart Association, New Jersey Affiliate.

REFERENCE

Irish, M. J., Hirabayashi, T., and Wilson, F. J. (1977). *Tissue and Cell 9*, 499-505.

Motility in Cell Function
Proceedings of the First John M. Marshall Symposium in Cell Biology

LOCALIZED CHAIN SEPARATION OF TROPOMYOSIN NEAR THE TROPONIN BINDING SITE

S. S. Lehrer
and
S Betcher-Lange

Department of Muscle Research
Boston Biomedical Research Institute
Boston, Massachusetts
and
Department of Neurology
Harvard Medical School
Boston, Massachusetts

Rabbit skeletal tropomyosin (Tm), a component of the Ca^{2+} control system of striated muscle contraction (Weber and Murray, 1973), is made up of two parallel α-helical chains interacting in register (Johnson and Smillie, 1975; Lehrer, 1975; Stewart, 1975a; McLachlan and Stewart, 1975). We have noted a preferential instability in the molecule near Cys 190 (Stone *et al.*,1975), which is close to the troponin binding site (Stewart, 1975b). This instability is detected by two different methods which measure the proximity of the adjacent Cys 190's of each chain. One method utilizes 5-5'-dithiobis (2-nitrobenzoate) (Nbs_2) to probe the formation of interchain S-S bonds (Lehrer, 1975). The other approach studies the degree of excited state dimer (excimer) formation between pyrene fluorophors covalently linked to the cysteines (Betcher-Lange and Lehrer, 1978). Both techniques indicated that localized chain separation near Cys 190 induced by increasing [GuHCl] is half complete when the molecule is less than 20% unfolded.

SDS-polyacrylamide gels were used to obtain the fraction of crosslinked chains produced by Nbs_2 treatment at different [GuHCl] (Fig. 1). In the absence of GuHCl the crosslinking reaction is nearly quantitative but, as reported previously, in 4 *M* GuHCl, where the chains are dissociated, the reaction with Nbs_2 does not produce S-S crosslinks (Lehrer, 1975).

ISBN 0-12-551750-5

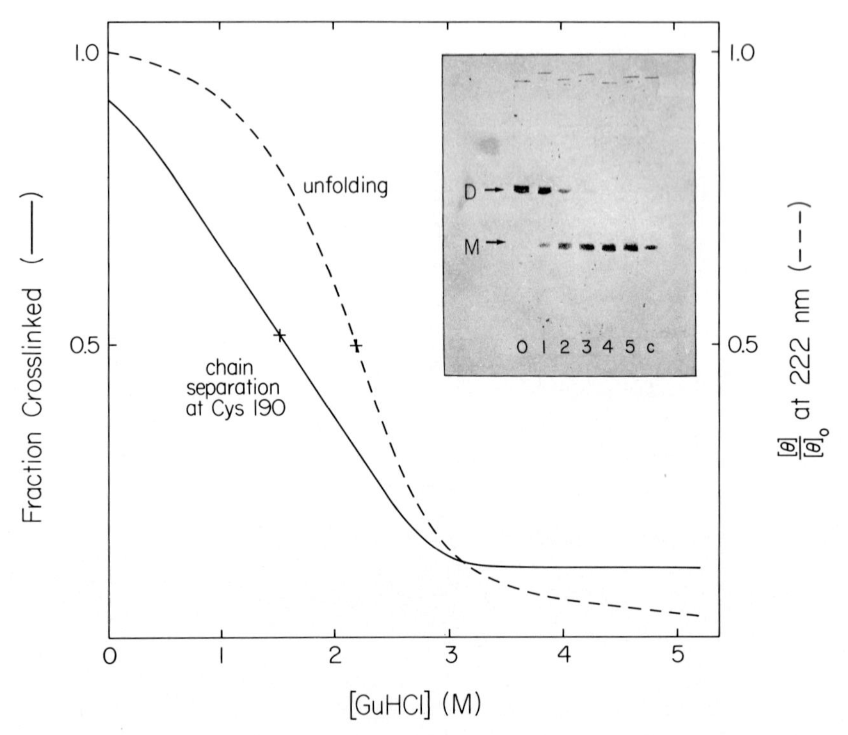

FIGURE 1. Comparison of the effect of GuHCl on the ability to crosslink Tm with the loss of ellipticity at 222 nm. The fraction crosslinked was obtained after reaction of Tm with 2 mM Nbs_2, by SDS-polyacrylamide gel electrophoresis and densitometry. Insert: photograph of stained gels of samples reacted at indicated [GuHCl]. See Lehrer (1977) for details.

The fraction of chains crosslinked at each [GuHCl] is also compared with the fractional ellipticity at 222 nm, which measures the degree of unfolding. It is clear that the chains separate near Cys 190 before there is a substantial degree of unfolding of the α-helix.

N-(1-pyrene)maleimide reacts with rabbit skeletal α-tropomyosin (ααTm) at pH 7.5 to form covalent bonds with Cys 190 and a neighboring lysine, probably Lys 189 (Betcher-Lange and Lehrer, 1978). The structure of the fluorescence probe and the spectra observed in the native and unfolded state are shown in Fig. 2. In the native state there are two fluorescence bands, a structured band with a major peak at 384 nm associated with emission from the pyrene excited monomer and a broad band near 480 nm corresponding to emission from an intramolecular excimer originating from an interaction between the pyrenes attached to each α chain. In the unfolded state

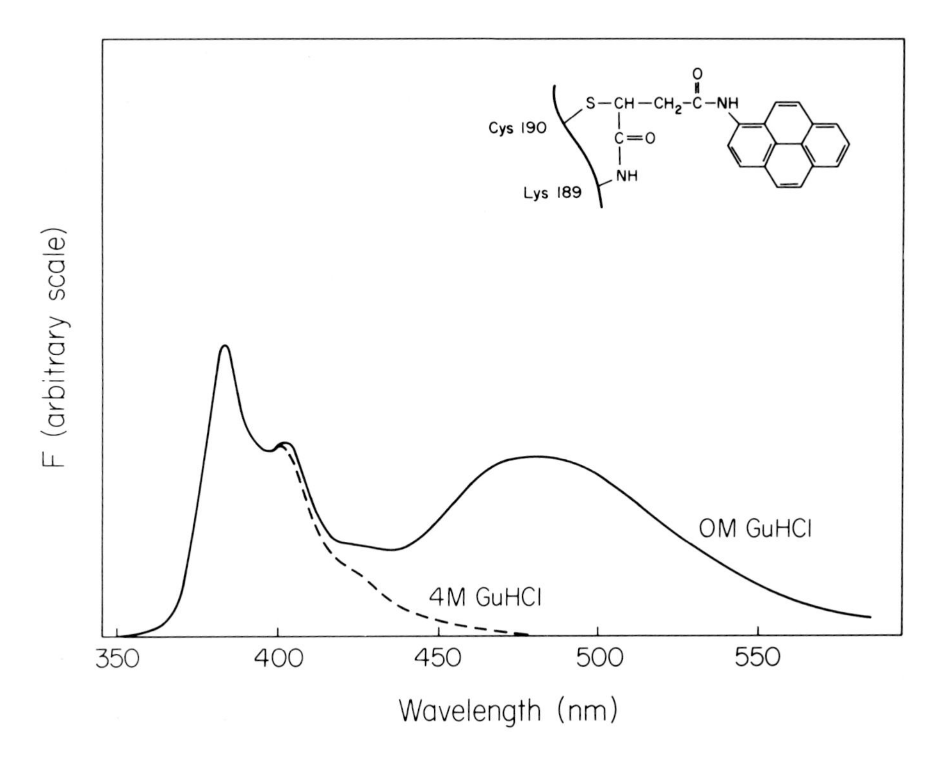

FIGURE 2. Fluorescence spectra of N-(1-pyrene)maleimide labeled αα-Tm (Pyr-ααTm), labeled and measured at pH 7.5, in the absence and presence of 4 mM GuHCl. Insert shows probe structure and probable mode of attachment to αα-Tm. λ_{ex}*=340 nm,* T = *22°, buffer = 5 mM Hepes, 1* M *NaCl, 1* mM *EDTA.*

(4 *M* GuHCl) the excimer does not form. The unfolding profiles obtained by measuring the excimer fluorescence and the ellipticity at 222 nm of Pyr-ααTm are compared in Fig. 3. The excimer fluorescence profile shows a transition midpoint at 1.5 *M* GuHCl, whereas the transition midpoint for the loss of secondary structure occurs at 2.1 *M* GuHCl, in agreement with values obtained from a control sample of unlabeled ααTm.

These results may explain why there is an unfolding pretransition, in the physiological temperature region, for S-S crosslinked Tm, whereas only a monotonic decrease in ellipticity is observed for uncrosslinked Tm (Lehrer, 1977). A small degree of strain introduced by the crosslink may be transmitted across the chains via the crosslink when the chains attempt to locally separate. This local chain separation may have relevance to the mechanism of regulation in view of its proximity to the troponin binding site.

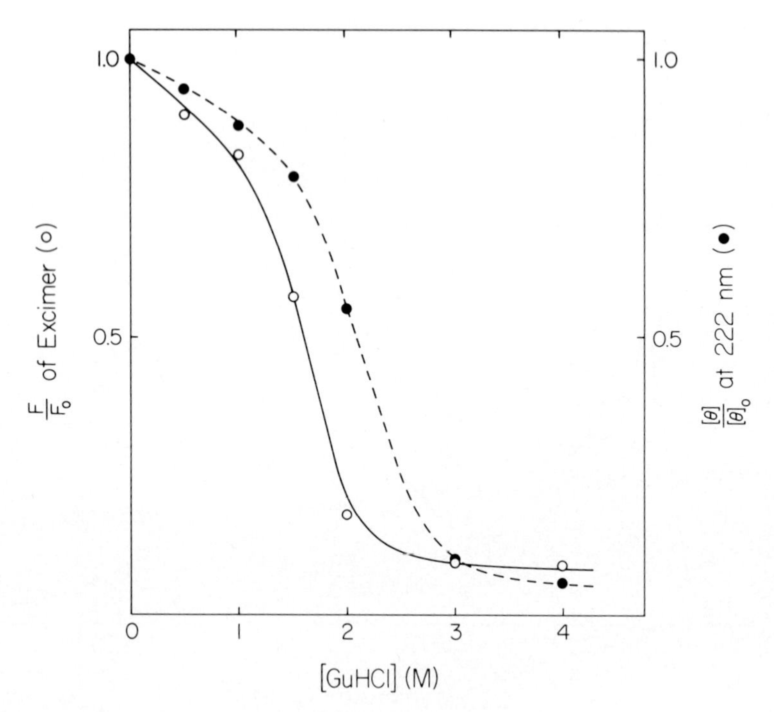

FIGURE 3. Comparison of the decrease in excimer fluorescence (o) with the decrease in ellipticity at 222 nm (●), of Pyr-ααTm as a function of GuHCl concentration. Same conditions as for Fig. 2.

ACKNOWLEDGMENTS

This work was supported by Grants AM 11677 from the National Institutes of Health, GB 24316 from the National Science Foundation, and by the Muscular Dystrophy Association of America.

REFERENCES

Betcher-Lange, S., and Lehrer, S. S. (1978). to be published.

Johnson, P., and Smillie, L. B. (1975). *Biochem. Biophys. Res. Comm. 64*, 1316-1322.

Lehrer, S. S. (1975). *Proc. Nat. Acad. Sci. U.S.A. 72*, 3377-3383.

Lehrer, S. S. (1977). *J. Mol. Biol. 117*, 000.

McLachlan, A. D., and Stewart, M. (1975). *J. Mol. Biol. 98*, 293-304.

Stewart, M. (1975a). *FEBS Lett. 53*, 5-7.

Stewart, M. (1975b). *Proc. Roy. Soc. B190*, 257-266.
Stone, D., Sodek, J., Johnson, P., and Smillie, L. B. (1975). Proc. IX FEBS Meeting, *"Proteins of Contractile Systems"* (Biro, E. N. A., ed.), Vol. 31, pp. 125-136. Akad. Kiado, Budapest.
Weber, A., and Murray, J. (1973). *Physiol. Rev. 53*, 612-673.

Motility in Cell Function
Proceedings of the First John M. Marshall Symposium in Cell Biology

TROPONIN C AND MODULATOR PROTEIN OF 3′, 5′-CYCLIC NUCLEOTIDE PHOSPHODIESTERASE: STRUCTURAL SIMILARITIES AND DISTRIBUTION IN VARIOUS TISSUES

W. Drabikowski

Department of Biochemistry of Nervous System and Muscle
Nencki Institute of Experimental Biology
Warsaw, Poland

Both troponin C (TN-C) and modulator protein of 3′, 5′-cyclic nucleotide phosphodiesterase (MP) are composed of four homologous regions, each consisting of two α-helical segments and a loop with a Ca^{2+}-binding site in between. Our recent studies (Drabikowski *et al.*, 1977b) have indicated that MP reveals several features common with TN-C induced by Ca^{2+}-binding, as, e.g., enhancement of the intrinsic tyrosine fluorescence, change of the mobility in urea polyacrylamide gel electrophoresis, and interaction with troponin I. Both proteins are cleaved by trypsin in the same area and Ca^{2+} ions have a pronounced effect on this process (Drabikowski *et al.*, 1977a,b,c). In the presence of Ca^{2+} only these peptide bonds are susceptible for tryptic attack which are out of both α-helical segments and the loops. These are the peptide bonds at the positions 8, 84, and 88 in TN-C (Drabikowski *et al.*, 1977a) and that at the position 77 in MP (Walsh *et al.*, 1977), homologous to the position 84 in TN-C. As the result of the cleavage of both proteins in the presence of Ca^{2+} two peptides appear, one containing Ca^{2+}-binding sites 1 and 2 (TR_1) and the second with the Ca^{2+}-binding sites 3 and 4 (TR_2), which are resistant to further splitting. In the absence of Ca^{2+}, pronounced conformational changes in the molecules of both proteins occur and the most susceptible for trypsin are those peptide bonds which were originally situated in the α-helical parts of region 3, i.e., the bonds at the residues 100 and 120 in case of TN-C (Vinokurov and Grabarek, unpublished) and at the residues 90 and 106 in case of MP (Walsh *et al.*, 1977).

The peptides TR_1 and TR_2 were isolated, purified to homogeneity, and their properties compared. Although the homologous peptides from both proteins are not identical in size and charge, they have several common features. The TR_2 peptides from both proteins preserve most of the properties of the parent molecules, i.e., they change mobility in urea polyacrylamide gel depending on Ca^{2+} and interact with troponin I. These peptides also show the Ca^{2+}-dependent spectral changes characteristic for the parent molecules (Drabikowski *et al.*, 1977b).

The Ca^{2+}-induced properties of the TN-C tryptic fragments were compared with those of fragments obtained by cleavage of TN-C with thrombin and cyanogen bromide (Leavis *et al.*, 1977a,b, 1978). All the results indicate that the Ca^{2+}-binding sites 1 and 2 are low affinity Ca^{2+}-binding sites, and sites 3 and 4 the high affinity, Ca^{2+}, Mg^{2+}, sites. However, when the regions 3 and 4 are separated from each other, the affinity toward Ca^{2+} considerably decreases (Leavis *et al.*, 1977b, 1978).

In view of the recent suggestions that other tissues than skeletal and cardiac muscle contain TN-C or TN-C-like proteins, an attempt has been made to elucidate whether the latter proteins are identical with TN-C or MP. The following properties have enabled us to distinguish between these two proteins: Different mobility in 15% SDS - polyacrylamide gel, different mobility of the complex with troponin I in urea polyacrylamide gel in the presence of Ca^{2+}, and the ability to stimulate phosphodiesterase activity (Drabikowski *et al.*, 1978). The results obtained indicate that smooth muscle (gizzard and uterus), brain, adrenal medulla (Drabikowski *et al.*, 1978), platelets (Muszbek *et al.*, 1977) slime mold Physarum polycephalum, and some protozoan cells (Euglena gracilis and Amoeba proteus) (Kuźnicki *et al.*, 1977, 1979) contain MP but not TN-C. The protein identical with MP is also present in skeletal and cardiac muscle (Drabikowski *et al.*, 1978). In all cases a part of MP is bound to the insoluble proteins. Subsequent investigations have indicated that MP is present in conventional preparations of actomyosin from smooth muscle (Kuźnicki *et al.*, 1978), brain, platelets, and Physarum (Kuźnicki *et al.*, 1979) and in preparations of skeletal muscle myosin (Barylko *et al.*, 1978). In all cases MP can be released by EDTA treatment (Fig. 1). As shown for skeletal and smooth muscle the EDTA-extracts contain Ca^{2+}-dependent protein kinases which phosphorylate light chain (P-LC) of EDTA-treated myosin (Barylko *et al.*, 1978, Kuźnicki *et al.*, 1978). After fractionation on DEAE-Sephadex column and separation of MP, the EDTA-extract is no more able to phosphorylate P-LC. The phosphorylating activity of the endogenous kinase from skeletal muscle can be restored by the addition of MP isolated from either skeletal or smooth or cytoplasmic actomyosin (Barylko *et al.*, 1978).

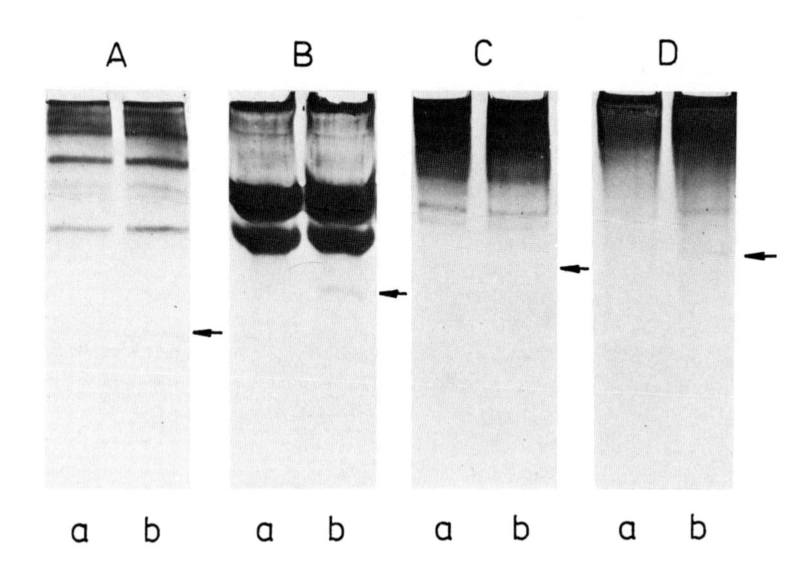

FIGURE 1. Urea polyacrylamide gel electrophoresis of EDTA-extracts from (A) rabbit skeletal muscle myosin, (B) chicken gizzard actomyosin, (C) human platelets actomyosin, (D) bovine brain actomyosin, a 0.1 mM $CaCl_2$ present; C - 0.1 mM EDTA present. Arrows indicate the position of MP. For details of procedure see Drabikowski et al. *(1978).*

In view of these results, as well as those of Dabrowska *et al.* (this volume) who showed that MP is a subunit of P-LC kinase from smooth muscle it is suggested that MP is a common subunit of all Ca^{2+}-dependent kinases which phosphorylate P-LC of skeletal, smooth and various cytoplasmic myosins.

REFERENCES

Barylko, B., Kuźnicki, J., and Drabikowski, W. (1978). *FEBS Lett. 90*, 301-304.

Drabikowski, W., Grabarek, Z., and Barylko, B. (1977a). *Biochim. Biophys. Acta 490*, 216-224.

Drabikowski, W., Kuźnicki, J., and Grabarek, Z. (1977b). *Biochim. Biophys. Acta 485*, 124-133.

Drabikowski, W., Kuźnicki, J., and Grabarek, Z. (1977c). *In* "Calcium Binding Proteins and Calcium Function" (R. H. Wasserman *et al.*, eds.), pp. 270-272. Elsevier, North-Holland, Amsterdam, New York.

Drabikowski, W., Kuźnicki, J., and Grabarek, Z. (1978). *Comp. Biochem. Physiol.*, *60*, 1-6.
Kuźnicki, J., Kuźnicki, L., and Drabikowski, W. (1977). *6th European Muscle Club*, Paris, Abstract of Commun.
Kuźnicki, J., Barylko, B., Górecka, A., and Drabikowski, W. (1978). *7th European Conference on Muscle and Motility*, Warsaw, Poland, Abst. Comm., pp. 100-101.
Kuźnicki, J., Kuźnicki, L., and Drabikowski, W. (1979). *Cell. Biol. Intern. Rep.* *3*, 17-23.
Leavis, P., Drabikowski, W., Rosenfeld, S., Grabarek, Z., and Gergely, J. (1977a). *Feder. Proc.* *36*, 831.
Leavis, P., Drabikowski, W., Rosenfeld, S., Grabarek, Z., and Gergely, J. (1977b). *In* "Calcium Binding Proteins and Calcium Function" (R. H. Wasserman *et al.*, eds.), pp. 274-276. Elsevier, North-Holland, Amsterdam, New York.
Leavis, P., Rosenfeld, S. S., Gergely, J., Grabarek, Z., and Drabikowski, W. (1978) *J. Biol. Chem.* *253*, 5452-5459.
Muszbek, L., Kuźnicki, J., Szabo, T., and Drabikowski, W. (1977). *FEBS Lett.* *80*, 300-313.
Walsh, M., Stevens, F. C., Kúźnicki, J., and Drabikowski, W. (1977). *J. Biol. Chem.* *252*, 7440-7443.

Motility in Cell Function
Proceedings of the First John M. Marshall Symposium in Cell Biology

SUBUNIT FUNCTION IN CARDIAC MYOSIN: ROLE OF CARDIAC LC_2

Ashok Bhan
Ashwani Malhotra
and
Shyuan Huang

Departments of Biochemistry and Medicine
Albert Einstein College of Medicine
and
Department of Medicine
Montefiore Hospital and Medical Center
Bronx, New York

A unifying feature of myosin from muscle and nonmuscle sources is the presence of the so called "regulatory subunit" with an approximate molecular weight of 18,000 - 20,000 daltons. In molluscan muscle myosin this subunit acts as an inhibitor of actin-myosin interaction in the absence of Ca^{2+} (Kendrick-Jones *et al.*, 1976). In smooth muscle myosin and platelet myosin the phosphorylation of this subunit by specific "kinases" results in the activation of Mg^{2+} ATPase of myosin by actin. Whether this subunit has a regulatory function in striated muscle myosin is not known. Partial removal of the 18,000 dalton subunit from rabbit skeletal myosin can be achieved by DTNB treatment under relatively mild conditions. It is impossible to remove the 18,000 dalton subunit of cardiac myosin without denaturing the molecule.

Based on the observation that this subunit in cardiac myosin is susceptible to proteolytic enzymes, (Bhan and Malhotra, 1976; Weeds and Frank, 1972) we have developed a technique for its removal under mild experimental conditions (Bhan *et al.*, 1978a; Bhan *et al.*, 1978b). Dog cardiac myosin in 0.4 *M* KCl (containing 1 m MEDTA and 10 m*M* imidazole, pH 6.8) is incubated with a myofibrillar protease (purified from the hearts of dystrophic syrian hamsters) for 20 hr at 4°C. The reaction is terminated by the addition of soybean trypsin inhibitor and myosin precipitated by dilution with water. SDS gel electro-

ISBN 0-12-551750-5

phoresis of protease treated myosin showed a complete loss of the 18,000 dalton subunit (Lc_2), with no apparent breakdown of the heavy chains (Bhan *et al.*, 1978b). The removal of Lc_2 did not affect the Ca^{2+} ATPase activity of myosin, however, the basal Mg^{2+} ATPase activity in 0.1 *M* KCl was increased 75% over that of native myosin. The actin-activated ATPase of light chain deficient myosin was increased threefold over that of native myosin at actin:myosin ratios greater than 1. Optimum substrate concentration for actin activated-ATPase of the light chain deficient myosin was 2.5×10^{-3} *M* in contrast to 0.5×10^{-3} *M* for that of native myosin. The values for actin activated-ATPase (at the optimum substrate concentrations in 0.1 *M* KCl, 30°C) were 0.042 μmoles Pi/mg/min for native myosin and 0.135 μmoles for the light chain deficient myosin. Addition of purified cardiac light chain (Lc_2) to the deficient myosin effectively reversed the changes in the basal Mg^{2+} ATPase and the actin-activated ATPase. DTNB light chain of rabbit skeletal myosin was equally effective in reversing these changes. Approximately 1 mole of cardiac Lc_2 or the DTNB light chains could be incorporated back per mole of the Lc_2 deficient myosin. Our results indicate a modulatory function of the Lc_2 in actin-activation of myosin ATPase of cardiac myosin.

ACKNOWLEDGMENTS

This work was supported by NIH Grants HL - 15498 and HL - 19458.

REFERENCES

Bhan, A., and Malhotra, A. (1976). *Arch. Biochem. Biophys. 174*, 27.

Bhan, A., Malhotra, A., and Hatcher, V. B. (1978a). *In* "Protein Turnover and Lysosomal Function" (H. L. Segal & D. Doyle, eds.). p. 607. Acadamic Press, New York.

Bhan, A., Malhotra, A., Hatcher, V. B., Sonnenblick, E. S., and Scheuer, J. (1978b). *J. Mol. Cell Cardiol. 10*, 769.

Kendrick-Jones, J., Szentkiralyi, E. M., and Szent-Györgyi, A. G. (1976). *J. Mol. Biol. 104*, 747.

Weeds, A. G., and Frank, G. (1972). *Cold Spring Harbor Symp. Quant. Biol. 37*, 9.

Motility in Cell Function
Proceedings of the First John M. Marshall Symposium in Cell Biology

MYOSIN PHOSPHORYLATION IN CULTURED AORTIC MEDIAL CELLS

Samuel Chacko[1]
Richard Mayne[2]
Stephen H. Blose[1]
and
Robert S. Adelstein[3]

[1]Department of Pathobiology
University of Pennsylvania
Philadelphia, Pennsylvania

[2]Department of Anatomy
University of Alabama Medical Center
and
[3]NIH
Bethesda, Maryland

Aortic smooth muscle cells can be grown in culture for several generations (Ross, 1971). These cells synthesize the same type and proportion of collagen as the aortic smooth muscle cells *in vivo* (Mayne *et al.*, 1977). However, unlike the muscle cells in the intact aortic media, the cultured medial cells lack the thick myosin filaments (Chacko *et al.*, 1977). Hence, their myosin appears to be similar to nonmuscle cells in their lack of ability to organize into thick filaments.

A specific kinase that phosphorylates the 20,000 dalton light chains (P-light chain) of myosin is present in the muscle as well as nonmuscle cells (Daniel and Adelstein, 1976; Perry *et al.*, 1975). This enzyme isolated from two nonmuscle cells: astrocytes, and proliferating skeletal myoblasts, is not dependent on Ca^{2+} for its activity (Scordilis and Adelstein, 1977). On the other hand, the light chain kinase isolated from muscle requires Ca^{2+} for activation (Perry *et al.*, 1975). Experiments were designed to determine whether or not the cultured aortic medial cells retain the calcium sensitivity of the kinase.

Supported by HL-22264 to S. C. and HL-21665 to R. M.

ISBN 0-12-551750-5

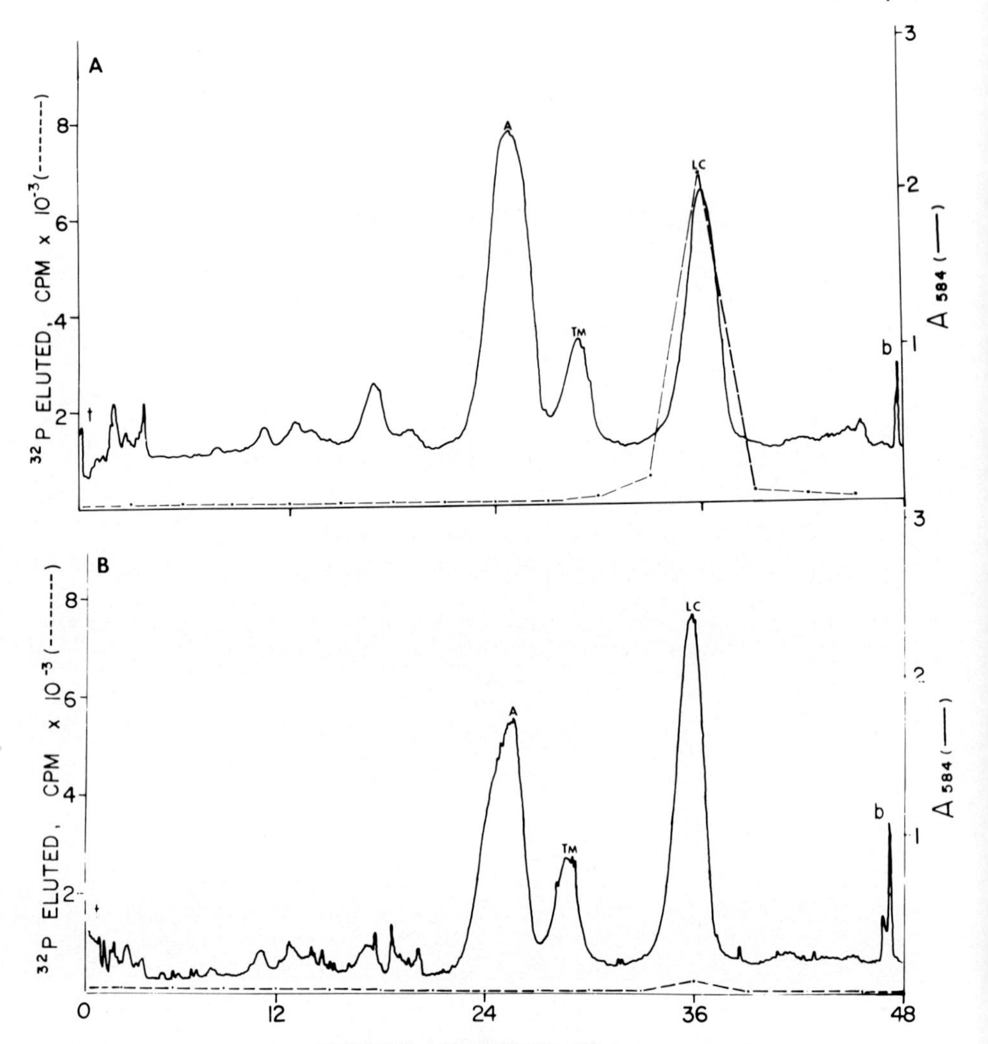

FIGURE 1. Gel scans and radioactivity elution profiles of exogenous smooth muscle light chain (LC) incubated with γ-AT^{32}P, partially purified light chain kinase isolated by Sepharose 4B chromatography and either 0.1 mM Ca^{2+}(A) or 2 mM EGTA (B). The 1% $NaDOdSo_4$/7.5% polyacrylamide gels stained with Coomassie brilliant blue were scanned at 584 nM and ^{32}P eluted from gel slices. Notice that the kinase separated from aortic smooth muscle actomyosin transferred the ^{32}P from γ-AT^{32}P to 20,000 dalton LC in the presence of Ca^{2+} (A) but not in the presence of EGTA (B) t, gel top; b, gel bottom; A, Actin; Tm, tropomyosin; LC, 20,000 dalton light chain.

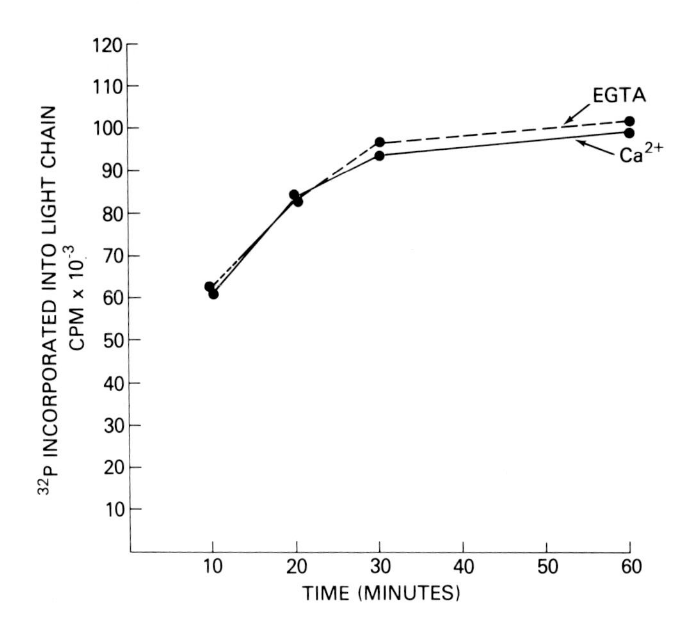

FIGURE 2. Time course for phosphorylation of exogenous smooth muscle light chain by partially purified kinase from cultured monkey aortic medial cells. The kinase from the cultured cells transferred ^{32}P from γ-AT^{32}P to 20,000 dalton LC both in the presence and absence of Ca^{2+}. The ^{32}P incorporation to LC was determined as reported previously (Chacko et al., *1977).*

CONCLUSION

These experiments demonstrate that the cultured aortic medial cells, in addition to lacking thick myosin filaments, possess a phosphorylation system that is similar to the nonmuscle cell, e.g., proliferating skeletal myoblasts, astrocytes.

REFERENCES

Chacko, S., Blose, S. H., and Adelstein, R. S. (1977). *In* "Excitation-Contraction Coupling in Smooth Muscle" (R. Casteels *et al.*, ed.). North-Holland Biomedical Press.

Chacko, S., Conti, M. A., and Adelstein, R. S. (1977). *Proc. Natl. Acad. Sci. U.S.A. 74,* 129-133.

Daniel, J. L., and Adelstein, R. S. (1976). *Biochemistry 15,* 2370-2377.

Mayne, R., Vail, M. S., Miller, E. J., Blose, S. H., and Chacko, S. (1977). *Arch. Biochem. Biophys. 181*, 462-464.
Perry, S. V., Cole, A. A., Morgan, M., Moir, A. J. G., and Pires. (1975). *Fed. Eur. Biochem. Soc. Meet. (Proc.) 31*, 163-176.
Ross, R. (1971). *J. Cell Biol. 50*, 172-186.
Scordilis, S. P., and Adelstein, R. S. (1977). *Nature 268*.

Motility in Cell Function
Proceedings of the First John M. Marshall Symposium in Cell Biology

THE INHIBITORY ROLE OF THE REGULATORY LIGHT CHAIN IN VASCULAR SMOOTH MUSCLE ACTOMYOSIN ATPase

U. Mrwa

Physiologisches Institute
Universität Heidelberg
Heidelberg, Germany

A simple system consisting of only actin from skeletal muscle purified, free of troponin and tropomyosin and heavy meromyosin subfragment 1 (HMM-S-1) from arterial muscle exhibits a Ca^{2+}-sensitive Mg-dependent ATPase activity which can be inhibited up to 60% by chelation of calcium ions with EGTA (Fig. 1A). The ATPase activity of HMM-S-1 alone is only about 30% of that of the Acto-HMM-S-1 system and is independent of trace amounts of calcium (Fig. 1C). Additionally the presence of a pseudo-ATPase, consisting of a contaminating phosphatase and kinase is ruled out by this experiment. When the regulatory light chain is removed from the HMM-S-1 by mild proteo-

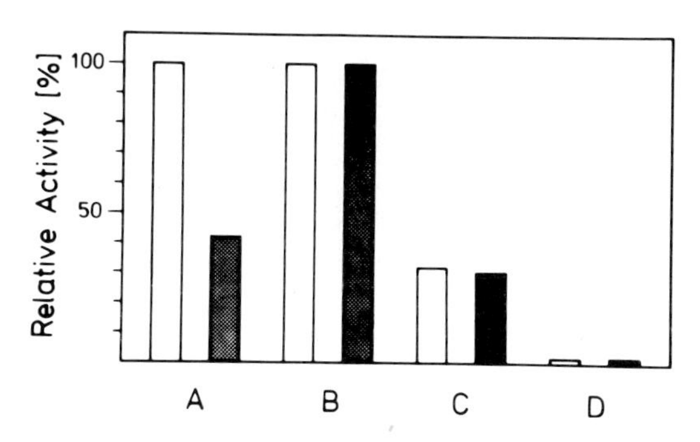

FIGURE 1. ATPase activity in percent of activity of acto S-1 in presence of Ca^{2+} (10^{-4} M) (open bars) or 10^{-8} M Ca^{2+} (closed bars). A: HMM-S-1 (arterial and actin), B: desensitized acto-HMM-S-1, C: HMM-S-1 alone, D: actin. Conditions: 50 mM KCl, 10 mM imidazole, 5 mM $MgCl_2$, 2 mM ATP, 0.1 mM Ca^{2+} or 2 mM EGTA pH 7, 20°C.

ISBN 0-12-551750-5

lytic digestion the ATPase activity of this desensitized acto-HMM-S1 system can no longer be inhibited by the removal of Ca-ions.

When the actin activation kinetics of the respective acto-HMM-S1 systems are analyzed in more detail (Fig. 2) the obtained double reciprocal plot indicates that the apparent dissociation constant of acto-HMM-S1 is unchanged for desensitized S1 (S1 lacking the regulatory light chain) at calcium concentrations of 10^{-5} and 10^{-8} *M* and in the calcium sensitive system (S1 containing the regulatory light chain) in presence of 10^{-5} *M* Ca^{2+} ions.

In the Ca-sensitive system at a Ca-concentration of 10^{-8} *M*, the affinity of actin to S1 seems to be reduced, whereas the Vmax for the actin activated ATPase activity seems to be unchanged.

From this data we conclude, that the 20,000 dalton light chain regulates vascular smooth muscle contraction by inhibiting actin-myosin interaction and thus the actin activated, Mg-dependent ATPase activity in the absence of calcium. In the presence of calcium this inhibitory effect is abolished.

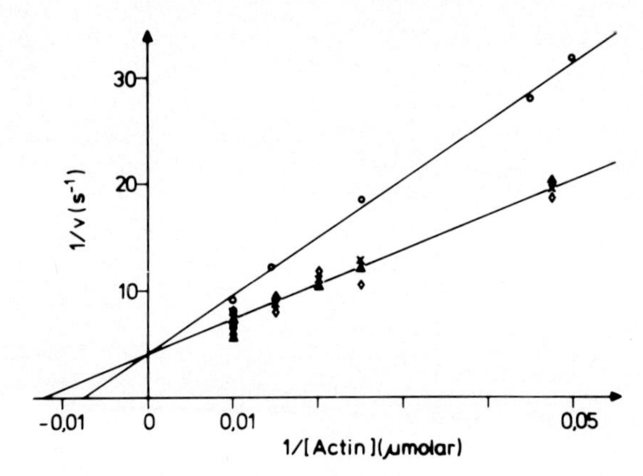

FIGURE 2. Actin activation of natural and desensitized S1. Double reciprocal plot of activity versus actin concentration. Conditions: 50 mM KCl, 10 mM imidazole, 5mM $MgCl_2$, 2 mM ATP, 0.6 μM S1, pH 7, 20°C.
▲ *natural S1 (10^{-4} M Ca^{2+}; × desensitized S1 (10^{-4} M Ca^{2+});*
◊ *desensitized S1 (10^{-8} M Ca^{2+}); O natural S1 (10^{-8} M Ca^{2+}).*

Motility in Cell Function
Proceedings of the First John M. Marshall Symposium in Cell Biology

RELATIONSHIP BETWEEN Ca^{2+} ACTIVATION AND THE PHOSPHORYLATION OF THE 20,000-DALTON LIGHT CHAIN IN SMOOTH MUSCLE

W. G. L. Kerrick
and
P. E. Hoar

Department of Physiology and Biophysics
University of Washington
Seattle, Washington

One of the major problems related to the Ca^{2+}-activation of muscle contraction is the role played by phosphorylation of the 20,000-dalton light chains of myosin. Considerable biochemical evidence has accumulated which indicates that phosphorylation of myosin light chains by themselves, or in combination with Ca^{2+} binding to other regulatory proteins, results in muscle contraction. A Ca^{2+}-independent light chain kinase from platelets has been shown to activate platelet actomyosin by phosphorylation of the 20,000-dalton light chains (Daniel, 1976). A Ca^{2+}-sensitive light chain kinase has been isolated from the rabbit skeletal muscle which has been shown to phosphorylate all 20,000 molecular weight light chains investigated (Perrie, 1973; Pires, 1974). A better characterized light chain kinase has been isolated from smooth muscle (chicken gizzard) which phosphorylates the light chains of smooth muscle myosin and is also calcium dependent (Dabrowska, 1978). This kinase has been shown to consist of two subunits: a 105,000-dalton subunit and a 17,000-dalton subunit. The 17,000-dalton subunit appears to be the calcium receptor site and similar in molecular weight, amino acid composition, and biologic activity to the TNC-like modulator protein. The gizzard muscle actomyosin in the presence of the light chain kinase shows a Ca^{2+}-sensitive actomyosin ATPase and Ca^{2+}-activated phosphorylation of the 20,000 molecular weight light chains both of which occur over the same concentration range of Ca^{2+}.

ISBN 0-12-551750-5

The purpose of this study was to determine whether a similar relationship exists between phosphorylation and the physiologic measure of contraction, "tension," in smooth muscle. Mechanically disrupted bundles of chicken gizzard muscle fibers were prepared by a homogenization procedure (Kerrick, 1975). These fibers were shown to be permeable to large proteins such as troponin-I and troponin-C and to be freely permeable to the nucleotide ATP (which cannot cross intact membranes) (Kerrick, to be published). Therefore, these fiber bundles are composed of functionally skinned (no functional sarcolemma) smooth muscle cells. The muscle bundles were mounted in a tension transducer similar to that of Hellam and Podolsky (1969) and immersed in solutions of varying concentrations of Ca^{2+} or Sr^{2+} while at the same time tension was monitored. Gizzard fibers were also immersed in similar solutions of identical ionic strength which contained γ-labeled P^{32}-ATP. These solutions differed from the tension solutions in that they contained no creatine phosphokinase or creatine phosphate. The fiber proteins were then dissolved in sodium dodocyl sulfate, electrophoresed on 10% SDS polyacrylamide gels, and stained with Coomasie blue. The gels were scanned with a spectrophotometer, bands cut out, and the amount of phosphate incorporated into the different proteins at various Ca^{2+} and Sr^{2+} concentrations was determined.

A comparison of the amount of ^{32}P-phosphate that is incorporated in the myofibrillar proteins of the gizzard fibers in the absence of calcium with that incorporated in its presence showed a large increase in the level of phosphorylation of the 20,000-dalton light chain in the presence of Ca^{2+}, but no significant change in phosphorylation levels of other proteins. This indicates that the only significant Ca^{2+}-sensitive kinase activity is associated with phosphorylation of the 20,000-dalton light chains of myosin. Approximately eight times higher Sr^{2+} concentration is required to activate tension in the fibers than Ca^{2+}. A comparison of the Ca^{2+}- and Sr^{2+}-activated phosphorylation of the 20,000-dalton light chains to the Ca^{2+} and Sr^{2+}-activated tension shows a good correlation.

These data show for the first time that activation of the physiological measure of contraction (tension) in smooth muscle by two different activators (Ca^{2+} and Sr^{2+}) shows a close correlation with phosphorylation of the myosin light chains in the divalent cation concentration range required for activation. They also demonstrate that the only proteins showing a significant change in phosphorylation during activation are the 20,000-dalton light chains. The evidence presented in this paper, therefore, is consistent with the hypothesis that phosphorylation of the light chains of myosin is a necessary requirement for activation of smooth muscle.

REFERENCES

Dabroska, R., Sherry, J. M. F., Aromatorio, D. K., and Hartshorne, D. J. (1978). *Biochemistry 17*, 253-258.

Daniel, J. L., and Adelstein, R. S. (1976). *Biochemistry 15*, 2370-2377.

Hellam, D. C., and Podolsky, R. J. (1969). *J. Physiol. (London) 200*, 807-819.

Kerrick, W. G. L., Hoar, P. E., Malencik, D. A., Stamps, L., and Fischer, E. H. (to be published).

Kerrick, W. G. L., and Krasner, B. (1975). *J. Applied Physiol. 39*, 1052-1055.

Perrie, W. T., Thomas, M. A. W., and Perry, S. V. (1973). *Biochem. Soc. Trans. 1*, 860-861.

Pires, E., Perry, S. V., and Thomas, M. A. W. (1974). *FEBS Lett. 41*, 292-296.

BIOCHEMICAL CHARACTERIZATION OF *ACANTHAMOEBA* MYOSINS I AND II AND *ACANTHAMOEBA* MYOSIN I HEAVY CHAIN KINASE

Edward D. Korn
Hiroshi Maruta
Jimmy Collins
and
Hana Gadasi

Laboratory of Cell Biology
National Heart, Lung, and Blood Institute
National Institutes of Health
Bethesda, Maryland

Acanthamoeba extracts contain at least three separable "myosins." *Acanthamoeba* myosin II, first identified by Maruta and Korn (1977a,b), is a two-headed molecule comprised of 170,000-dalton, 17,500-dalton, and 17,000-dalton polypeptides (molar ratio 2:1.5:2). We have now resolved *Acanthamoeba* myosin I (Pollard and Korn, 1973a) into two proteins. Myosin IA is a single-headed enzyme comprised of 130,000-dalton, 17,000-dalton, and 14,000-dalton polypeptides (1:1:0,5). Myosin IB is also single-headed but contains 125,000-dalton, 27,000-dalton, and 14,000-dalton polypeptides (1:1:0.5). The molecular weights and the stoichiometry of the constituent polypeptides are approximate values but the differences among the three myosins are real. Most importantly, myosin II is a two-headed enzyme while myosins IA and IB are single-headed.

Myosins IA and IB are enzymatically very similar to each other and very different from myosin II (Maruta and Korn, 1977b). For example, the ratio of (K^+,EDTA)-ATPase to Ca^{2+}-ATPase activity is 7.7 for myosin IA, 5.6 for myosin IB, and 0.14 for myosin II. All three *Acanthamoeba* myosins have very low Mg^{2+}-ATPase activities that are activated only about two-fold by F-actin. However, a major difference between myosins IA and IB and myosin II is that the addition of a cofactor protein (Pollard and Korn, 1973b) to the incubation mixtures

ISBN 0-12-551750-5

allows considerable actin-activation of myosins IA and IB but has no effect on myosin II (Maruta and Korn, 1977b). We (Maruta and Korn, 1977c) have purified cofactor protein to two major components on gel electrophoresis (95,000 and 58,000 daltons), although minor bands are still present. The purified protein is fully active in a molar ratio to myosin I of no more than 1:1000 when added directly to the assay mixture.

When incubated with myosin and [γ-^{32}P]ATP, cofactor protein catalyzes the phosphorylation of myosins IA and IB to the extent of about 0.7 mol/mol but does not phosphorylate myosin II significantly. Phosphorylation requires Mg^{2+}, but neither Ca^{2+}, cAMP, nor cGMP has any effect. Phosphorylated myosins IA and IB are fully actin-activated in the absence of added cofactor protein. Removal of the phosphate moiety by partially purified platelet myosin phosphatase (gift of R. S. Adelstein) eliminates the actin-activated Mg^{2+}-ATPase activities of the previously phosphorylated myosins. None of these treatments affects the (K^+,EDTA)- or Ca^{2+}-ATPase activities of myosins IA or IB. Finally, gel electrophoresis demonstrates that all of the radioactivity of phosphorylated myosins IA and IB is located on their heavy chains (130,000-daltons for IA and 125,000-daltons for IB). Thus, cofactor protein is an *Acanthamoeba* myosin I heavy chain kinase. All other kinases known to affect the actin-activated Mg^{2+}-ATPase activities of myosins are light chain kinases. The specific activities of *Acanthamoeba* myosins IA and IB are very high: about 3 $\mu mol \cdot min^{-1} \cdot mg^{-1}$ for the (K^+,EDTA)-ATPase, 0.4 for the Ca^{2+}-ATPase and 2 for the actin-activated Mg^{2+}-ATPase of the phosphorylated myosins.

Because of their similarities in size, enzymatic properties and as substrates for myosin I heavy chain kinase, we assume that myosins IA and IB are products of the same gene. However, neither myosin IA nor IB can be the precursor of the other if the polypeptides of the isolated molecules are, in each case, a single heavy chain and two light chains (the heavy chain of IA is larger than the heavy chain of IB, but one of the light chains of IB is larger than either light chain of IA). In that case, IA and IB could be derived from a common precursor with a heavy chain of 130,000 daltons and light chains of 27,000 and 14,000 daltons. However, if the 27,000-dalton chain of myosin IB is a degradation product of the heavy chain, and not a true light chain, myosin IB could be derived from IA.

Our assumption that myosins IA and IB are related proteolytically makes it necessary for us to consider seriously the possibility that both myosins I may be related to myosin II, despite the inability of Maruta and Korn (1977b) and Pollard (this volume) to obtain evidence for the conversion of myosin II to myosin I in extracts or homogenates of *Acanthamoeba*. If the 27,000-dalton chain of myosin IB is a true light chain, myosin II cannot be its precursor. However, an undetected myosin minimally consisting of two heavy chains of 170,000 daltons, two light chains of 27,000 daltons, and two light chains of 17,000 daltons could, hypothetically, be a common precursor to *Acanthamoeba* myosins IA, IB, and II. Such a putative precursor would be expected to have a phosphorylation site for myosin I kinase that would be unavailable when the precursor is converted to myosin II but accessible upon degradation to myosins IA and IB. The cysteine residues of the hypothetical parent molecule would have to be retained in myosin II (Pollard, this volume) but lost (Pollard and Korn, 1973a) in the formation of myosins IA and IB. The several myosin products would also have to be so modified as to reverse their relative (K^+,EDTA)- and Ca^{2+}-ATPase activities. (Of course, if the 27,000-dalton polypeptide of myosin IB were not a true light chain, it would be physically possible for myosin II to be the precursor of myosins IA and IB but all of the above chemical and functional considerations would still apply. Thus far, however, all attempts to convert myosin II to a myosin I have failed.) Although we believe the aforementioned possible transformations to be unlikely, peptide maps and sequence data will probably be required to determine unequivocally whether myosin I and myosin II are related or independent gene products.

REFERENCES

Maruta, H., and Korn, E. D. (1977a). *J. Biol. Chem. 252*, 399-402.

Maruta, H., and Korn, E. D. (1977b). *J. Biol. Chem. 252*, 6501-6509.

Maruta, H., and Korn, E. D. (1977c). *J. Biol. Chem. 252*, 8329-8332.

Pollard, T. D. (This volume).

Pollard, T. D., and Korn, E. D. (1973a). *J. Biol. Chem. 248*, 4682-4690.

Pollard, T. D., and Korn, E. D. (1973b). *J. Biol. Chem. 248*, 4691-4697.

Motility in Cell Function
Proceedings of the First John M. Marshall Symposium in Cell Biology

THE Mg^{2+}-DEPENDENT MYOSIN GUANOSINE TRIPHOSPHATASE MECHANISM

John F. Eccleston

Department of Biochemistry and Biophysics
School of Medicine
University of Pennsylvania
Philadelphia, Pennsylvania

As part of a study involving the spectral characterization of the intermediates of the Mg^{2+}-dependent myosin ATPase using absorption, circular dichroism, and fluorescence measurements (Eccleston and Trentham, 1977), the mechanisms of the hydrolysis of thioGTP and GTP catalyzed by myosin subfragment 1 were investigated. Both showed marked differences from the mechanism of ATP hydrolysis.

When a 10 μM solution of subfragment 1 was rapidly mixed with either 50 μM [γ-^{32}P] thioGTP or 50 μM [γ-^{32}P]GTP, no transient production of [^{32}P]P_i could be detected at pH 8 and 0.1 M KCl, (Fig. 1) indicating that unlike the myosin ATPase the steady-state intermediate consists predominantly of a myosin-substrate complex.

The binding of GTP or GDP to subfragment 1 could not be measured directly since it was not accompanied by any detectable change in the intrinsic protein fluorescence or release of protons. However, it could be followed indirectly by measuring the effect of GTP or GDP on the rate of binding to subfragment 1 of ATP (which enhances subfragment 1 fluorescence) or thioITP (which quenches subfragment 1 fluorescence). Both ATP and ADP bind to subfragment 1 at an identical observed second-order rate constant of $1.2 \times 10^6\ M^{-1}c^{-1}$ (Bagshaw *et al.*, 1974) but the rate constants of binding of GTP and GDP to subfragment 1 were $1.1 \times 10^5\ M^{-1}c^{-1}$ and $1.5 \times 10^4\ M^{-1}c^{-1}$, respectively. The dissociation rate of GDP from subfragment 1 was measured by displacing it with either ATP or thioITP, and found to be 0.06 $1c^{-1}$. Since this is slower than the steady-state rate of subfragment 1 catalyzed hydrolysis of GTP (0.5 $1c^{-1}$) it is clear that GDP bound to subfragment 1 forms a

ISBN 0-12-551750-5

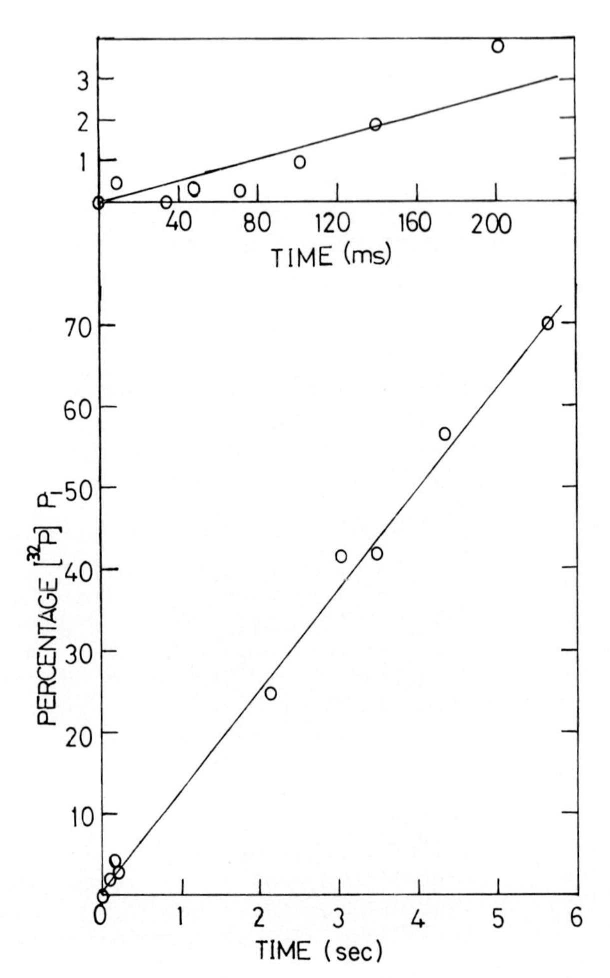

FIGURE 1. Rate of transient production of [^{32}P]P_i *in the reaction of 10* µM *subfragment 1 with 50* µM [γ-^{32}P]*GTP in the quenched-flow apparatus at 20°C and pH 8.0. The line in the upper graph is the steady-state rate shown in the lower graph. Both syringes also contained 0.1* M *KCl, 5* mM $MgCl_2$*, 0.1* mM *dithiothreitol and 0.1* M *tris adjusted to pH 8.0 with HCl.*

complex which is not kinetically competent to be an intermediate of the GTPase mechanism. Calculation of the dissociation constant of GDP from subfragment 1 from the association and dissociation rate constants gives a value of 4 µ*M*.

A further unusual feature of the interaction of GDP with myosin is that hydrolysis of GDP to GMP and P_i occurs in the presence of subfragment 1. Purified preparations of subfragment 1 containing either A1 or A2 light chains (Weeds and Taylor, 1975) hydrolyzed GDP at identical rates and the presence of ADP inhibited this hydrolysis. Although not definitely excluding the presence of a contaminating enzyme in the subfragment 1 preparations, the results strongly suggest that subfragment 1 can catalyze the hydrolysis of GDP although the biological and chemical significance of this remains obscure.

Although these results show that the mechanism of GTP hydrolysis by myosin is markedly different from that of ATP hydrolysis it shows several similarities to the GTPase of elongation factor Tu of the protein biosynthetic process of E. coli. In this system, the energy of GTP hydrolysis is apparently utilized in order to increase the specificity of codon-anticodon binding over that predicted from binding studies of the appropriate bases (Thompson and Stone, 1977). The steady-state intermediate of the EF-Tu GTPase is also a protein-substrate complex and GDP binds tightly to the protein. A similarity between the nucleotide binding sites of myosin and EF-Tu has also been detected using electron paramagnetic resonance techniques. (Wilson and Cohn, 1977).

ABBREVIATIONS

ATPase, adenosine triphosphatase; GTPase, guanosine triphosphatase; thioGTP, 2-amino-6-mercapto-9-β-ribofuranosylpurine 5'-triphosphate; thioITP, 6-mercapto-9-β-ribofuranosylpurine 5'-triphosphate.

REFERENCES

Bagshaw, C. R., *et al.* (1974). *Biochem. J. 141,* 351-364.
Eccleston, J. F., and Trentham, D. R. (1977). *Biochem. J. 163,* 15-29.
Thompson, R. C., and Stone, P. J. (1977). *Proc. Natl. Acad. Sci. U.S.A. 74,* 198-202.
Weeds, A. G., and Taylor, R. S. (1975). *Nature 257,* 54-56.
Wilson, G. E., and Cohn, M. (1977). *J. Biol. Chem. 252,* 2004-2009.

Motility in Cell Function
Proceedings of the First John M. Marshall Symposium in Cell Biology

MECHANOCHEMISTRY OF ACTIN-MYOSIN INTERACTION

M. Kawai
and
P. W. Brandt

Departments of Neurology and Anatomy
Columbia University
New York, New York

Fenn (1923) discovered that the chemical energy mobilized in muscle contraction depends on the external load. Today, we believe that this energy conversion and force production takes place in the myosin to actin cross-bridges. In accord with Fenn's results, we assume that some reaction rates in the ATP hyprolysis cycle which resides in the cross-bridges are strain sensitive (Huxley, 1957; Hill, 1975). Two complementary techniques have evolved to observe this strain sensitivity on single fiber preparations. One analyzes tht length transients following step load changes; the other analyzes the tension changes following a step or sinusoidal length perturbation.

In an effort to optimize signals from chemomechanical transitions which are strain sensitive, we use the technique of sinusoidal length perturbation at various frequencies (0.25-133 Hz) and record the tension amplitude and phase. This technique is a mechanical equivalence of spectrophotometry, and kinetics in cross-bridges can be probed for rate constants ranging from 1.5 to 800 sec^{-1} during activation. We have applied this technique to intact and skinned crayfish and frog muscle preparations (Kawai *et al.*, 1977; Kawai and Brandt, 1975) and to skinned rabbit psoas (Kawai, 1978; Kawai and Orentlicher, 1976). Data is displayed as a Nyquist plot (Fig. 1) or in amplitude/phase versus frequency plots (Fig. 2). After appropriate data reduction procedures, three exponential rate processes are readily identifiable in these plots. The smooth curve is theoretically fit to the data (open symbols). These processes are A), a slow exponential lead (indicated by letter *a* in the figure), B), a middle frequency lag (*b*, oscillatory work), and C), a fast lead (*c*). The data in the figs (solid points and

ISBN 0-12-551750-5

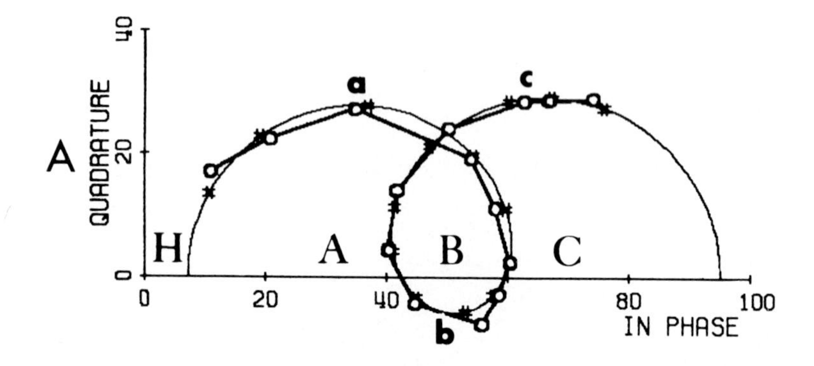

FIGURE 1.

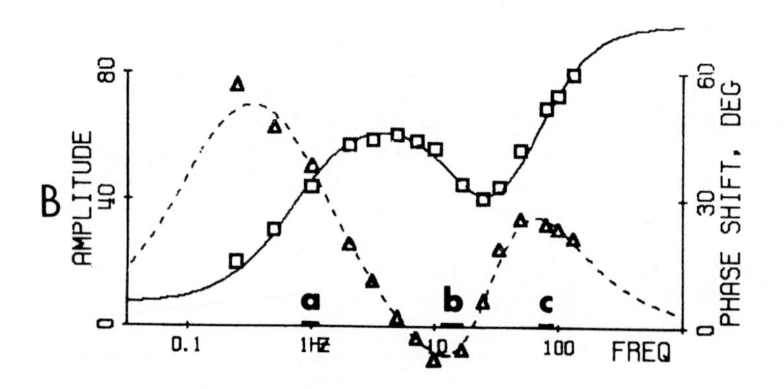

FIGURE 2.

lines connecting them) was obtained on chemically skinned rabbit preparations (5 m*M* MgATP, 2.4 CaATP, 5 ATP, 7 P_i, pCa 4, pH 7.00, μ = 225, 3 fiber bundle, 20°C), but the appearance of the data plots is essentially the same for all the muscle types tested. Apparent rate constants from these studies are listed in Table 1.

While some progress toward relating cross-bridge kinetics to actomyosin biochemistry will result from experiments with intact muscle, major advances will most likely come from skinned fiber studies. With this preparation we can experimentally control the chemistry surrounding the contractile proteins while monitoring the mechanical parameters. Our preliminary results show that process A) relates to rearrangements of sarcomere segments, B) to chemomechanical energy transduction, and C) to MgATP binding to myosin head and dissociation of actomyosin (Kawai. 1978; Kawai and Orentlicher, 1976).

TABLE 1. *Apparent Rate Constant (Approximate) (units in* sec^{-1}*)*

Symbol	Polarity	Nomenclature of Huxley (1974)	Intact preparations		Skinned preparation		
			Frog 8°C	Crayfish 20°C	Rabbit 20°C	Frog 10°C	Crayfish 20°C
$2\pi a$	+	Phase 4	3	6	8	5	1.5
$2\pi b$	-	Phase 3	40	140	120	40	30
$2\pi c$	+	Phase 2	300	570	600	220	200

Skinned preparations: The data depends upon solution conditions (Kawai and Orentlicher, 1976; Kawai, 1978). In general, the solutions contain saturating concentrations of MgATP and Ca^{++}. In addition they contain 7 mM P_i and 5 Imidizol at pH 7.0.

Intact preparations: Frog, one head of semitendinosus, high K^+ activation. Crayfish, single fiber activated with high K^+ or caffeine.

ACKNOWLEDGMENTS

This work was supported by grants from NSF (PCM 76-00441) and NIH (NS 05910, AM21530).

REFERENCES

Fenn, W. O. (1923). *J. Physiol. (Lond.) 58*, 175.
Hill, T. L. (1975). *Prog. Biophys. 29*, 105.
Huxley, A. F. (1957). *Prog. Biophys. 7*, 255.
Huxley, A. F. (1974). *J. Physiol. (Lond.) 243*, 1.
Kawai, M. (1978). *Biophys. J. 22*, 97.
Kawai, M., and Brandt, P. W. (1975). *Biophys. J. 15*, 154a (Abstr.).
Kawai, M., and Orentlicher, M. (1976) *Biophys. J. 16*, 152a (Abstr.).
Kawai, M., Brandt, P. W., and Orentlicher, M. (1977). *Biophys. J. 18*, 161.

Motility in Cell Function
Proceedings of the First John M. Marshall Symposium in Cell Biology

REMOVAL OF THE M-LINE BY TREATMENT WITH FAB' FRAGMENTS OF ANTIBODIES AGAINST MM-CREATINE KINASE

Theo Wallimann
*Gudrun Pelloni**
*David C. Turner**
and
*Hans M. Eppenberger**

Department of Biology
Brandeis University
Waltham, Massachusetts
and
*Institute for Cell Biology
Swiss Federal Institute of Technology
Zurich, Switzerland

The M-line of vertebrate striated muscle has recently been shown to consist of a) longitudinal elements, called M-filaments, b) transverse elements, called (primary) M-bridges, that link thick filaments to M-filaments and that are responsible for the principal M-line substriations seen in longitudinal sections, and c) additional transverse elements, called secondary M-bridges, interconnecting the M-filaments (Luther and Squires, 1977). The protein composition of these different elements is unknown.

Low ionic strength extraction of myofibrils removes all traces of the M-line structure. The structural proteins of the M-line must, therefore, be present in low ionic strength extracts. Such extracts, however, presumably also contain proteins not integral to the M-line structure. Among the proteins found in low ionic strength extracts is the MM isoenzyme of creatine kinase (CK). There is strong evidence that a fraction of the MM-CK in skeletal muscle cells is indeed a structural component of the M-line (Wallimann *et al.*, 1977). The amount of CK firmly bound to myofibrils is roughly that expected if CK molecules were to form primary or secondary M-bridges.

ISBN 0-12-551750-5

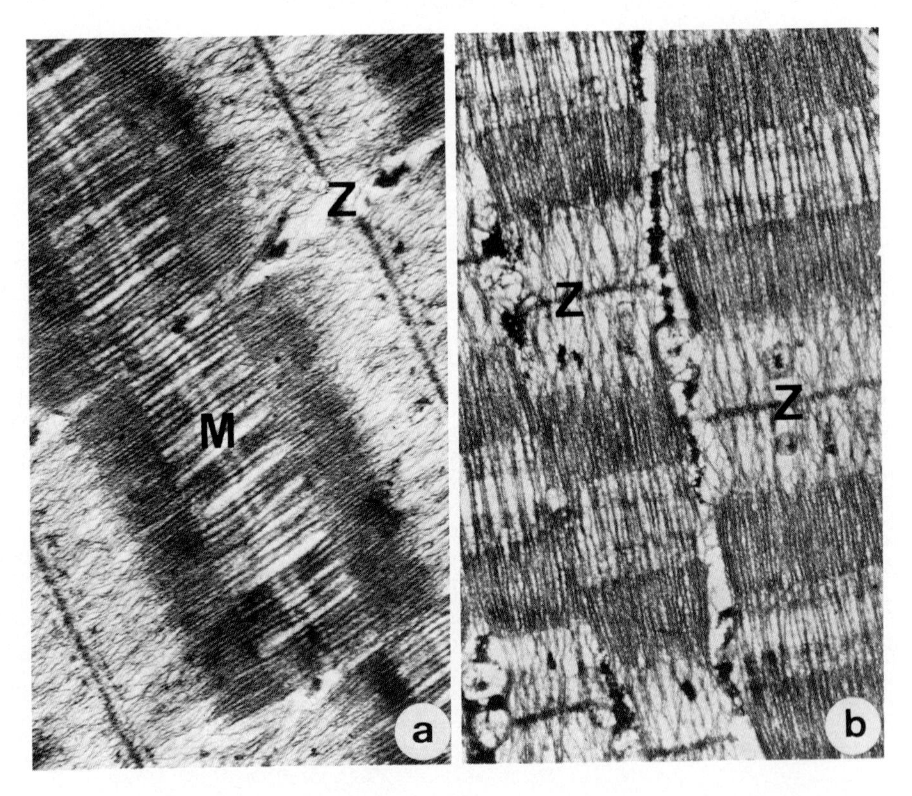

FIGURE 1. (a) Fiber bundle treated with Fab' fragments of IgG from nonimmune serum; M-line is intact. (b) Fiber bundle treated with Fab' fragments of affinity-purified antibodies against MM-creatine kinase; M-line appears to have been completely and specifically removed. Magnification: x37600.

Using specific antiserum against CK, we found specific immunofluorescent staining at the middle of the A-band (Wallimann *et al.*, 1977). We then showed that incubation of fiber bundles with affinity-purified antibody against CK led to a thickening of the M-line detectable in the electron microscope (Wallimann *et al.*, 1978). Moreover, the increased electron density appeared to correspond to M-line substriations, as would be expected if CK molecules were to contribute to the structure of transverse elements (Wallimann *et al.*, 1978). In an attempt to achieve better antibody penetration and better resolution, we prepared Fab' fragments of affinity-purified anti-CK antibody and incubated them with fiber bundles (Wallimann *et al.*, 1978). To our surprise, we found that the M-line was no longer detectable in electron micrographs (Fig. 1). This M-line removal was not due to residual papain activity in the Fab' preparation (Wallimann *et al.*, 1978); Fab' fragments of control IgG failed to remove the M-line (Fig. 1). Immunofluorescence studies confirmed that myofibrils pretreated with anti-CK-Fab' no longer contained detectable CK in the M-line (Wallimann *et al.*, 1978). Treatment with anti-CK-Fab' releases a protein from myofibrils that comigriates with MM-CK in SDS gel electrophoresis and that can be unequivocally identified as MM-CK by the immunoreplica technique (Wallimann *et al.*, 1978). This protein is not released by treatment with Fab' fragments of control IgG (Wallimann *et al.*, 1978).

The apparently specific and complete removal of the M-line by treatment with anti-CK-Fab' is the best evidence yet that CK is a structural protein of the M-line. Myofibrils from which the M-line has been removed by this procedure offer promising possibilities for studies of the reconstitution of the M-line structure. Finally, we hope to identify proteins associated with CK in the M-line by examining the proteins removed along with CK by Fab' treatment.

REFERENCES

Luther, P., and Squire, J. (1977). *Proc. of 6th Meeting of Europ. Muscle Club.* p. 61.

Wallimann, T., Turner, D. C., and Eppenberger, H. M. (1977). *J. Cell Biol. 75*, 297.

Wallimann, T., Pelloni, G., Turner, D. C., and Eppenberger, H. M. (1978). *Proc. Natl. Acad. Sci. U.S.A. 75*, 4296.

Motility in Cell Function
Proceedings of the First John M. Marshall Symposium in Cell Biology

M-BAND PROTEINS:
EVIDENCE FOR MORE THAN ONE COMPONENT

Prokash K. Chowrashi
and

Frank A. Pepe

Department of Anatomy
School of Medicine
University of Pennsylvania
Philadelphia, Pennsylvania

Different proteins have been identified as being present in the M-band of skeletal muscle myofibrils (Trinnick and Lowey, 1977; Masaki and Takaiti, 1974; Turner *et al.*, 1973; Eaton and Pepe, 1972; Morimoto and\Harrington, 1972). It is not known whether the M-band is made up of one or more protein components. Stromer *et al.* (1969) extracted M-band protein from myofibrils and were able to reconstruct the M-band by treating the extracted myofibrils with the crude M-band extract.

We have found that a fraction of the crude M-band extract (obtained using 5 m*M* Tris pH 8 buffer), which precipitates between pH 5.7 and 5.1, will reconstruct the M-band in myofibrils from which the M-band has previously been extracted. This fraction contains primarily components with chain weights of about 160,000, 42,000, and 38,000 with smaller amounts of a component with a chain weight of about 22,000 (Fig. 1a). Addition of 18% ammonium sulfate to this fraction gives a precipitate containing the 160,000 and 42,000 components (Fig. 1b) and a supernatant containing some 42,000 and the 38,000 and 22,000 chain weight components (Fig. 1c). Creatine kinase activity is present in both of these fractions (Fig. 1b and 1c) and creatine kinase migrates with the 42,000 component (Fig. 1d), therefore, indicating that creatine kinase is present in the 42,000 chain weight bands. Using conditions described by Stromer *et al.* (1969), the preparation in Fig. 1b does not reconstruct the M-band. The preparation in Fig. 1c gives only

ISBN 0-12-551750-5

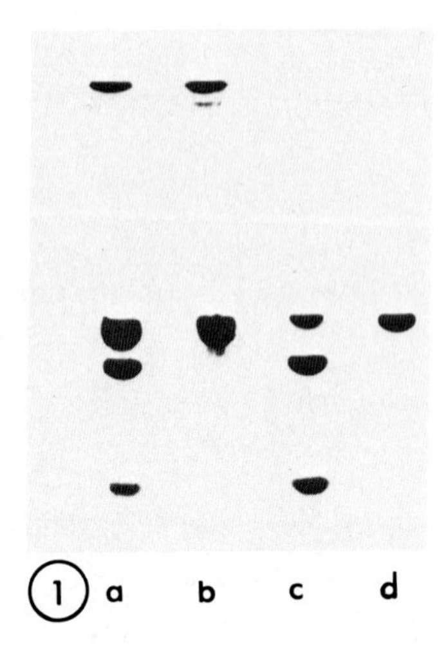

FIGURE 1. SDS polyacrylamide gel electrophoresis of fractions of crude M-band extract (obtained with 5 mM Tris pH 8 buffer) using 6% polyacrylamide gels. (a) Fraction precipitating between pH 5.7 and 5.1. (b) Precipitate obtained from (a) on adding 18% ammonium sulfate. (c) Supernatant from (b). (d) Creatine kinase.

a small increase in density along the myosin filaments in the M-band region (Fig. 2b). On recombination of the separated fractions in Fig. 1b and 1c reconstruction of the M-band equivalent to that obtained before separation (Fig. 1a) is observed (Fig. 2a). In Fig. 2a M-bridges between filaments can be seen. Creatine kinase alone (Fig. 1d) will not reconstruct the M-band. This suggests that at least one component from each of the preparations in Fig. 1b and 1c is required for satisfactory reconstruction of the M-band (Fig. 2a) with only the lower chain weight components being required for partial reconstruction of the M-band (Fig. 2b). Further it suggests that the lower chain weight components (Fig. 1c) are binding to the myosin filaments (Fig. 2b) whereas the 160,000 chain weight component may be responsible for bridging between the filaments.

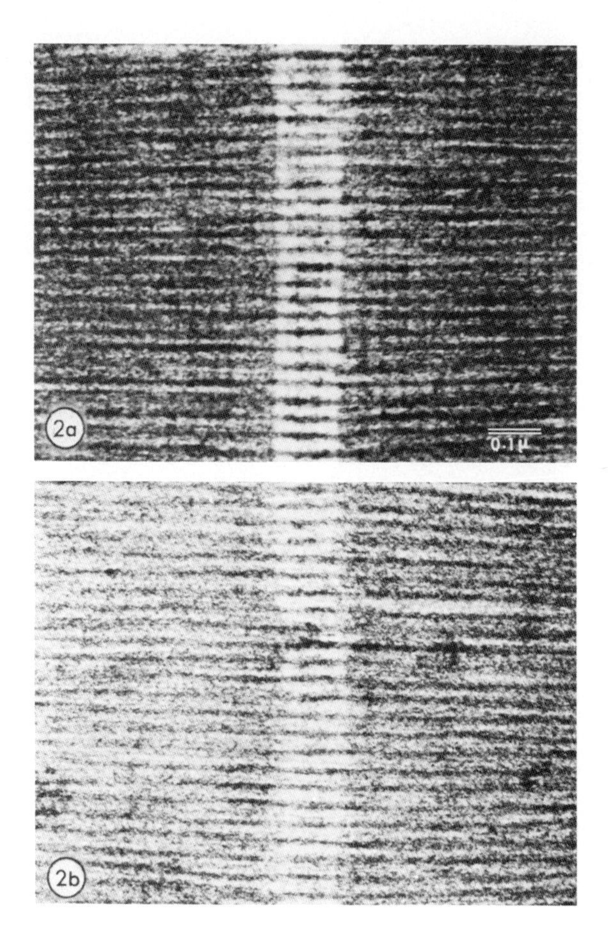

FIGURE 2. M-band reconstruction. (a) Reconstruction (Stromer et al.*, 1969) using a recombination of the preparations in Fig. 1b and 1c. This reconstruction is equivalent to that obtained with the preparation in Fig. 1a before separation. M-bridges can be seen extending between the myosin filaments. (b) Reconstruction using the preparation in Fig. 1c. An increase in density along the myosin filaments in the M-band region is observed. No bridges between filaments are evident.*

ACKNOWLEDGMENTS

This work was supported by USPHS Grant HL 15835 to the Pennsylvania Muscle Institute.

REFERENCES

Eaton, B. L., and Pepe, F. A. (1972). *J. Cell Biol.* *55*, 681-695.

Masaki, T., and Takaiti, O. (1974). *J. Biochem. (Tokyo)* *75*, 367-380.

Morimoto, K., and Harrington, W. F. (1972). *J. Biol. Chem.* *247*, 3052-3061.

Stromer, M. H., Hartshorne, D. J., Mueller, H., and Rice, R. V. (1969). *J. Cell Biol.* *40*, 167-178.

Trinick, J., and Lowey, S. (1977). *J. Mol. Biol.* *113*, 343-368.

Turner, D. C., Wlllimann, T., and Eppenberger, H. M. (1973). *Proc. Nat. Acad. Sci., U.S.A.* *70*, 702-705.

Motility in Cell Function
Proceedings of the First John M. Marshall Symposium in Cell Biology

CREATINE KINASE DURING GROWTH AND DEVELOPMENT OF THE UTERINE SMOOTH MUSCLE

M. R. Iyengar
and
C. W. L. Iyengar

Laboratories of Biochemistry
School of Veterinary Medicine
University of Pennsylvania
Philadelphia, Pennsylvania

Differentiation of the skeletal muscle cell is acccompanied by a characteristic change in the isoenzyme form of creatine kinase (CK) from the embryonic or brain type BB to the muscle type MM (Turner *et al.*, 1974). The amount of CK synthesized by growing muscle closely parallels its actomyosin content (Eppenberger *et al.*, 1963; Hauschka, 1968). The uterine smooth muscle (SM) undergoes marked growth (30 to 40 times in larger mammals) in response to stimulation by ovarian hormones and by physical stretch (Reynolds, 1965). Increase in the contractile proteins of the myometrium in pregnancy has been reported (Csapo, 1971). We have studied the activity and distribution of CK in the bovine myometrium in early pregnancy (60 to 120 day gestation).

The enzyme was obtained by exhaustive extraction with low and high ionic strength buffers to ensure solubilization of any structure-bound enzyme. CK activity was determined spectrophotometrically in the direction of ATP synthesis coupling ATP to NADP reduction with the use of hexokinase and glucose-6-phosphate dehydrogenase (Oliver, 1955) in the presence of AMP to inhibit adenylate kinase. Endogenous reduction of NADP in the absence of phosphoryl creatine (PCr) was routinely checked. The CK activity of the muscle from the nonpregnant animals was 30.4 (S.E.M.± 4.8) units/g wet tissue (1 unit = 1 μmole NADPH/min at 30°C whereas the hormone dominated tissue contained 65.5 (± 8.8) units/g wet weight. This represents a twofold increase in specific activity and five to sixfold increase in the tissue content. It is noteworthy that the in-

ISBN 0-12-551750-5

crease in muscle mass (two to three times) at this stage is less than 10% of the weight at term.

In order to determine how the increased enzyme was distributed, two types of tissue fractionation were done: a) Isolation of subcellular organelles by differential centrifugation in 0.3 *M* sucrose, and b) separation of sarcoplasmic proteins and actomyosin by sequential extraction with a low ionic strength buffer (0.05 *M* KCl, 0.01 *M* imidazole-HCl, pH 7.0, 0.001 *M* β-mercaptoethanol) and an actomyosin extractant (0.5 *M* KCl, 0.01 *M* imidazole-HCl, pH 7.0, 0.001 *M* β-mercaptoethanol). Among the cell fractions in sucrose, activity was found only in the cytosol (97% of the total) and in the microsomes (3%) in the nonpregnant tissue, whereas with the pregnant tissue about 10 to 12% of the total sucrose soluble activity was found in the microsomes and the rest in the cytosol. Actomyosin in this medium sediments with nuclei and other insoluble fragments. Estimation of CK in the actomyosin extracts demonstrated approximately a 100% increase in this fraction (from 4% in the nonpregnant to 9% in the pregnant). Association of CK with actomyosin and microsomes in SM resembles skeletal muscle (Baskin and Deamer, 1970; Turner *et al.*, 1973) in contrast to cardiac muscle in which significant amounts of CK are bound to the mitochondria in addition to the myofibrils (Sobel *et al.*, 1972). The increase of CK in the actomyosin fraction is of the same order of magnitude as the change in total cell enzyme, whereas the enzyme localized in the microsomal fraction increases 300 to 400% of basal value. The role of CK in the membrane is unknown.

Whether the enzyme found in actomyosin extracts was bound to the contractile proteins was studied by further fractionation of actomyosin. The actomyosin from uterine muscle was subjected to two dilution-precipitation cycles followed by dissociation with MgATP (5 m*M*) at high ionic strength. The actin pellet had little activity. The enzyme activity was demonstrable in myosin thus purified (0.1 unit per mg protein). Recently Eppenberger and co-workers (Wallimann *et al.*, 1977) have reported that the M-line region of the myosin filaments in chicken breast muscle is the site of association of CK (0.21 unit/mg protein) with skeletal myosin. Cardiac myosin from beef and dog (Iyengar, unpublished) also contains associated CK activity (0.06 to 0.1 unit/mg protein). Whether these latter associations are site specific is being investigated.

CK from uterine SM has been purified (Iyengar and Iyengar, 1978). Kinetic characterization of the pure enzyme showed a marked cooperativity in the binding of the substrate. Thus the binding of PCr to the binary complex E-MgADP was facilitated five to sixfold when compared to direct binding to the free enzyme. Since nearly all the ADP in muscle is actin-bound or in the mitochondria, initial accumulation of ADP during contraction of SM observed in smooth muscle (Butler *et al.*, 1977) might be involved in the activation of CK.

REFERENCES

Baskin, R. J., and Deamer, D. W. (1970). *J. Biol. Chem. 245*, 1345-1347.

Butler, T. M., Siegman, M. J., Mooers, S. U., and Davies, R. E. *In* "Excitation-contraction Coupling in Smooth Muscle" (Casteels *et al.*, eds.), pp. 463-469. Elsevier/N.Holland Biomed. Press, Amsterdam.

Csapo, A. I. (1971). *In* "Contractile Proteins and Muscle Contraction" (Laki, ed.), pp. 413-481. Marcel Dekker, New York.

Eppenberger, H. M., von Fellenberg, R., Richterich, R., and Aebi, H. (1963). *Enzyme Biol. Clin. 2*, 139-174.

Hauschka, S. D. (1968). *In* "The Stability of the Differentiated State" (Ursprung, ed.), pp. 38-57. Springer Verlag, New York.

Iyengar, M. R., and Iyengar, C. W. L. (1978). AIBS Meeting, June, Atlanta.

Oliver, I. T. (1955). *Biochem. J. 61*, 116-122.

Reynolds, S. R. M. (1965). *In* "The Physiology of the Uterus," 2nd ed., pp. 236-348. Haffner, New York.

Sobel, B. E., Shell, W. E., and Kelin, M. S. (1972). *J. Mol. Cell. Cardiol. 4*, 367-380.

Turner, D. C., Williams, T., and Eppenberger, H. M. (1973). *Proc. Natl. Acad. Sci. U.S.A. 70*, 702-705.

Turner, D. C., Maier, V., and Eppenberger, H. M. (1974). *Dev. Biol. 37*, 63-89.

Wallimann, T., Turner, D. C., and Eppenberger, H. M. (1977). *J. Cell Biol. 70*, 297-317.

Motility in Cell Function
Proceedings of the First John M. Marshall Symposium in Cell Biology

OBSERVATIONS ON THE THREE-DIMENSIONAL ORGANIZATION OF MYOFIBRILS IN VERTEBRATE SMOOTH MUSCLE

Roland M. Bagby
and
Robert K. Abercrombie

Department of Zoology
University of Tennessee
Knoxville, Tennessee

It is convenient to think of all muscles as having the same basic organization of contractile machinery, but most studies of smooth muscle structure have provided little comfort to generalists. Studies with isolated cell preparations have added observations which, at first sight, seemed to add to the confusion. One of the most perplexing observations has been the change in orientation of myofilaments from parallel to the cell's axis to increasingly oblique or random as the cell shortened (Fay and Delise, 1973; Fisher, 1974; Fisher and Bagby, 1974, 1977). Models to account for the oblique arrangement have usually been two-dimensional (Rosenbluth, 1965; Fay and Delise, 1973). Our laboratory noted a helical arrangement which was most evident in shortened cells (Fisher and Bagby, 1972, 1977). A three-dimensional model (Fig. 1) was proposed to account for the oblique arrangement, the helical appearance, and reorientation of the myofilaments (Fisher, 1974). The basic assumption was made that bundles of myofilaments remained straight during shortening and merely changed their orientation. The helical appearance especially evident in the contracted cell model (Fig. 1b) was dependent upon the attachment sites for the filament bundles being arranged in a helix around the inner surface of the plasma membrane. The evidence for this model is contained in a recent publication (Fisher and Bagby, 1977). A model recently published by Small (1977) is substantially the same as this model; and, as far as we can determine, is based on the same assumptions.

ISBN 0-12-551750-5

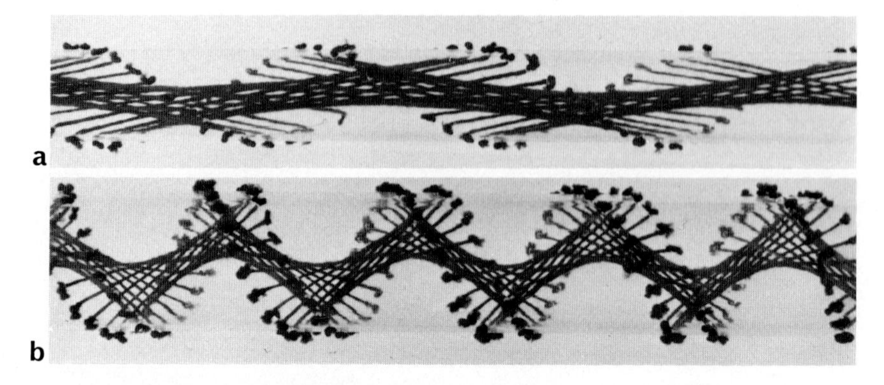

FIGURE 1. A model developed by Fisher (1974) showing the extended cell (a) and the contracted cell (b). Strands of yarn (bundles of myofilaments) are attached at either end to opposite sides of a plexiglas tube (plasma membrane). Attachment sites are arranged helically, leading to an overall helical appearance and the illusion that the strands are curved. The individual strands in "b" are ½ their length in "a". Cell diameter and the angle filaments make with the cell axis are increased proportionately.

Studies with isolated cells (Bagby and Pepe, 1977), isolated myofibrils (Bagby and Pepe, unpublished data), and cultured smooth muscle cells (Gröschel-Stewart *at al.*, 1975) stained with fluorescent antimyosin have provided strong evidence fór myofibrils, and occasionally these myofibrils showed banding patterns reminiscent of those in skeletal muscle. However, gizzard muscle cells (Fig. 2) often had myofibrils which undulated or coiled around the cell axis. The undulating myofibrils were seen most consistently in the shorter cells. The possibility then occurred to us that undulations could account for the observations of more oblique orientation of myofilaments in shortened cells. We decided to make further observations with live cells.

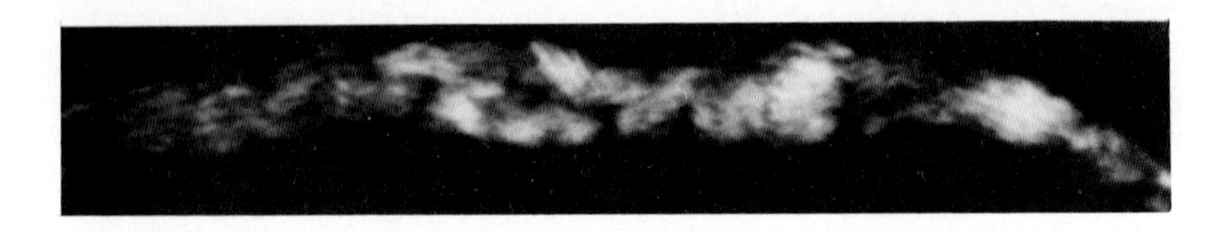

FIGURE 2. Glycerinated chicken gizzard cell stained with fluorescent anti-gizzard myosin. Note the curved myofibrils (X1900).

Isolated living cells were prepared from *Bufo marinus* stomach according to Bagby *et al.* (1971). To follow the course of individual myofibrils in these cells, cinemicrography with polarizing optics was performed while focusing through the cell. The movies, studied extensively with an editor, showed many myofibrils which followed curved paths. Sequences from these movies are reproduced in Figs. 3 and 4. Although separate frames are not as satisfactory as a movie, some myofibrils appear curved, and myofibrils seem to be oriented differently at different levels. There is a distinct possibility that the helical organization formerly observed in contracting cells (Fisher and Bagby, 1977) was due to curved paths of myofibrils rather than to straight myofibrils with helically arranged attachment sites. Although curved myofibrils in smooth muscle would create additional questions as to mode of force generation, etc., their presence would do much to explain why myofibrils have been so elusive in thin-sectioned material.

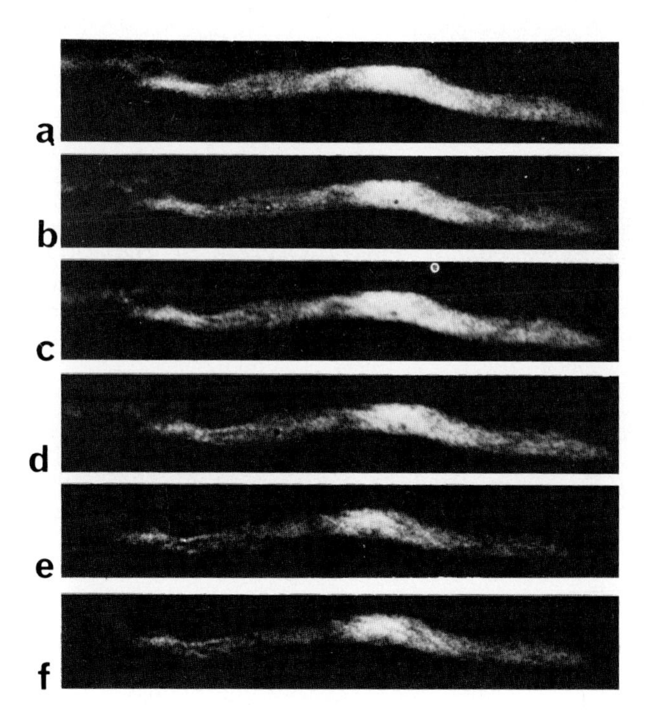

FIGURE 3. A sequence taken from a movie of Bufo marinus *cells filmed with polarizing optics (crossed polars) while focusing through a cell. Myofibrils appear as light streaks on the dark background. Note the apparent curved path of groups of myofibrils around the central portion of the cell (X620).*

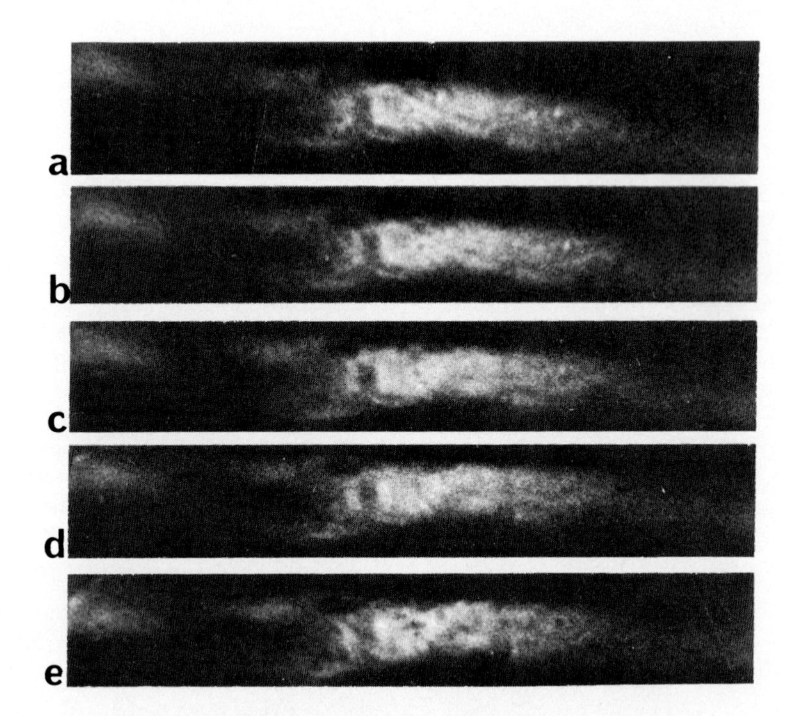

FIGURE 4. Same as Fig. 3 except a higher magnification was used to allow better discrimination of individual myofibrils. Note the myofibril in the top frames of the sequence which appears to wrap around myofibrils located deeper in the cell (X1300).

ACKNOWLEDGMENTS

The authors thank Dr. Bruce Fisher for his permission to reproduce the model which he constructed.

This investigation was supported by NIH Grant HL18077 and by a grant from the Tennessee Heart Association.

REFERENCES

Bagby, R. M., and Pepe, F. A. (1977). *Fed. Proc. 36*, 602.

Bagby, R. M., Young, A., Dotson, R., Fisher, B., and McKinnon, K. (1971). *Nature 234*, 351-352.

Fay, F. S., and Delise, C. S. (1973). *Proc. Natl. Acad. Sci. U.S.A. 70*, 641-645.

Fisher, B. A. (1974). Ph.D. dissertation, University of Tennessee, Knoxville.
Fisher, B. A., and Bagby, R. M. (1972). *Am. Zoologist 12,* xl.
Fisher, B. A., and Bagby, R. M. (1974). *Fed. Proc. 33,* 435.
Fisher, B. A., and Bagby, R. M. (1977). *A. J. P. Cell Physiol. 1,* C5-C14.
Gröschel-Stewart, U., Schreiber, J., Mahlmeister, C. H., and Weber, K. (1975). *Histochem. 43,* 215-224.
Rosenbluth, J. (1965). *Science 184,* 1337-1339.
Small, J. V. (1974). *Nature 249,* 327.
Small, J. V. (1977). *J. Cell Sci. 24,* 327-349.

Motility in Cell Function
Proceedings of the First John M. Marshall Symposium in Cell Biology

MYOSIN LOCALIZATION IN CULTURED EMBRYONIC CARDIAC MYOCYTES

Robert R. Kulikowski
and
Francis J. Manasek

Department of Anatomy
The University of Chicago
Chicago, Illinois

Embryonic chick cardiac myocytes suffer reversible disruption of their myofibrils when myocardia are trypsinized in Ca^{++}, Mg^{++}-free medium (Cedergren and Harary, 1964). Upon culturing, typical cross-banded myofibrils reform in these cells (Cedergren and Harary, 1964; Fischman and Zak, 1971; Gross and Müller, 1971; Kasten, 1971). We are using this system to explore events in early cardiac myofibrillogenesis. In the present paper we utilize immunofluorescent techniques employing anti-skeletal muscle myosin antibody to demonstrate some of the various morphologies assumed by myosin-containing structures in these cells during reassembly of cross-banded myofibrils. We show that myosin-containing structures in muscle cells are not limited to myofibrils, but that a variety of structures can be formed, some similar to those found in nonmuscle cells (Fujiwara and Pollard, 1976; Lazarides, 1975, 1976; Lazarides and Weber, 1974; Weber and Gröschel-Stewart, 1974).

In the myocytes, myosin assumes a variety of morphological forms prior to its reassembly into overt myofibrils. In cases where the cells are not yet well spread upon the substrate, myosin appears as a nonregular, granular accumulation fairly evenly dispersed throughout the cytoplasm (Fig. 1A). As cells spread out more on the substrate, linearity evolves in the myosin-containing structures (Fig. 1B,C), which often assume a stress fiberlike morphology (Fig. 1C). The linear elements, seen only after myocytes have begun to spread, appear to form in the direction of cell elongation (strain). This seems to be a prerequisite for their formation. In older, well-spread

ISBN 0-12-551750-5

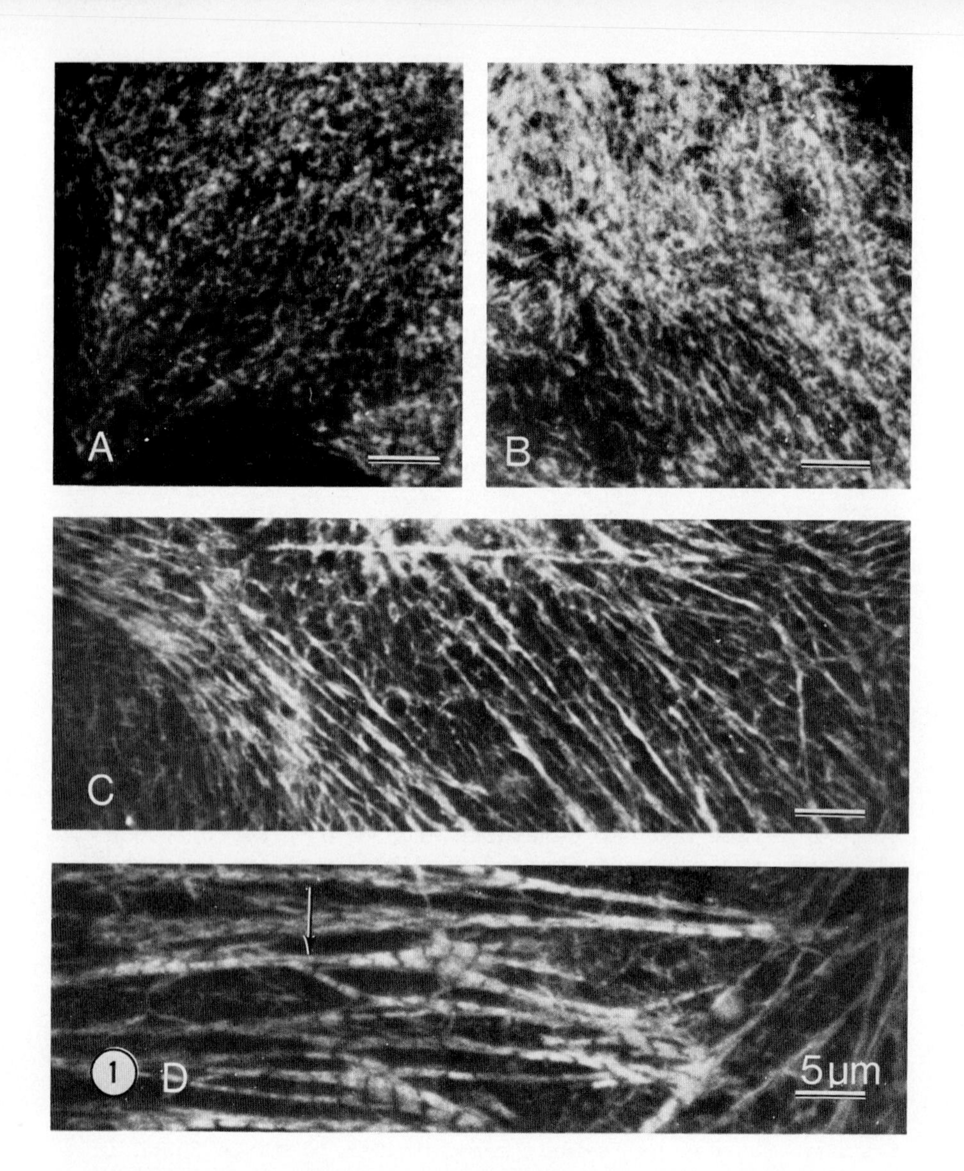

FIGURE 1. Indirect immunofluorescence staining of cultured embryonic cardiac myocytes using anti-skeletal muscle myosin antibody. Representative areas of cytoplasm have been depicted. With progressive cell spreading the dispersed myosin (A) assumes a more linear distribution (B). Noncross-banded fibers, similar to stress fibers, are formed (C). Typical cross-banded myofibrils can occur concomitantly (D). The bars indicate 5 µm.

cultures typical mature myofibrils are seen (Fig. 1D), often continuous (arrow) with fibers which closely resemble stress fibers described in nonmuscle cells. In nearly all cells, a heavy accumulation of myosin is detected in the perinuclear region.

The observed patterns show that myosin, in cardiac muscle, can be incorporated into structures that are morphologically similar to contractile protein-containing structures present in nonmuscle cells. Indeed, newly reassembled myofibrils and stress fiberlike stuctures often appear within the same cell and at times are seen to be continuous. While it has not yet been ascertained that the stress fiberlike structures also contain actin, it would appear safe to assume that once a typical myofibril has reassembled, it contains the full complement of contractile proteins. Clearly, striated muscle myosin can form continuous fibers indicating that under appropriate conditions the contractile proteins in both muscle and nonmuscle cells can form similar structures. The observed relationship between these fibers and myofibrils suggests that myofibrils can reassemble via stress fiberlike intermediates in cultured cardiac myocytes.

ACKNOWLEDGMENT

This work was supported by USPHS HL 13831.

REFERENCES

Cedergren, B., and Harary, I. (1964). *J. Ultrastruct. Res. 11*, 443.

Fischman, D. A., and Zak, R. (1971). *J. Gen. Physiol. 57*, 245A.

Fujiwara, K., and Pollard, T. D. (1976). *J. Cell Biol. 71*, 848.

Gross, W. O., and Müller, C. (1977). *Cell and Tissue Res. 178*, 483.

Kasten, F. (1971). *Acta Histochem. Suppl. 9*, 775.

Lazarides, E. (1975). *J. Histochem. Cytochem. 23*, 507.

Lazarides, E. (1976). *J. Cell Biol. 68*, 202.

Lazarides, E., and Weber, K. (1974). *Proc. Nat. Acad. Sci. U.S.A. 71*, 2268.

Weber, K., and Gröschel-Stewart, U. (1974). *Proc. Nat. Acad. Sci. U.S.A. 71*, 4561.

Motility in Cell Function
Proceedings of the First John M. Marshall Symposium in Cell Biology

ISOTONIC SPECTRIN SOLUBILIZATION

Michael P. Sheetz
and
David Sawyer

Department of Physiology
The University of Connecticut Health Center
Farmington, Connecticut

The major peripheral protein of the erythrocyte membrane, spectrin, is thought to be organized in a network along with actin on the cytoplasmic face of the membrane. The crosslinking of spectrin by anti-spectrin antibodies promotes the ATP dependent shape change in ghosts (Sheetz and Singer, 1977) which is the initial step of the ghost endocytic process (Penniston and Green, 1968). Erythrocyte membrane shape changes have been correlated with the dephosphorylation of membrane proteins including spectrin (Birchmeier and Singer, 1977; Shohet and Greenquist, 1977). Further Pinder *et al.* (1978) have reported that spectrin phosphorylation increases the actin polymerizing activity of spectrin. An explanation consistent with these results is that crosslinking and interaction in a spectrin-actin complex underlying the erythrocyte membrane is dependent upon spectrin phosphorylation. Therefore, it is interesting to note that spectrin is solubilized from Triton shells of intact erythrocytes with its dephosphorylation.

As described previously (Sheetz *et al.*, 1978) Triton shells are prepared in isotonic solutions from intact erythrocytes by a modification of the procedure of Yu *et al.* (1973). Washed erythrocytes are mixed with a Triton solution and are centrifuged into a 20-60% sucrose gradient in an isotonic buffer to separate the shells from the cytoplasm, solubilized membrane proteins, and Triton. The light-scattering band from the sucrose gradient which contains the shells (at $\rho \simeq 1.20$ gm/cc for ½% Triton extraction) is washed with an isotonic buffer to remove the sucrose. The shells are composed of spectrin (~70%), actin (~5%), band 3 (~9%), band 4 (~7%), and

ISBN 0-12-551750-5

several other minor proteins but the shells do not contain any material that reacts with periodic acid Schiff reagent (a carbohydrate stain). Spectrin alone is solubilized from the Triton shells upon incubation in a divalent cation-containing isotonic buffer for 24 hr at 0° or for 30 min at 20°C (Fig. 1). After such incubations, 20-40% of the protein in the shells is solubilized. The soluble spectrin coelectrophoreses with spectrin prepared by hypotonic extraction and contains less than one mole of Triton per mole of spectrin. The treatment of the membrane with nonionic detergent is required since isotonic lysis alone will not result in spectrin solubilization. When the spectrin is prelabelled by incubation of the intact cells with $^{32}P_i$, the soluble spectrin always has 1/2 or less of the phosphate specific activity of spectrin in

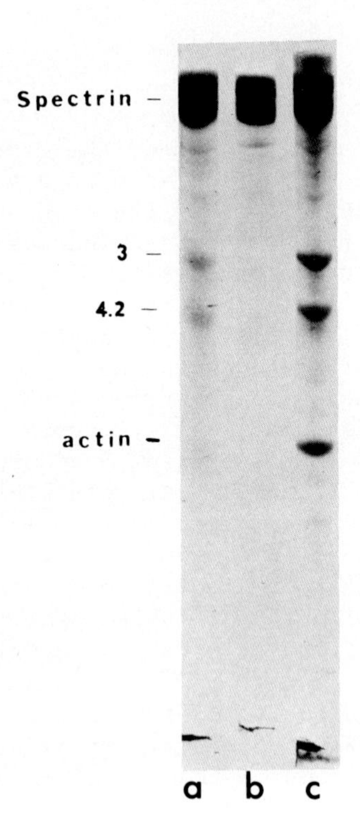

FIGURE 1. SDS polyacrylamide electrophoresis gels stained with Coomassie blue of (a) the Triton shells prepared by 0.5% Triton X-114 extraction in 140 mM KCl, 20 mM HEPES, 5 mM reduced glutathione, 5 mM $MgCl_2$ (pH 7.0), (b) spectrin solubilized from the shells by dialysis against 146 mM NaCl, 20 mM Tris, 0.5 mM $MgGl_2$, and 0.2 mM dithiothreitol (pH 7.4) for 24 hr on ice, and (c) pelleted shells after spectrin solubilization.

the original shell or in hypotonically prepared spectrin (Table 1). The absence of Mg^{2+} inhibits both solubilization and dephosphorylation (Table 1). Ca^{2+} accelerates spectrin solubilization more than Mg^{2+}.

We believe that there are two steps in isotonic spectrin solubilization. First, the membrane is perturbed by the non-ionic detergent. Bennett and Branton (personal communication) have observed that the spectrin binding to inside out vesicles is destroyed by Triton since no membrane protein-spectrin complex can be formed. Further, the portions of components 3 and 4 in the Triton shells dissociate from spectrin and actin with time and appear as a low density light-scattering band on the sucrose gradient. This suggests that Triton X-114 destroys the associations of other membrane proteins with spectrin and actin. The second step of the solubilization is divalent cation dependent and correlates with spectrin dephosphorylation. It is difficult to rule out the possibility that cation-dependent proteolysis is causing the solubilization; however, co-electrophoresis of the isotonically prepared spectrin with spectrin prepared by low ionic strength extraction shows no difference in mobility even on 3.25% acrylamide gels where the two spectrin components are well resolved. The phosphatase activity is not washed away from the shells during the removal of the sucrose and that activity, like the solubilization of spectrin, is stimulated by divalent cations. There is a clear suggestion that spectrin solubilization is caused by dephosphorylation and further experiments are underway to establish a causal relationship. The proposal which we favor to explain these results is that spectrin solubilization follows upon the dissociation of spectrin from the membrane binding protein(s) and the disaggregation of spectrin from other spectrin molecules or actin with dephosphorylation.

This proposal explains the correlation of membrane shape changes and dephosphorylation. If the maintainance of the biconcave disc morphology in red cells depends upon the extent of crosslinking in the actin-spectrin complex (see Sheetz and Singer, 1977; Birchmeier and Singer, 1977) and spectrin dephosphorylation results in spectrin disaggregation, then a change in shape would follow upon spectrin dephosphorylation.

TABLE 1. Relative ^{32}P Specific Activity of Spectrin[a]

Ionic conditions[b]	%shell protein solubilized	Solubilized spectrin	Pelletable spectrin	Before incubation
EGTA	13%	0.48	0.70	1.18
Mg^{2+} + EGTA	23%	0.32	0.54	1.18

[a]The specific activities of spectrin component 2 were determined for each sample as described in Sheetz and Singer (1977) and were normalized to the activity of spectrin in ghosts prepared by lysis of red cells in 10 m Tris pH 7.4.

[b]Shells were prepared according to Sheetz et al. (1978) with ½% Triton X-114 extraction. The material was diluted, pelleted and resuspended in 146 mM NaCl, 20 mM Tris (pH 7.4), and 0.5 mM EGTA with or without 0.5 mM Mg^{++}.

ACKNOWLEDGMENT

This work was supported by NIH grant HL-18317.

REFERENCES

Birchmeier, W., and Singer, S. J. (1977). *J. Cell Biol. 73,* 647-659.

Fairbanks, G., Steck, T. L., and Wallach, D. F. H. (1971). *Biochemistry 10,* 2606-2616.

Penniston, J. T., and Green, D. E. (1968). *Arch. Biochem. Biophys. 128,* 339-350.

Pinder, J. C., Bray, D., and Gratzer, W. B. (1978). *Nature 270,* 752-754.

Sheetz, M. P., Sawyer, D., and Jackowski, S. (1978). *In* "The Red Cell" Alan R. Liss, Inc. pp. 431-450.

Sheetz, M. P., and Singer, S. J. (1977). *J. Cell Biol, 73,* 638-646.

Shohet, S. B., and Greenquist, A. C. (1977). *Blood Cells 3,* 115-133.

Yu, J., Fischman, D. A., and Steck, T. L. (1973). *J. Supramol Struct. 1,* 233-241.

Motility in Cell Function
Proceedings of the First John M. Marshall Symposium in Cell Biology

CYTOCHEMICAL STUDY OF TRANSPORT ATPASE IN RHESUS MONKEY CORNEA ENDOTHELIUM

K. C. Tsou

School of Medicine
University of Pennsylvania
Philadelphia, Pennsylvania

B. Stuck
S. Schuschereba
and
E. Beatrice

Letterman Army Institute of Research
California

Little is known concerning the laser effect on the cornea because of the limitation of material amenable for such study. Recently, the use of new electron cytochemical methods (Tsou *et al.*, 1974, 1975, 1976) has made possible such investigations. This report illustrates a study of the alteration of morphology and Na^+-K^+ ATPase on Rhesus monkey cornea after CO_2 laser treatment at threshold safety level.

MATERIALS AND METHODS

Laser treatment of the Rhesus monkey cornea at 3-6 W/cm^2, 100 msec with a CO_2 laser has been carried out. Tissue preparation and incubation condition, using the new 5-nitroindoxyl phosphate method, has been reported elsewhere (Tsou and Khatami, 1977).

ISBN 0-12-551750-5

RESULTS AND DISCUSSION

Normal Rhesus monkey localization of Na^+-K^+ ATPase has been recently described by us (Tsou and Khatami, 1977). It is usually on the plasma membrane of endothelium cells, and is extremely weak in epithelial cells of the cornea except the wing cells and some necrotic cells on the surface layer. The laser treated cornea usually showed vacuolization in endothelial cells in the vicinity of the center of lasing. The plasma membrane ATPase (Fig. 1) activity is often in patches, but is mostly on the inner side of the membrane. Fusion at intercellular membrane can sometimes be seen between cells in the center of lasing, which show higher activity than the adjacent cells. While lasing produces heat, the alteration of enzyme activity could also take place. Na^+-K^+ ATPase controls both the ion flow and the hydration process of the cornea.

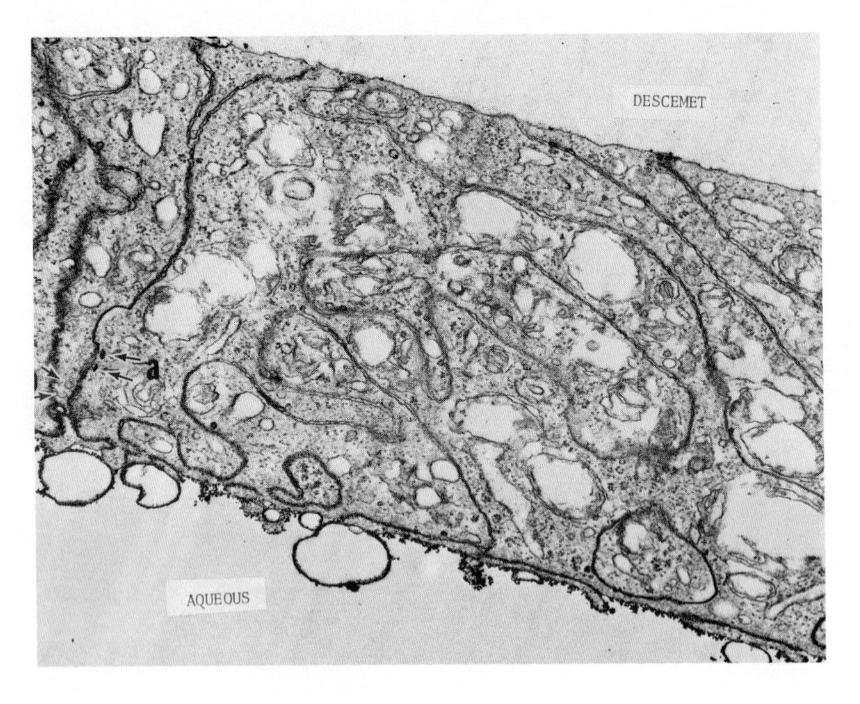

FIGURE 1. Na^+-K^+ ATPase in Rhesus monkey cornea endothelium - CO_2 laser treated, 3 W/cm^2, 100 msec. Note plasma membrane activity is heterogeneous. Stain vacuoles numerous on aqueous side not in control. Note also two stain particles (a) versus unstain particles (b) on juxtaposition of two cells (29,000X).

This function is therefore important in the sudden alteration of the cellular environment by laser irradiation. Based on these experimental results, CO_2 lasing on cornea endothelium initially activates the ATPase "pump" near the aqueous side and may inactivate the ATPase pump near the Descemet membrane side. The net result then brings water into the cells as manifested by the vacuolization process.

REFERENCES

Tsou, K. C., Morris, J., Shawaluk, P., Stuck, B., and Beatrice, E. (1974). *Proc. EMSA, 32nd,* pp. 224-225.

Tsou, K. C., Lo, K. W., and Yip, K. F. (1975). *J. Histochem. Cytochem. 23,* 307.

Tsou, K. C., Severdia, J., Stuck, B., and Beatrice, E. (1976). *Proc. EMSA, 34th,* p. 100.

Tsou, K. C., and Khatami, M. (1977). *Proc. EMSA, 35th,* p. 444.

Motility in Cell Function
Proceedings of the First John M. Marshall Symposium in Cell Biology

ANESTHETIC EFFECTS ON AMEBOID CELLS

David L. Bruce

Department of Anesthesiology
University of California
Irvine, California

John Marshall contributed, indirectly, to the basic knowledge of anesthetic action by taking a postdoctoral student into his laboratory to learn about the giant ameba *Chaos chaos*. The results of their studies (Bruce and Marshall, 1965) formed the foundation for an investigation of anesthetic effects on that ameba (Bruce and Christiansen, 1965). Both ether and halothane caused *C. chaos* to undergo a striking change in morphology, in which the cytoplasm adjacent to the plasmalemma became clear and the granular inclusions in the cytosol moved centrally. Between these two intracellular phases, there was a 65 mV potential difference with the peripheral, "clear" zone less negative than the normal potential recorded from the center of the cell. The peripheral part was also relatively rigid, and this led to the formation of several testable hypotheses.

If the ameba is transformed by anesthetics, ameboid mammalian cells should also be affected. Leukocytes would be the obvious cell to study, and they should be inhibited, under anesthesia, from fulfilling functions requiring plasticity of the subplasmalemmal cytoplasm. Tests were made in mice anesthetized with halothane to see if this agent inhibited transvascular diapesis of, and phagocytosis by, neutrophils. As predicted by analogy to the ameba, both functions were found to be inhibited (Bruce, 1966, 1967).

Cell division requires cytoplasmic cleavage, and so would also be inhibited by a phase change in the cytosol. Evidence that this happens under halothane anesthesia was furnished from results in rat bone marrow (Bruce and Koepke, 1966, 1971).

ISBN 0-12-551750-5

It was later postulated that halothane might cause similar inhibition of the *in vitro* dedifferentiation and division of cultured human lymphocytes treated with phytohemagglutinin. Again, the prediction proved to be accurate (Bruce, 1972).

Other workers, more sophisticated cell biologists, have since become interested in anesthetic effects on cells. Hinkley, working with cultured mouse neuroblastoma cells, showed reversible dissolution of 40-80 Å microfilaments when the cells were treated with halothane (Hinkley and Telser, 1974). Subsequently, the same investigators reported negative results in studies of halothane effect on protein and RNA synthesis in these cells and concluded, "halothane affects the morphology and growth rate of cultured mouse neuroblastoma cells by disrupting cytoplasmic actinlike microfilaments" (Telser and Hinkley, 1977).

Fourteen years ago, Marshall told his student, an anesthesiology resident, that if he learned about amebas he might some day understand why anesthetics inhibit the contractility of cardiac muscle. That made no sense to the student at that time. Now, it does, and he only wishes John were here so he could tell him so.

REFERENCES

Bruce, D. L. (1972). *Anesthesiology 36*, 201-205.
Bruce, D. L. (1966). *J. Cell. Physiol. 68*, 81-84.
Bruce, D. L. (1967). *J. Surg. Res. 7*, 180-185.
Bruce, D. L., and Christiansen, R. (1965). *Exp. Cell Res. 40*, 544-553.
Bruce, D. L., and Koepke, J. A. (1971). *Anesthesiology 34*, 573-576.
Bruce, D. L., and Koepke, J. A. (1966). *Anesthesiology 27*, 811-816.
Bruce, D. L., and Marshall, J. M., Jr. (1965). *J. Gen. Physiol. 49*, 151-178.
Hinkley, R. E., and Telser, A. G. (1974). *J. Cell Biol. 63*, 531-540.
Telser, A., and Hinkley, R. E. (1977). *Anesthesiology 46*, 102-110.

Motility in Cell Function
Proceedings of the First John M. Marshall Symposium in Cell Biology

COMPARISON OF A 95,000 MOLECULAR WEIGHT PROTEIN FROM *LIMULUS* SPERM WITH MUSCLE α-ACTININS

Issei Mabuchi

Department of Biology
University of Tokyo
Komaba, Meguro-ku
Tokyo 153, Japan

A protein which has the same subunit molecular weight as muscle α-actinin (95,000) has been found in isolated "false discharges" of *Limulus polyphemus* sperm and it has been suggested that this protein plays some role in the actin bundle organization (Tilney, 1975). To elucidate the role of this protein (sperm 95K protein), I have compared its properties with those of muscle α-actinins.

The false discharges were isolated as previously described (Tilney, 1975), and the 95K protein was purified by means of preparative gel electrophoresis in the presence of sodium dodecyl sulfate (SDS) (Stephens, 1975). A 95K protein, presumably α-actinin, was purified from a water-extract of *Limulus* myofibrils by precipitation with 15% saturation of ammonium sulfate followed by preparative SDS gel electrophoresis. Chicken gizzard α-actinin was a gift from Dr. T. D. Pollard. Proteins ran as single bands on SDS-gel electrophoresis (Fig. 1), and comigrated with each other. An antiserum against the sperm 95K protein was produced in a rabbit and the specificity of the antibodies was assayed by Ouchterlony's test and immunoelectrophoresis (Mabuchi and Okuno, 1977).

ISBN 0-12-551750-5

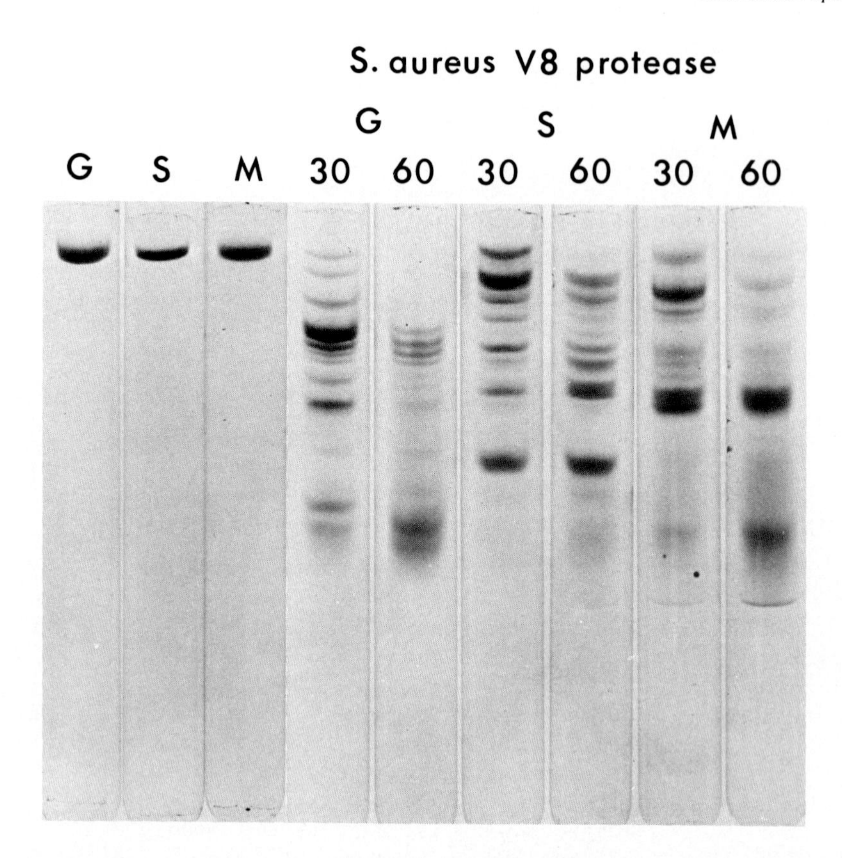

FIGURE 1. One-dimensional peptide mapping. Electrophoresis was carried out in 10% acrylamide gels containing 0.05% SDS, 5% glycerol, and 25 mM Tris/glycine (pH 8.5). The left three gels are purified proteins while the six gels to the right contain proteins digested by 25 µg/ml of S. aureus *V8 protease in 0.05% SDS, 1 mM EDTA, 0.5 mM dithiothreitol, and 20 mM Tris/HCl (pH 8.0) at 35 C° for 30 or 60 min. G, chicken gizzard α-actinin; S, sperm 95K protein; M,* Limulus *muscle α-actinin.*

One-dimensional peptide mapping (Cleveland *et al.*, 1977) using *S. aureus* V8 protease digests showed no similarity among these proteins (Fig. 1). Papain digests gave only two, three, and four peptides of molecular weights ranging 17,500-12,000 for chicken gizzard, *Limulus* sperm, and *Limulus* muscle proteins, respectively, with the two higher molecular weight peptides apparently similar in molecular weight among these proteins. *Limulus* proteins were considerably more resistant to both trypsin and chymotrypsin digestions compared to chicken gizzard α-actinin.

Two-dimensional peptide mapping obtained after a prolonged digestion by TPCK-trypsin gave 62, 54, and 75 peptides for chicken gizzard, *Limulus* sperm, and *Limulus* muscle proteins, respectively (not shown). Among the peptides of the sperm 95K protein digest, only 31% were found to be similar to those of chicken gizzard α-actinin digest and 48% were similar to *Limulus* muscle 95K protein.

Both immunoelectrophoresis (not shown) with crude protein preparations and Ouchterlony's test with purified proteins (Fig. 2) demonstrated that the antiserum against the sperm 95K contained antibodies specific to this protein but did not contain antibodies against other constituents of the false discharges, *Limulus* muscle 95K protein or chicken gizzard α-actinin.

The techniques employed here to compare the 95K proteins are capable of resolving both differences and similarities in protein primary structure. All of these methods suggest that there are major differences among the three proteins. Therefore, the primary structure of these proteins may differ significantly from each other. However, the possibility remained that the sperm 95K protein corresponds phylogenically to α-actinin, since it retains the ability to bind actin (Tilney, 1975) and also has the same molecular weight as α-actinin. Biological properties and the mode of interaction of this protein with actin remain to be elucidated.

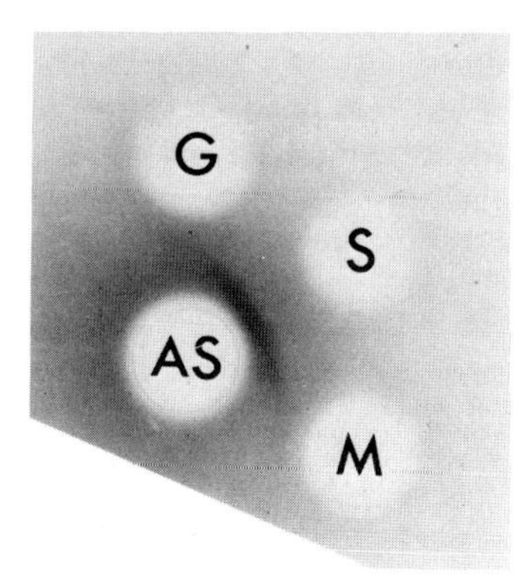

FIGURE 2. Ouchterlony's test of the anti-sperm 95K protein. The 1% agarose gel contained 0.02% SDS, 50 mM NaCl, and 10 mM sodium phosphate buffer (pH 7.0). AS, antiserum against sperm 95K protein; G, chicken gizzard α-actinin; S, sperm 95K protein; M, Limulus *muscle α-actinin. The protein concentration of each antigen fraction was 1 mg/ml.*

ACKNOWLEDGMENTS

I am grateful to Lew Tilney for encouragement and discussion and to Tom Pollard for the gift of chicken gizzard α-actinin. This research was supported by an NSF grant to L. G. Tilney.

REFERENCES

Cleveland, D. W., Fischer, S. G., Kirschner, M. W., and Laemmli, U. K. (1977). *J. Biol. Chem. 252,* 1102-1106.
Mabuchi, I., and Okuno, M. (1977). *J. Cell Biol. 74,* 251-263.
Stephens, R. E. (1975). *Anal. Biochem. 65,* 369-379.
Tilney, L. G. (1975). *J. Cell Biol. 64,* 289-310.

Motility in Cell Function
Proceedings of the First John M. Marshall Symposium in Cell Biology

SECOND MESSENGER CONTROL OF SPERM MOTILITY

Leonard Nelson

Department of Physiology
Medical College of Ohio
Toledo, Ohio

John M. Marshall prepared and generously provided the antibodies to skeletal muscle contractile proteins enabling us in 1958 to demonstrate that the site of binding of anti-myosin in frozen-dried rat sperm flagella coincided with the intracellular locus of Ca-activated ATPase (Nelson, 1958a,b). Electron microscopic examination established that the enzymatic-contractile protein site occupies the core of each of the nine coarse longitudinal fibers, while the purported, anti-actin antibody reacted with the cortical region partially surrounding each of the "cores." Subsequent studies have shown that in the presence of ATP and Mg^{2+} the contractile proteins extracted from both mammalian and invertebrate spermatozoa may be induced to undergo the conformational change sequences referred to as "precipitation" and "super-precipitation" (Young and Nelson, 1968; Nelson, 1966); furthermore, so-called glycerinated sperm models have been reactivated by ATP to propagate flagellar waves. We have recently observed that the motile performance of living sperm cells depends in large measure on the status of putative, intracellular, cholinergic neuroeffector systems. That is, the swimming rate, determined objectively, quantitatively, and reproducibly, of the sperm cells responds to a number of agents which operate directly or indirectly on cholinergic receptors. When acetylcholine is introduced into sperm cells or reversible depolarizing agents such as succinylcholine (both in the presence of dimethyl sulfoxide), or the cholinomimetic, nicotine, the swimming speed exhibits a biphasic response, nearly doubling at lower concentrations, and slowing down at higher concentrations (Nelson, 1972a,b). However, irreversible depolarizers, such as decamethonium, postsynaptic receptor blockers such as d-tubocurarine (Nelson, 1973) and α-bungarotoxin (Nelson, 1976),

ISBN 0-12-551750-5

and such ACh synthesis inhibitors as hemicholimium, cause only partial or complete depression of sperm motility. The abrupt and total inhibition of motility of all cells in the sperm suspension at less than micromolar quantities of α-bungarotoxin, and the complete and virtually instantaneous cessation of motility in over 50% of the cells at picomolar concentrations points specifically to the controlling presence of a nicotinic cholinergic receptor in the sperm cell surface membrane.

Since the divalent cations, calcium and magnesium (Young and Nelson, 1969, 1974), also appear to play a critical role in modulating motility of the living cells, a conceptual model has been devised which can account for the overall control of motility by integrating the two processes, the one centering on acetylcholine regulation of plasma membrane permeability, the second relating to the translocation of calcium ions through the cell membrane and between the intracellular compartments (Nelson, 1978).

In the fluid, lipid-globular protein, mosaic model, the cholinergic receptor extends through both the cytoplasmic and external surfaces of the plasma membrane and is depicted as undergoing a conformational change in response to binding with acetylcholine. Thus the cholinergic receptor may function as a Ca-ionophore, the ion channel opening as a consequence, facilitating transport of enough Ca into the cell to trigger simultaneous Ca release from intracellular sequestration sites. The released Ca is then available to mediate the coupling of the excitation process to the propagation of the propulsive contractile wave along the coarse, longitudinal, flagellar fibers. The acetylcholine which binds to the receptor and activates the ionophore mechanism is synthesized intracellularly by choline acetyltransferase (the presence of which in sperm has been reported independently in other laboratories, one at Vanderbilt (Bishop *et al.*, 1976) and the other at the University of Otago, New Zealand (Stewart and Forrester, 1976). While it remains to be determined whether acetylcholine is synthesized continuously or "on demand," inhibitory concentrations of hemicholinium, which interfere with acetylcholine biosynthesis, cause very rapid depression of motility, suggesting that the sperm cell contains little, if any, stored reserves of acetylcholine.

Closure of the Ca channels follows on the catalytic hydrolysis of ACh by the receptor-associated ACh-esterase, permitting reversal of the ACh induced conformational change. Restoration of the intracellular free Ca levels would be accomplished partially by extrusion and partly by resequestration of the Ca, all of which would bring about reversal of the contractile phase.

Changes in the cAMP content of spermatozoa which precede initiation of motility also participate in control of the intracellular Ca^{2+} traffic by regulation of the activity of the sequestration organelles (viz., mitochondria, intracellular membrane systems).

In summary, agents which interfere with ACh biosynthesis or hydrolysis, with ACh binding or depolarization of the cholinergic receptor, all affect sperm motility at physiologic concentrations. Concomitantly, factors such as lanthanum, procaine, chelating agents which alter the extracellular or intracellular $[Ca^{2+}]$, or which block its combination or release from binding sites, also cause alteration in motility.

REFERENCES

Bishop, M. R., Sastry, B. V., Schmidt, D. E., and Harbison, R. D. (1976). *Biochem. Pharmacol. 25*, 1617.

Nelson, L. (1958a). *Biophysical Soc. Abstracts*, p. 18.

Nelson, L. (1958b). *Biochim. Biophys. Acta 27*, 634.

Nelson, L. (1966). *Biol. Bull. 130*, 376.

Nelson, L. (1972a). *Biol. Reprod. 6*, 319.

Nelson, L. (1972b). *Exp. Cell Res. 74*, 269.

Nelson, L. (1973). *Nature 242*, 401.

Nelson, L. (1976). *Exp. Cell Res. 101*, 221.

Nelson, L. (1978), *Fed. Proc., 37*, 2543.

Stewart, T. A., and Forrester, I. T. (1976). *Proc. U. Otago Med. School, 54*, 53.

Young, L. G., and Nelson, L. (1968). *Exp. Cell Res. 51*, 34.

Young, L. G., and Nelson, L. (1969). *J. Cell Physiol. 74*, 315.

Young, L. G., and Nelson, L. (1974). *J. Reprod. Fertil. 41*, 371.

Motility in Cell Function
Proceedings of the First John M. Marshall Symposium in Cell Biology

A MOTILE SYSTEM ACTING ON THE CELL NUCLEUS*

John F. Aronson

The Wistar Institute of Anatomy and Biology
Philadelphia, Pennsylvania

An association between the cell nucleus and that region of the cytoplasm which acts as a center for microtubules and contains the centriole is common in animal cells and may be instrumental in relating events at the cell surface to the nucleus.

In a previous study (*J. Cell Biol. 51*, 579, 1971) with sea urchin eggs at the two cell stage, we were able to demonstrate that the nucleus, when separated from the center by centrifugation, is returned to the center by a Colcemid sensitive motile system. In this work we consider that movement of the egg nucleus to the sperm nucleus following fertilization is a comparable and experimentally more accessible system. This view is based on experiments in which movement of the female nucleus was delayed by Colcemid pretreatment until after the centers had doubled and under conditons where only one center was associated with the sperm nucleus. In such instances the egg nucleus moved to the nearest center and not necessarily to the center associated with the sperm nucleus.

These observations and those to be described were made primarily on eggs of the sea urchin *Lytechinus variagatus* which were treated with 5×10^{-7} *M* Colcemid for 30-60 min before fertilization. When fertilized, these eggs did not show nuclear fusion and a birefringent aster was not apparent by polarized light. When irradiated with 366 n*M* light (20-60 sec, high pressure mercury arc), a birefringent center appeared, and after a lag period, the egg nucleus began to move and eventually fused with the sperm nucleus.

**This work is dedicated to John Marshall in memory of a stimulating and all too brief association.*

ISBN 0-12-551750-5

In experiments involving approximately 100 eggs, we were able to show a linear correlation between the lag period and the separation between nuclei. In addition, the extrapolated intercept was near zero. On the basis of this data, the rate of development of the motile system is 10 μm/min at 23°C.

Movement of the nucleus following release of the Colcemid block was studied in photographic sequences taken at 1 sec intervals with Nomarski differential interference optics. Rates of movement as high as 30 μm/min for distances of over 20 micra were frequently seen.

The initial movement of the female nucleus was not an abrupt transition to the maximal rate but rather was a slow movement increasing to the maximal rate over about 20 sec. Since the nuclear face is deformable and deformation is seen only when the nucleus is moving rapidly, it is possible to say that this slow movement is a characteristic of the motile system and does not reflect resistance to movement. There was also some evidence of discontinuity in the rate of nuclear movement over a time scale of tens of seconds which again, on the basis of changes in axial ratio of the nucleus, appears to be a property of the motile system.

Deformation of the nucleus during rapid movement is often seen as a blunt projection of roughly 15 square micra in cross section suggesting that the force to move the nucleus is applied over a broad area and not by one or a few filaments of molecular dimensions. The site of attachment is the face nearest the center and the attachment site does not shift since nuclei with small birefringent crystals attached did not rotate as they moved to the center.

Movement of the female nucleus once initiated was stopped by high concentrations of Colcemid or by near 0° temperatures, and there was no gross evidence of a transient increase in the rate of movement during this process. Redevelopment of movement after cold treatment involves a lag period and the reformation of a birefringent aster. These observations suggest that the motile system is dependent on the presence of microtubules or has a lifetime of less than about 30 sec in their absence.

Motility in Cell Function
Proceedings of the First John M. Marshall Symposium in Cell Biology

MEASUREMENT OF THE NUCLEOTIDE EXCHANGE RATE AS A DETERMINATION OF THE STATE OF CELLULAR ACTIN

J. L. Daniel
L. Robkin
L. Salganicoff
and
H. Holmsen

Specialized Center for Thrombosis Research
Temple University
Philadelphia, Pennsylvania

Extensive study of adenine nucleotide metabolism in a variety of cells has indicated a central role of ATP as the direct energy source for many cellular functions. Several methods for nucleotide extraction have been employed. In blood platelet studies, one of two extraction methods is commonly used: perchloric acid or ethanol-EDTA extraction. These two methods produce slightly different results (Holmsen, 1972). When platelets are extracted with ethanol, a fraction consisting mainly of ADP is not solubilized, but remains bound to insoluble proteins. We have fairly conclusive evidence to suggest that this protein is actin (French and Wachowicz, 1974; Daniel and Holmsen, unpublished).

The metabolic adenine nucleotide pool can be labeled by incubation of platelets with ^{14}C-adenine (the large amounts of ATP and ADP in the dense granules are not labeled). The actin-bound (ethanol insoluble) ADP is also labeled and attains the same specific radioactivity as the metabolic nucleotide pool. This fact indicates that ADP of actin turns over rapidly in the intact cell. We have been able to measure this turnover rate using a double isotope procedure.

ISBN 0-12-551750-5

Platelets are incubated with ^{14}C-adenine under conditions which allow the ethanol-soluble and insoluble fractions to be labeled to the same specific radioactivity. These platelets are then incubated with ^{3}H-adenine, and the rate of incorporation of ^{3}H into ethanol soluble and insoluble nucleotides is determined. The results are expressed as a $^{3}H/^{14}C$ ratio.

The rate of incorporation of ^{3}H into the ethanol-soluble and -insoluble fractions is shown in Fig. 1. The rate of incorporation into each is indistinguishable at this low time resolution.

We can resolve this process if the kinetics of the first minute is examined in greater detail (Fig. 2) using a mathematical model based on the transfer of ^{3}H from the medium to the soluble nucleotides and then to the actin-bound ADP, developed with the help of Dr. R. Tallerida, the rate constant for this turnover can be calculated, and yields a rate constant of 0.1 sec^{-1} ± .025 for seven determinations. Assuming that ATP is the precursor of actin-ADP, and that there is 0.3 μ moles actin-ADP/10^{11} platelets, this rate constant means that at least 30% of the ATP utilized in resting platelets is consumed by this process (Holmsen and Akkerman, unpublished).

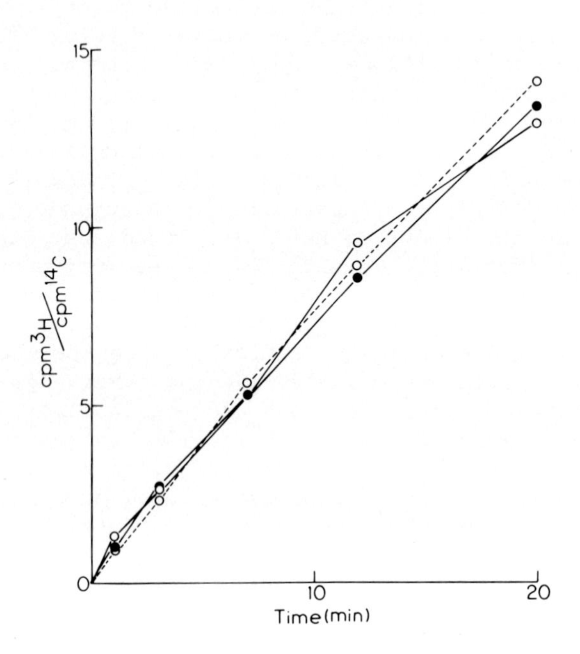

FIGURE 1. Incorporation of ^{3}H-adenine into adenine nucleotides of intact platelets. Ethanol soluble ATP (-O-), ethanol soluble ADP (-O-), ethanol insoluble ADP (---O---).

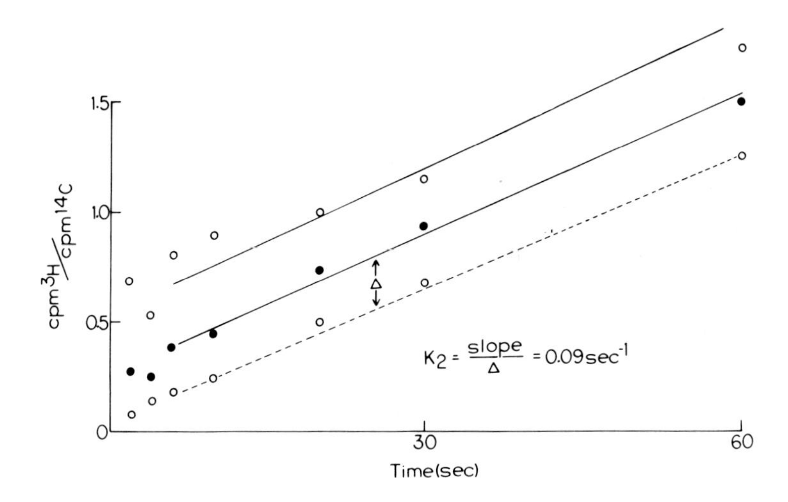

FIGURE 2. Early time course of labeling by ^{3}H-adenine of platelets bound and soluble mucleotides. Symbols are the same as for Fig. 1.

Furthermore, this rate is many orders of magnitude faster than expected for F-actin and indicates that actin in resting platelets is in a nonfilamentous form. This aggrees with morphologic studies on resting platelets (Behnke *et al.*, 1971).

Microfilaments are seen in activated platelets (Hovig, 1968). We studied the kinetics in a model of activated platelets with well-described morphology (Russo and Salganicoff, unpublished). In this model, large filament bundles are easily seen. Platelets were incubated with ^{14}C-adenine in plasma, subsequently activated by thrombin (1 U/ml) and 2 m*M* $CaCl_2$ on a nylon mesh, and finally incubated with ^{3}H-adenine. The rate of incorporation into soluble and actin-bound pools is given in Fig. 3. The rate of uptake of the actin-bound ADP is depressed by 70%, which indicates, according to our working hypothesis, that actin is polymerized consistent with the morphological appearance of such platelets.

Study of the turnover rate of actin-bound nucleotide has allowed us to approach both the polymerization state and the energetic requirements of actin in intact living cells. We feel this technique will allow further insights in the role of actin in cell motility.

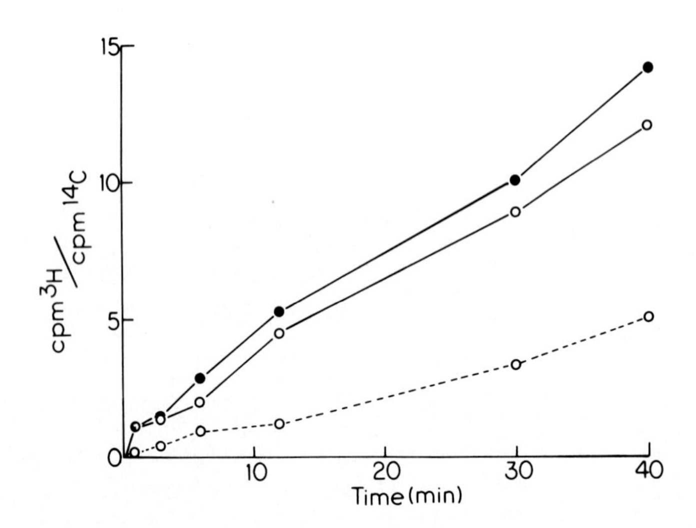

FIGURE 3. Rate of ^{3}H-adenine incorporation into adenine nucleotide in thrombin-Ca^{2+} activated platelets. Symbols are the same as Fig. 1.

REFERENCES

Behnke, O., Kristensen, B. I., and Nielsen, L. E. (1971). *In* "Platelet Aggregation" (J. Caen, ed.), pp. 3-31. Masson and Cie, Paris.

French, P. C., and Wachowicz, B. (1974). *Haemostasis 3*, 271-281.

Holmsen, H. (1972). *Ann. New York Acad. Sci. 201*, 109-121.

Hovig, T. (1968). *Series Haematologica 1*, 3-64.

Motility in Cell Function
Proceedings of the First John M. Marshall Symposium in Cell Biology

CESSATION OF PROTOPLASMIC STREAMING DURING MITOSIS IN PLASMODIA OF *PHYSARUM POLYCEPHALUM*

D. Kessler
and
M. J. Lathwell

Department of Biology
Haverford College
Haverford, Pennsylvania

Microplasmodia of the myxomycete *Physarum polycephalum* may be fused at 26°C to produce a single multinucleate plasmodium in which all nuclei periodically undergo mitosis simultaneously (Guttes and Guttes, 1964). Sachsenmaier *et al.* (1973) indicate that shuttle streaming stops during mitosis, although contrary observations have been reported (Miller and Anderson, 1971). With Haverford College students John Ahrens, Ira Kelberman, and Peter Steenbergen, we have examined streaming in channels of small plasmodia (about 1.5 to 2 cm in diameter) under the compound microscope during the third mitosis after fusion. At the time that streaming is observed, small samples from the same plasmodium are taken at intervals and immediately fixed and stained in aceto-lactic-orcein stain for identification of stage of mitosis. Thus, we can correlate qualitative observations of streaming velocity with the morphological events of mitosis in a plasmodium. We have found that streaming velocity begins to slow in all channels throughout the plasmodium in late prophase, and actually stops for a short time in late metaphase or anaphase (Fig. 1). By late telophase and early interphase the velocity of streaming has increased back to normal interphase levels. We have never observed all channels throughout a plasmodium ceasing motion in this fashion during interphase.

ISBN 0-12-551750-5

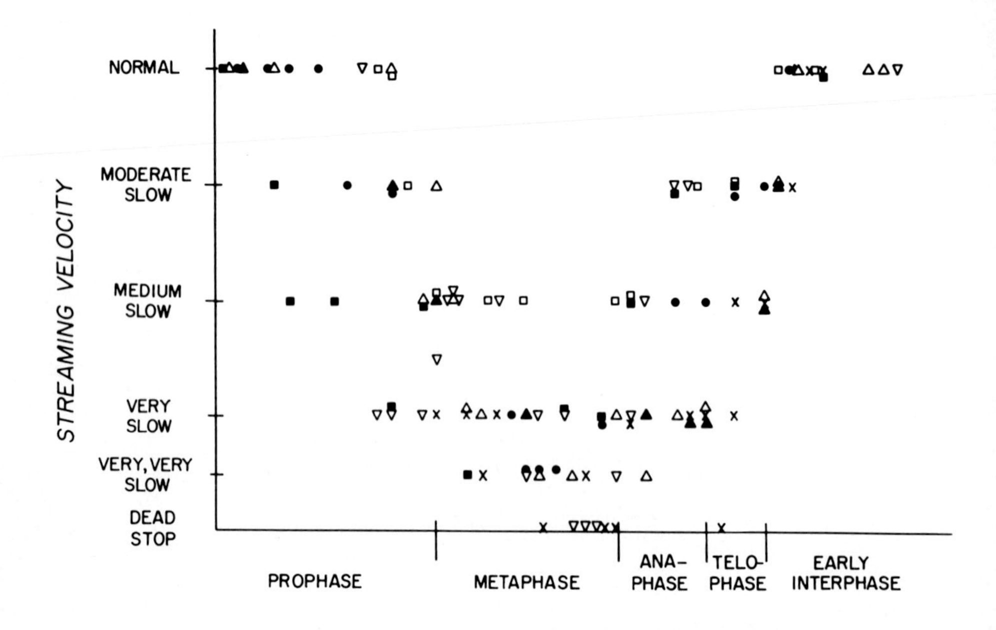

FIGURE 1. Microscopic observations of streaming velocity in seven different plasmodia of Physarum polycephalum *during mitosis.*

Since mitosis is not accompanied by cytokinesis in the plasmodium at this stage in the life cycle, an obvious morphological explanation for the cessation of streaming during mitosis---that the actomyosin filaments are redistributed in the cytoplasm in preparation for cytoplasmic cleavage---would seem improbable here. However, the distribution of cytoplasmic filaments during mitosis in *Physarum* has not been systematically studied. Therefore, redistribution or depolymerization of cytoplasmic filaments during mitosis has not been excluded as a regulatory mechanism in this system. We have studied this question recently by using a combined osmium tetroxide---glutaraldehyde solution to fix samples of mitotic plasmodia in which streaming had stopped or samples from vigorously streaming interphase cultures, embedding them in epon-araldite, and examining thick and thin sections of this material by phase-contrast microscopy and by electron microscopy. We have found that interphase plasmodia which had been streaming vigorously exhibit large channels in which invaginations of the plasma membrane produce a complex network of elongated vesicles in the outer portion of the cytoplasm (ectoplasm), while the center of the channel (endoplasm) is free of this tubular network. Cytoplasmic filaments form fibrils which are often seen associated with the ectoplasmic invaginations of the plasmalemma. Bundles of filaments are not seen in the endoplasm. In mitotic cultures in which streaming had ceased before fixation, invaginations of the plasmalemma into the channel ectoplasm are very pronounced while the endoplasm is reduced in volume or even absent. Cytoplasmic filaments are still evident, usually in association with the cell membrane invaginations. The nuclei are undergoing an intranuclear mitosis at this time. Anaphase figures may be seen in which the nuclear envelope around the spindle pole region has disintegrated, exposing the mitotic spindles to the cytoplasm.

These observations disprove the idea that a major portion of the cytoplasmic actomyosin filaments have depolymerized or radically redistributed themselves during mitosis. We are currently examining several other hypotheses which might explain the regulation of actomyosin interactions in this system.

ACKNOWLEDGMENT

This work was supported by National Institutes of Health Grant 5 RO1 GM22924.

REFERENCES

Guttes, E., and Guttes, S. (1964). *In* "Methods in Cell Physiology" (D. M. Prescott, ed.), Vol. I, pp. 43-54. Academic Press, New York.

Miller, D. M., and Anderson, J. (1971). *In* "Experiments in Physiology and Biochemistry," Vol. 4, pp. 183-202. Academic Press, New York.

Sachsenmaier, W., Blessing, J., Brauser, B., and Hansen, K. (1973). *Protoplasma* 77, 381-396.

Motility in Cell Function
Proceedings of the First John M. Marshall Symposium in Cell Biology

EVIDENCE FOR DIRECT LARGE FILAMENT INTERACTIONS IN MOLLUSCAN CATCH MUSCLES

William H. Johnson
and
Van P. Thompson

Rensselaer Polytechnic Institute
Troy, New York
and
University of Maryland Dental School
Baltimore, Maryland

Without question the ability of molluscan muscles to maintain tension with little or no expenditure of energy is a unique mode of action of muscles. Enough evidence exists to require consideration of mechanisms which differ from those responsible for contraction of skeletal muscles (Ruegg, 1971).

One such suggestion, which has been proposed by several authors (see Ruegg, 1971, for review), is that large filaments interact with one another to form a kind of rigid framework, perhaps as part of a more extensive cytoskeleton. This framework would passively bear tension which is actively developed by a bifilament actomyosin system. We have tested this possibility by studying interactions of large filaments isolated from the white adductor of *Mercenaria mercenaria*, and have also attempted to isolate filament aggregates from glycerinated preparations of the Anterior Byssus retractor muscle (ABRM) of *Mytilus edulis*. Both are known to be catch muscles.

Results are outlined in Fig. 1. In this figure, part A shows a phase micrograph of isolated *Mercenaria* filaments in buffer solution at pH 7.8; part B shows the same preparation after titration with gentle stirring to pH 6.2. The filaments clearly form bundles at the lower pH, and aggregation begins at a pH comparable to the pH at which paramyosin forms paracrystals. This aggregation occurs equally in the presence or absence of myosin. In Fig. 1C and 1D, scanning electron micrographs of this preparation at pH 7.8 and 6.2, respectively,

ISBN 0-12-551750-5

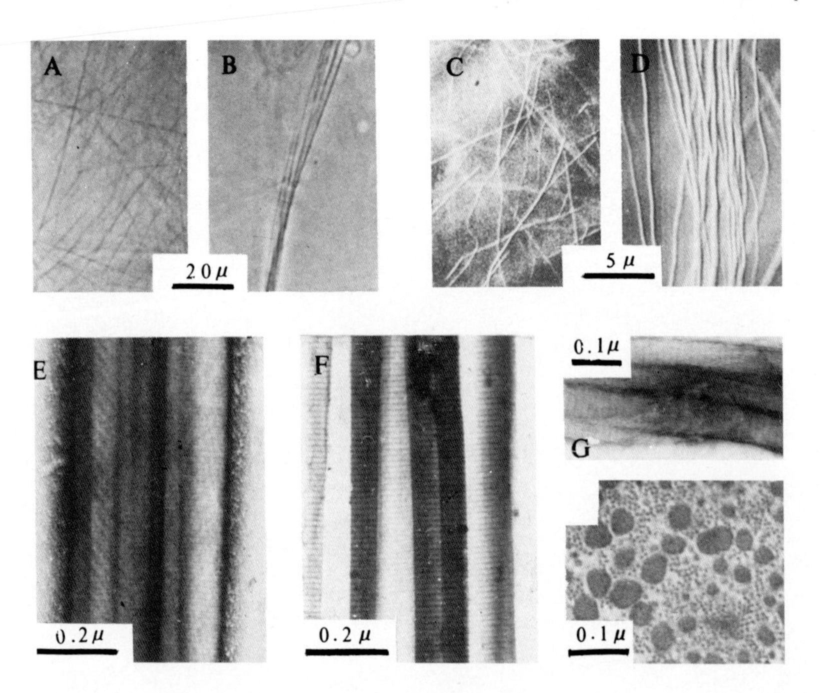

FIGURE 1. Micrographs of various preparations from molluscan muscles and paramyosin preparations, as detailed in the text.

are shown with the same results. The bundles formed at pH 6.2 are composed of laterally aggregated native filaments, as seen in electron micrographs of negatively stained preparations (Fig. 1E).

It is significant that α-paramyosin will spontaneously form bundles of paracrystals (Merrick and Johnson, 1977). An example of such a negatively stained bundle of paracrystals is seen in Fig. 1F. Note that in both Fig. 1E and 1F, the 14.5 nm repeat is in register across the bundle, indicating a structurally specific interaction between the filaments or paracrystals.

Glycerinated ABRM fibers were blendored at pH 6.2, where such fibers exhibit a high, catchlike elastic modulus. When such preparations are negatively stained and examined in the electron microscope, aligned bundles of large filaments are commonly observed (Fig. 1G). These bundles are not present at a pH higher than 7.0; however, in the high pH range, only individual filaments were seen. The filaments are very fragile and tend to fall apart in the presence of the negative stain.

This evidence indicates that the large filaments of catch muscles have the potential for side-by-side interaction. Are large filaments in living catch muscles close enough for such interactions to occur? Morrison and Odense (1974) suggested this on the basis of their comparative work, which included the *Mercenaria* adductor muscle used by us. In Fig. 1H, a micrograph of a cross section of the living ABRM is shown. Aggregates of large filaments without interdigitating small filaments are very common. They are also common in glycerinated preparations (see also Sobieszek, 1973). According to preliminary findings in our laboratory, such large filament bundles or aggregates are not found in noncatch, paramyosin containing muscles such as the red adductor of *Mercenaria*. Clearly, then, large filament interaction is possible in catch muscles and should be considered as playing a possible role in catch.

REFERENCES

Merrick, J. P., and Johnson, W. H. (1977). *Biochemistry 16*, 259-264.

Morrison, C., and P. H. Odense. (1974). *J. Ultrastruct. Res. 49*, 228-240.

Ruegg, J. C. (1971). *Physiol. Rev. 51*, 201.

Sobieszek, A. (1973). *J. Ultrastruct. Res. 43*, 313.

Index

T

U

V

W

X